Mechanisms and Machine Science

Volume 86

This book series establishes a well-defined forum for monographs, edited Books, and proceedings on mechanical engineering with particular emphasis on MMS (Mechanism and Machine Science). The final goal is the publication of research that shows the development of mechanical engineering and particularly MMS in all technical aspects, even in very recent assessments. Published works share an approach by which technical details and formulation are discussed, and discuss modern formalisms with the aim to circulate research and technical achievements for use in professional, research, academic, and teaching activities.

This technical approach is an essential characteristic of the series. By discussing technical details and formulations in terms of modern formalisms, the possibility is created not only to show technical developments but also to explain achievements for technical teaching and research activity today and for the future.

The book series is intended to collect technical views on developments of the broad field of MMS in a unique frame that can be seen in its totality as an Encyclopaedia of MMS but with the additional purpose of archiving and teaching MMS achievements. Therefore, the book series will be of use not only for researchers and teachers in Mechanical Engineering but also for professionals and students for their formation and future work.

The series is promoted under the auspices of International Federation for the Promotion of Mechanism and Machine Science (IFToMM).

Prospective authors and editors can contact Mr. Pierpaolo Riva (publishing editor, Springer) at: pierpaolo.riva@springer.com

Indexed by SCOPUS and Google Scholar.

More information about this series at http://www.springer.com/series/8779

Eusebio E. Hernandez ·
Sajjad Keshtkar · S. Ivvan Valdez
Editors

Industrial and Robotic Systems

LASIRS 2019

Springer

Editors
Eusebio E. Hernandez
ESIME Ticoman
Instituto Politecnico Nacional
Mexico City, Mexico

Sajjad Keshtkar
School of Engineering and Science
Tecnologico de Monterrey
Altamira, Mexico

S. Ivvan Valdez
CONACYT-CENTROGEO
Mexico City, Mexico

ISSN 2211-0984 ISSN 2211-0992 (electronic)
Mechanisms and Machine Science
ISBN 978-3-030-45401-2 ISBN 978-3-030-45402-9 (eBook)
https://doi.org/10.1007/978-3-030-45402-9

This Springer imprint is published by the registered company Springer Nature Switzerland AG
The registered company address is: Gewerbestrasse 11, 6330 Cham, Switzerland

Preface

This book compiles selected and reviewed papers of the Latin American Symposium on Industrial and Robotic Systems 2019 (LASIRS 2019), from October 30 to November 01, 2019, Tecnologico de Monterrey, Campus Tampico, Mexico. This conference is a second calling for the community interested in robotics, in particular, for the Latin American community interested in robotics engineering applications from algorithms, simulations, and design, to prototypes and implementations. A first meeting took place at Guanajuato, México, in 2017, under the name of School of Optimization, Simulation and Numerical Methods in Robotics, and as the name suggests, this first experience aimed to bring together researchers and students with a computer science profile but devoted to solving robotics problems; nevertheless, the experience was so enriching that make us noticed that the frontiers in robotics problems are not strict, nor in the nature of the problems either in the necessary perspectives to solve them. Hence, for this second edition, we convoke researchers in a more broad spectrum of robotics, keeping in mind several goals: to focus on the applicability of the methods and proposals presented in the Symposium, the aim to present different perspectives and complementary disciplines around robotics, the aim to include introduction and tutorial talks for undergraduate students as an essential part of the symposium in order to promoting this beautiful area of knowledge to the new generations, and the last but not least, and possibly our main challenge for future editions, to promote the meeting among the large community, we know, exists in Latin America. In these last two points, we are aware of the talented community we have; for instance, very often Latin American teams are participants and winners of international robotics contest, so our future vision of this meeting is an academic festivity around robotics with Latin American students as the special guest.

In this compilation, the reader can find several papers, with topics from new algorithms, design, simulation, and implementation of industrial and robotics systems; many of them are valuable by their tutorial content even if they introduce new material, other are new rigorous research papers, and some others are descriptions of novel robotics implementation. We expect that the reader finds this compilation useful, that is, at the same time, an invitation to enjoy the broad and fascinating industrial–robotics world.

We would like to take this opportunity to extend our recognition to the scientific and organizing committee members, session chairs, reviewers, and conference participants for the scientific success of this event. Their active participation in the discussions and the numerous interesting presentations were vital to achieving the scientific meeting objectives. We are grateful to Tecnologico de Monterrey, Campus Tampico, for hosting the event. We would also like to thank the auspices and support of The International Federation for the Promotion of Mechanisms and Machine Science (IFToMM) and the financial support of the National Council of Science and Technology (CONACyT), Mexico, through grant 299374.

Eusebio E. Hernandez
Sajjad Keshtkar
S. Ivvan Valdez

Organization

LASIRS 2019 Organizing Committee

Manuel Martínez Martínez	School of Engineering and Science, Tecnologico de Monterrey, Altamira, Mexico
Irandi Gutierrez Carmona	School of Engineering and Science, Tecnologico de Monterrey, Altamira, Mexico
Sajjad Keshtkar	School of Engineering and Science, Tecnologico de Monterrey, Altamira, Mexico
Juan Gabino Díaz Martínez	School of Engineering and Science, Tecnologico de Monterrey, Altamira, Mexico
Roberto Abraham Valdivia Beutelspacher	School of Engineering and Science, Tecnologico de Monterrey, Altamira, Mexico
Leopoldo Morales Gómez	School of Engineering and Science, Tecnologico de Monterrey, Altamira, Mexico
León Felipe Guevara Chávez	School of Engineering and Science, Tecnologico de Monterrey, Altamira, Mexico
Jorge Álvarez Ramírez	School of Engineering and Science, Tecnologico de Monterrey, Altamira, Mexico
Miguel Ángel Contreras Enríquez	School of Engineering and Science, Tecnologico de Monterrey, Altamira, Mexico

LASIRS 2019 Scientific Committee

Alexander Poznyak	CINVESTAV of National Polytechnic Institute, Mexico
Eusebio Hernández	National Polytechnic Institute, Mexico
Rafiq Ahmad	University of Alberta, Canada
Hirohisa Kojima	Tokyo Metropolitan University, Japan
Mario Acevedo Gómez	Panamerican University, Mexico
Juan Carlos Jáuregui Correa	Autonomous University of Querétaro, Mexico
Héctor Becerra Fermín	Mathematics Research Center, Mexico

Edgar Alfredo Portilla Flores	National Polytechnic Institute, Mexico
Sajjad Keshtkar	Tecnologico de Monterrey, Mexico
Ivvan Valdez Peña	CentroMet-CONACyT, Mexico
Adriana Vargas Martínez	Tecnologico de Monterrey, Mexico
Jorge Lozoya	Tecnologico de Monterrey, Mexico
Armando Román	Tecnologico de Monterrey, Mexico
Irandi Gutierrez Carmona	Tecnologico de Monterrey, Mexico

Contents

Nanotechnology Applications in Industry and Heat Transfer

Sina Razvarz[1] (✉), León Felipe Guevara Chavez[2], and Cristóbal Vargas-Jarillo[1]

[1] Departamento de Control Automático CINVESTAV, IPN (National Polytechnic Institute), 07360 Mexico City, Mexico
srazvarz@yahoo.com, cvargas@ctrl.cinvestav.mx

[2] School of Engineering and Science, Tecnológico de Monterrey, Altamira, 89600 Tampico, Mexico
Leon.guevara@tec.mx

Abstract. Nanotechnology is stated as the utility of nanomaterials for human profit. Nanomaterials contain distinct attributes because of their physical as well as chemical features at the nanoscale. Nanotechnology has a vast domain of usages and has been utilized in providing new products in the industrial section, electronic as well as energy. In this paper, the properties of fluids with nanoparticles are reviewed. Furthermore, the developed models with nanoparticle consortium are studied.

Keywords: Nanotechnology · Nanomaterials · Industry

1 Introduction

Industrial engineering is one of the fields that artificial neural network, fuzzy logic, and nano technology have found an extensive implementation area [1–19]. Nanoscale refers size dimensions between approximately 1 and 100 nm (or more appropriately, 0.2 and 100 nm) because at this scale the properties of materials differ with respect to their physical properties from a larger scale.

The size dimensions of Nanoscale are among nearly 1 and 100 nm (or more suitably, 0.2 and 100 nm), since at this scale the attributes of materials change concerning their physical attributes from a greater scale. Every material which contains one or more dimensions in the nanoscale is stated as nanomaterial [20]. Nanoparticles contain distinct biological attributes such as tiny size, a great surface-area-to-volume ratio that permits them to paste, attract as well as to transfer composites like tiny-molecule drugs, proteins as well as probes having great effectiveness [21].

The field of nanotechnology is built itself as an essential technology for a vast domain of implementations, so being a first-class preference for science as well as technology policy growth. Nanotechnology is the implementation of scientific knowledge for manipulating and control issues at the nanoscale. Nevertheless, there is no definite answer regarding the usage of nanoparticles and nanomaterials, mostly due to the shortage of information regarding their attributes and applicability.

E. E. Hernandez et al. (Eds.): LASIRS 2019, MMS 86, pp. 1–8, 2020.
https://doi.org/10.1007/978-3-030-45402-9_1

The attributes of materials are not the same at the nanoscale due to two major reasons. The first reason is that the superficial region of nanomaterials is great and the second reason is that the quantum effects dominate the treatment of the nanoscale. The physical nature of the nanoparticles with size less than 20 nm changes more extremely when is compared with the conventional size materials [22]. The basic attribute for characterizing nanomaterials are [23]: size propagation, surface region and porosity, and techniques of synthesis. It is on the basis of their inherent attributes or in the ones acquired using connecting them with other materials as if possible usages are generated for achievement, for instance [24]: reduction; weight diminution, and amended functionalities concerning to the materials.

2 Nanofluids Preparation and Characterization

2.1 Particle Size Distribution

The nanomaterial is a material having an exterior dimension in the nanoscale or with interior construction at the nanoscale [25] that can reveal new properties in comparison with the equal material without nanoscale characteristics. The attributes of the materials may not be the same at the nanoscale due to two major reasons. The first reason is that the superficial region of nanomaterials is great and the second reason is that the quantum effects dominate the treatment of the nanoscale.

Applying a standard from 2007, the nanoparticle is stated as a particle having a nominal diameter of less than nearly 100 nm [22]. On a standard from 2008, the nanoparticle is stated as a particle as if it contains all its three dimensions approximately 100 nm or minor, also can be mentioned as nano-object [26].

The size dimensions of Nanoscale are among nearly 1 and 100 nm (or more suitably, 0.2 and 100 nm), since at this scale the attributes of materials change concerning their physical attributes from a greater scale. Every material which contains one or more dimensions in the nanoscale is stated as nanomaterial. Nanoparticles contain distinct biological attributes such as tiny size, a great surface-area-to-volume.

The physical attributes of nanoparticles having sizes less than 20 nm can vary more greatly when are compared with the conventional size materials.

The other perception which is utilized is nanostructured nanoparticles, composed of particles with structural characteristics less than 100 nm that can impact their physical, chemical as well as biological characteristics.

2.2 Kind of Nanomaterial

Nanomaterials can cover a heterogeneous domain of materials using classification by species not entirely clear of controversy.

The European committee demonstrates the below-mentioned classification, on the basis of different Sustainable and Responsible Investment (SRI) deliberated records [27]:

- **Inorganic non-metallic nanomaterials:** titanium dioxide (TiO2), silicon dioxide (SiO2), zinc oxide (ZnO), strontium titanate (SrTiO3), strontium carbonate (SrCO3),

indium tin oxide (ITO), antimony tin oxide (ATO), disbismuth trioxide (Bi2O3), nickel monoxide (NiO) disilver oxide (Ag2O), aluminium nitride (AlN), silicon nitride (Si3O4), titanium nitride (TiN), titanium carbonitride (TiCN), tungsten carbide (WC), tungsten sulphide (WS2), etc.;
- **Metals and Metal alloys:** gold (Au), silver (Ag), platinum (Pt) and palladium (Pd) alloy, copper (Cu) nanopowders, iron (Fe) nanoparticles, nickel (Ni), cobalt (Co), aluminium (Al), zinc (Zn), manganese (Mn), molybdenum (Mo), tungsten (W), lanthanum (La), lithium (Li), rhodium (Rh), etc.
- **Carbon-based nanomaterials:** fullerenes (C60), carbon nanotubes (CNT, SWCNT, MWCNT), carbon nanofibers (CNF), carbon black and graphene flaks;
- **Nanopolymers and dendrimers:** polymer nanoparticles, polymer nanotubes, nanowires and nanorods, polyglycidylmethacrylate (PGMA) fibres, nanocellulose, nanostructured polymer-films, polyacrylonitrile, nanostructures (PAN), dendrimers;
- **Quantum dots**; are inorganic fluorescent semiconductor nanoparticles made of 10–50 atoms having a diameter domain of 2–10 nm [28].
- **Nanoshells** are nanoparticles made up of a gold Shell encircling a semiconductor [29].
- **Liposomes** are the easiest types of nanovectors that are composed of lipids enclosing the water core.

Nevertheless, there also exist novel kinds of nanomaterials in progress that are mostly mentioned as the second generation, third generation nanomaterials [27].

2.3 Volume Concentration of Nanofluid

Nanofluids technology can be considered as a novel challenge for the heat transfer fluid as the thermal conductivity of nanofluids unusually increases at an extremely less volume concentration [30].

The quantity of nanoparticles needed to prepare nanofluids is computed utilizing the law of mixture formula. The volume concentration for nanofluids is defined as below [31]

$$\Phi = \frac{V_{nanoparticle}/\rho_{nanoparticle}}{V_{nanoparticle}/\rho_{nanoparticle} + V_{basefluid}/\rho_{basefluid}} \tag{1}$$

such that $V_{nanoparticle}$ is considered as the volume of nano particle, $\rho_{nanoparticle}$ is considered as the density of nano particle, $V_{basefluid}$ is taken to be the volume of base fluid, and $\rho_{basefluid}$ is taken to be the density of base fluid.

2.4 Nano Fluid

A nanofluid is a fluid containing nanometer-sized particles, called nanoparticles. These fluids are engineered colloidal suspensions of nanoparticles in a base fluid. The nanoparticles used in nanofluids are typically made of metals, oxides, carbides, or carbon nanotubes.

Nanofluids have been constructed by utilizing two techniques: one step technique and two step technique. The principal aim of these techniques is to provide a homogeneous as well as stable solid-liquid mixture, also avoiding the agglomeration, feasible erosion and clogging etc. In one step technique, the nanoparticles have been constructed and dispersed into the foundation fluid at the same time. Whereas in a two-step technique, the nanoparticle has been generated in the first step and dispersed into host fluids in the second step. Investigation reveals that nanofluids with oxide nanoparticles and carbon nanotubes can be produced utilizing two-step technique [32]. Two-step technique is not appropriate for metallic nanoparticles. Under particular circumstances, nanofluids represent excellent thermo physical features that result in better effect of the thermal usage.

2.5 Nanofluids Viscosity Characterization

The density of nanofluids increases with an increment in concentration also reduces with an increment in temperature. The width of nanofluids for every volume concentration is calculated using Hygrometer. The acquired density information is in contradiction form also, the qualities are obtained using the thickness connection status developed by Pak and Choi in 1998 [33–35] for nanofluids.

The density data for water-propylene glycol mix like a function of temperature, which is extracted from ASHRAE (2009) Hand Book [36].The density of nanofluids is defined using a blending rule as below

$$\rho_{nanofluid} = (1 - \Phi)\rho_{basefluid} + \emptyset\rho_{basefluid} \tag{2}$$

Based on the outcomes demonstrated in Fig. 1, the density of nanofluids can forecast experimental data of density precisely.

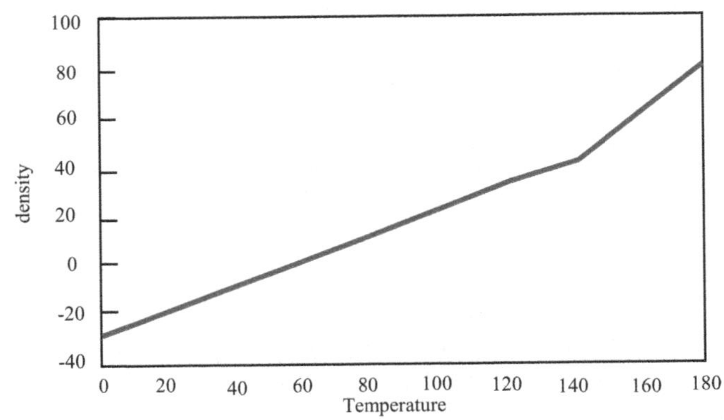

Fig. 1. Relation among temperature and density.

2.6 Nanofluid Particular Heat Rate Measurements

The nanofluid particular heat rate can be considered as one of the main characteristics and also has a major role in impacting heat transfer rate of nanofluids.

Particular heat is the rate of heat needed for increasing the temperature of one gram of nanofluids by one degree centigrade. The heat transfer rate is stated as below

$$Q = mc_p \Delta T \tag{3}$$

such that Q is taken to be the heat transfer rate, m is considered as the mass flow rate, also ΔT is taken to be the temperature variation of the cooling liquid.

The nanofluid particular heat changes with the nanoparticle concentration in the base fluid based on Eq. (3) which is generated by Buongiorno in [37].

$$\left(c_p\right)_{nanofluid} = \frac{(1-\varphi)\left(\rho c_p\right)_{fluid} + \varphi\left(\rho c_p\right)_{nanofluid}}{(1-\varphi)(\rho)_{fluid} + \varphi(\rho)_{nanofluid}} \tag{4}$$

such that c_p is taken to be the heat capacity of the nanofluid or base fluid.

3 Applications of Nanotechnology in Industry

3.1 Biomedical Applications

Nanoparticles have been utilized for disease recognition, drug transfer, gene therapy of cancer, and also to prevent other infections.

There exist various nanoparticle-based delivery systems accessible for distinguishing utilizations like Liposomes (Controlled and targeted drug delivery, targeted gene delivery); Dendrimers (Targeted drug delivery); Nanoshells (Tumor targeted); Nanowires, (fullerenes-based derivatives As targeting and imaging agents); Quantum dots (targeting and imaging agents) Gold nanoparticles (Targeted delivery and imaging agents); Carbon nanotube (Drug gene and DNA delivery, tumor targeting).

3.2 Nanotechnology in Agricultural and Food Production

The diversity of nanomaterials, mainly metal-based nanomaterials as well as carbon-based nanomaterials, are used because of their displacement, agglomeration as well as affect on formation and evolution of crop plants [38].

3.3 Nanotechnology in the Food Industry

Nanotechnology usages in foods proposed the technologies, which generate added values, which are greatest required, also contain the most effect on consumers.

Four main kinds of usages of nanoscience in the food industry, are agriculture, food dealing food wrapping as well as supplements.

Silver nanoparticles have been considered as an efficient concept opposed to vancomycin-resistant enterococcus as well as extended Extended-spectrum beta-lactamases [39].

4 Conclusion

Nanoscience has been broadly utilized in different areas like agricultural industry because of the specific properties of nanomaterials. In this paper, the properties of fluids with nanoparticles are reviewed. Furthermore, the developed models with nanoparticle consortium are studied. The attributes of the nanofluid can be estimated precisely applying the mixing rule. Viscosity are estimated wisely applying the Einstein model at low concentrations.

References

1. Razvarz S, Jafari R, Yu, W, Khalili A (2017) PSO and NN modeling for photocatalytic removal of pollution in wastewater, vol 20, pp 1–6
2. Jafari R, Yu W (2015) Artificial neural network approach for solving strongly degenerate parabolic and burgers-fisher equations. In: 12th international conference on electrical engineering, computing science and automatic control
3. Jafari R, Razvarz S, Gegov A (2018) A new computational method for solving fully fuzzy nonlinear systems. In: Computational collective intelligence, ICCCI 2018. LNCS, vol 11055. Springer, Cham, pp 503–512
4. Razvarz S, Jafari R (2017) ICA and ANN modeling for photocatalytic removal of pollution in wastewater. Math Comput Appl 22:38–48
5. Razvarz S, Jafari R, Gegov A, Yu W, Paul, S (2018) Neural network approach to solving fully fuzzy nonlinear systems. In: Fuzzy modeling and control methods application and research. Nova science Publisher Inc., NewYork, pp 45–68. ISBN 978-1-53613-415-5
6. Razvarz S, Jafari R (2017) Intelligent techniques for photocatalytic removal of pollution in wastewater. J Electr Eng 5:321–328
7. Jafari R, Razvarz S (2017) Solution of fuzzy differential equations using fuzzy Sumudu transforms. In: IEEE international conference on innovations in intelligent systems and applications, pp 84–89
8. Jafari R, Razvarz S, Gegov A, Paul S (2018) Fuzzy modeling for uncertain nonlinear systems using fuzzy equations and Z-numbers. In: Advances in computational intelligence systems: contributions presented at the 18th UK workshop on computational intelligence, advances in intelligent systems and computing, 5–7 September 2018, Nottingham, UK, vol 840. Springer, Cham, pp 66–107
9. Jafari R, Razvarz S (2018) Solution of fuzzy differential equations using fuzzy Sumudu transforms. Math Comput Appl 23:1–15
10. Jafari R, Razvarz S, Gegov A (2019) Solving differential equations with Z-numbers by utilizing fuzzy Sumudu transform. In: Arai K, Kapoor S, Bhatia R (eds) Intelligent systems and applications, IntelliSys 2018, vol 869. Advances in intelligent systems and computing. Springer, Cham, pp 1125–1138
11. Yu W, Jafari R (2019) Modeling and Control of Uncertain Nonlinear Systems with Fuzzy Equations and Z-Number. In: IEEE Press Series on Systems Science and Engineering. Wiley-IEEE Press, Wiley, Hoboken
12. Razvarz S, Jafari R (2018) Experimental study of Al_2O_3 nanofluids on the thermal efficiency of curved heat pipe at different tilt angle. J Nanomater 2018:1–7
13. Razvarz S, Vargas-Jarillo C, Jafari R (2019) Pipeline monitoring architecture based on observability and controllability analysis. In: IEEE international conference on mechatronics (ICM), vol 1, pp 420–423

14. Razvarz S, Vargas-Jarillo C, Jafari R, Gegov A (2019) Flow control of fluid in pipelines using PID controller. IEEE Access 7:25673–25680
15. Razvarz S, Jafari R (2016) Experimental study of Al2O3 nanofluids on the thermal efficiency of curved heat pipe at different tilt angle. In: 2nd international congress on technology engineering and science (ICONTES), Malaysia
16. Jafari R, Yu W, Li X (2016) Solving fuzzy differential equation with Bernstein neural networks. In: IEEE international conference on systems, man, and cybernetics, Budapest, Hungary, pp 1245–1250
17. Jafari R, Yu W (2017) Uncertain nonlinear system control with fuzzy differential equations and Z-numbers. In: 18th IEEE international conference on industrial technology, canada, vol 1, pp 890–895
18. Jafarian A, Measoomynia S, Jafari R (2012) Solving fuzzy equations using neural nets with a new learning algorithm. J Adv Comput Res 3:33–45
19. Jafari R, Razvarz S, Gegov, A (2019) Neural network approach to solving fuzzy nonlinear equations using Z-numbers. IEEE Trans Fuzzy Syst 1:1-12
20. Sekhon B (2014) Nanotechnology in agri-food production: an overview. Nanotechnol Sci Appl 7:31–35
21. Khan MY, Saroj BK, Roy M, Aziz I (2015) A review—emerging use of nano-based carriers in diagnosis and treatment of cancer novel approaches. Int J Adv Multidiscip Res 2(1):38–54
22. ISO/TR 27628:2007 (2007) Workplace atmospheres - Ultrafine, nanoparticle and nano-structured aerosols -Inhalation exposure characterization and assessment
23. Stone V, Nowack B, Baun A, van den Brink N, von der Kammer F (2010) Nanomaterials for environmental studies: classification, reference material issues, and strategies for physico-chemical characterization. Sci Total Environ 408:1745–1754
24. European Agency for Safety and Health at Work, E-fact, Nanomaterials in maintenance work (2013) occupational risks and prevention
25. Keshavarz Moraveji M, Razvarz S (2012) Experimental investigation of aluminum oxide nanofluid on heat pipe thermal performance. Int Commun Heat Mass Trans 39(12):1444–1448
26. ISO/TS 27687:2008(en) (2008) - Nanotechnologies — Terminology and definitions for nano-objects — Nanoparticle, nanofibre and nanoplate
27. European Commission (2012) Types and uses of nanomaterials, including safety aspects, Brussels
28. Santos CSC, Gabriel B, Blanchy M, Menes O, García D, Blanco M (2015) Industrial applications of nanoparticles – A prospective overview, materials today, proceedings, vol 2, pp 456 – 465
29. Park JW (2002) Liposome-based drug delivery in breast cancer treatment. Breast Cancer Res 4(3):95–99
30. Tavman T, Chirtoc T, Schuchmann HP (2008) Experimental investigation of viscosity and thermal conductivity of suspensions containing nano sized ceramic particles. Arch Mater Sci Eng 34:99–104
31. Pak BC, Cho YI (1998) Hydrodynamic and heat transfer study of dispersed fluids with submicron metallic oxide particles. Exp Heat Transf 11:151–170
32. Eastman JA, Choi SUS, Li S, Yu W, Thompson LJ (2001) Anomalously increased effective thermal conductivities of ethylene glycol-based nanofluids containing copper nanoparticles. Appl Phys Lett 78(6):718–720
33. Choi SUS, Wang X, Xu W (1999) Thermal conductivity of nano-particle-fluidmixture. J Thermophys Heat Transf 13(4):474–480
34. Choi SUS (1995) Enhancing thermal conductivity of fluids with nano-particles. ASME 231:99–105
35. Choi SUS (1998) Nanofluid technology: current status and future research. In: Energy, vol 26

36. American Society of Heating, Refrigerating and Air-Conditioning Engineers (2009): 2009 Ashrae Handbook: fundamentals. ASHRAE, Atlanta
37. Buongiorno J (2005) Convective transport in nanofluids. J Heat Trans 128:240–250
38. Nair R, Varghese SH, Nair BG, Maekawa T, Yoshida Y, Kumar DS (2010) Nanoparticulate material delivery to plants. Plant Sci 179(3):154–163
39. Phan TD, Debeaufort F, Voilley A (2008) Moisture barrier, wetting and mechanical properties of shellac/agar or shellac/cassava starch bilayer bio-membrane for food applications. J Membr Sci 325:277–283

Applications of Fuzzy Logic, Artificial Neural Network and Neuro-Fuzzy in Industrial Engineering

Raheleh Jafari[1(✉)], Miguel A. Contreras[2], Wen Yu[3], and Alexander Gegov[4]

[1] School of Design, University of Leeds, Leeds LS2 9JT, UK
r.jafari@leeds.ac.uk
[2] School of Engineering and Science, Tecnológico de Monterrey, 89600 Altamira, Tampico, Mexico
macontr@tec.mx
[3] Departamento de Control Automático, CINVESTAV-IPN (National Polytechnic Institute), 07360 Mexico City, Mexico
yuw@ctrl.cinvestav.mx
[4] School of Computing, University of Portsmouth, Buckingham Building, Portsmouth PO1 3HE, UK
alexander.gegov@port.ac.uk

Abstract. Artificial intelligence methods namely artificial neural network, fuzzy logic, and neuro-fuzzy have been effectively utilized in different applications like business, marketing, control engendering, health care, and social services. To demonstrate the usage of fuzzy set theory, artificial neural network, as well as neuro-fuzzy in industrial engineering and also for providing a basis for future investigation, a literature review of the artificial neural network, fuzzy logic, and neuro-fuzzy in industrial engineering is conducted in this paper.

Keywords: Fuzzy logic · Artificial neural network · Neuro-fuzzy · Industrial engineering

1 Introduction

Industrial engineering is one of the fields that artificial neural network, fuzzy logic, and neuro-fuzzy have found an extensive implementation area. Artificial neural networks can be considered as the most effective technique over the last few decades that are extensively utilized in a wide variety of implementations in different fields [1–6]. Artificial neural networks have been considered as efficient and versatile tools. They have learning ability as well as model-free characteristics.

Fuzzy systems are appropriate for estimated reasoning, mainly for the system with a mathematical design which is hard to obtain [7–11]. Fuzzy set theory can be considered as a major problem designing as well as a solution approach. The main contribution of fuzzy logic is its ability to present vague data. Fuzzy logic has been implemented

E. E. Hernandez et al. (Eds.): LASIRS 2019, MMS 86, pp. 9–14, 2020.
https://doi.org/10.1007/978-3-030-45402-9_2

to design systems, which are difficult to define accurately. In recent years, successful implementations of fuzzy logic in industrial engineering have been reported. Industrial engineers encounter with numerous problems having incomplete as well as uncertain information [12–15]. Fuzzy logic theory can be considered as a capable tool for solving these kinds of problems. Fuzzy logic presents an effective tool to facilitate investigation in industrial engineering when the dynamics of the decision environment restrict the accurate evaluation of model parameters.

Neuro-fuzzy has been used in an extensive range of domains [16–18]. It is the combination of artificial neural networks and fuzzy logic. The cause of utilizing artificial neural networks with fuzzy logic is that artificial neural networks never make a presumption on the probability distribution functions of data [19]. Generally, the neuro-fuzzy model explains solutions better than artificial neural networks [20].

This paper represents details of the application of artificial neural network, fuzzy logic, and neuro-fuzzy in industrial engineering. In this paper, the most recent researches in the field of artificial neural network, fuzzy logic, and neuro- fuzzy are covered. Since some industries have successfully used these techniques, detailed discussions are supplied to stimulate future investigations. This article remaining sections are organized into four Sections. In Sect. 2 the applications of fuzzy logic in industrial engineering are given. In Sect. 3 the applications of artificial neural network in industrial engineering are given. Section 4 presents the applications of neuro-fuzzy in industrial engineering. Conclusions are included in Sect. 5.

2 Applications of Fuzzy Logic in Industrial Engineering

Fuzzy logic systems have been effectively applied in different industrial fields like automobile speed control [21], robot arm control [22], water quality control [23], and automatic train operation systems [24]. Fuzzy logic can be utilized for improving the efficiency of the system. Even though fuzzy logic is considered as an approach for presenting inaccurate and vague information, in [25] the Universal Approximation Theorem states that the fuzzy logic system can uniformly approximate any nonlinear function to any degree of preciseness.

The fuzzy inference system utilizes fuzzy set theory for mapping inputs to outputs. The common fuzzy inference system is Mamdani and Sugeno type. Fuzzy inference system consists of a fuzzification interface, a rule database, a decision making unit as well as defuzzification interface, see Fig. 1.

Fuzzy logic systems are extremely effective in highly complicated and nonlinear processes, and also in the lack of any simple mathematical model. In [26], fuzzy set theory is utilized for generating a novel pulse discriminator in electric dis- charge machining procedure. In electric discharge machining procedure, cutting efficiency indexes like material elimination rate as well as surface harshness are in direct relation with the electric discharge machining discharge pulses.

In [27], the fuzzy logic is utilized for control of the fluid catalytic cracking unit. In that paper, the fuzzy logic control as a control method is efficiently applied for improved procedure control of fluid catalytic cracking in the refinery process industry.

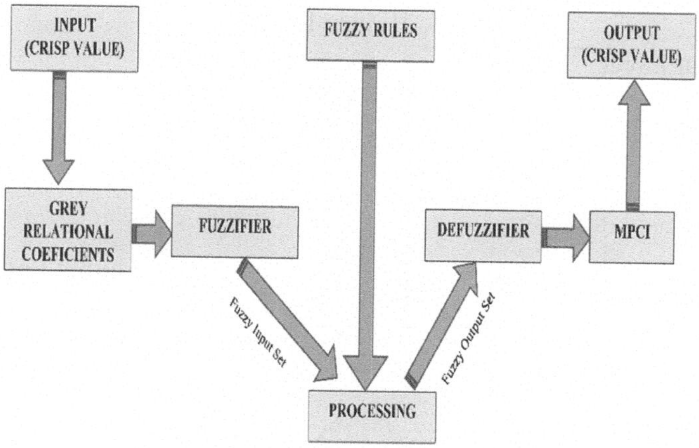

Fig. 1. The flowchart of a fuzzy inference system.

3 Applications of Artificial Neural Network in Industrial Engineering

In contrast with the fuzzy system, the neural network is desirable at predicting. This potential originates from the learning ability as well as model-free characteristics. The design of the artificial neural network is based on a group of interconnected artificial neurons along with linear or nonlinear transfer functions. Neurons are placed in various layers such as input layer, hidden layer, as well as output layer. The artificial neural network learns the relevance between input and output of the system employing an iterative procedure named training. For every input into the neuron, there is a weight. Weight is a tune able number that is defined while training the network. A simple model of the artificial neural network is shown in Fig. 2.

One of the applications of the artificial neural network is in condition monitoring. Artificial neural network provides a reliable procedure to condition monitoring. Condition monitoring can be considered as an effective device in maintenance planning, also it can be utilized for avoiding the unpredicted failures. In [28], artificial neural network technique is used for experimentally recognizing gears as well as bearings faults related to a typical gearbox system.

The artificial neural network has been used for evaluating vibrations as well as recognizing fault existence. Fault detection is considered as a crucial issue for gas turbine owners to move from preventative repair to predictive repair and hence to decrease the repair expenses. In [29], artificial neural network technique is used for fault detection of an industrial gas turbine.

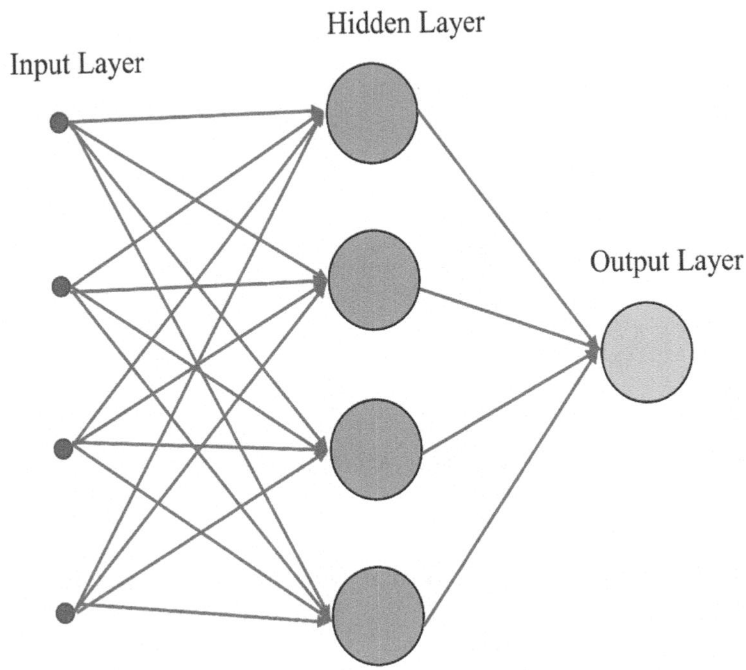

Fig. 2. A simple model of the artificial neural network.

4 Applications of Neuro-Fuzzy in Industrial Engineering

Artificial neural network and fuzzy logic are both considered as model-free numerical techniques. Each technique uses an uncomplicated algorithmic procedure instead of a complex mathematical analysis, and also the parameters are tune-able [30]. These resemblances make it possible to combine the two techniques.

Neuro-fuzzy system is a multi-layer feedforward adaptive network which figures out the fundamental elements as well as functions of a standard fuzzy logic system. As fuzzy logic systems are proved to be universal approximators, and since neuro-fuzzy systems are isomorphic to standard fuzzy logic systems concerning their functions hence neuro-fuzzy systems are likewise universal approximators [31]. For a special method in neuro-fuzzy, we can refer to the adaptive neuro-fuzzy inference system which is considered as one of the initial integrated hybrid neuro-fuzzy models [32].

One of the applications of the neuro-fuzzy system is in polymerization systems. Over the last few decades solubility of gases in polymers has been of interest to chemical engineers. In [33], a hybrid grid partitioning adaptive neuro-fuzzy inference system is utilized to predict carbon dioxide solubility in polymers.

The major aim of current manufacturing industries is to manufacture low cost, superb quality products in less time [34]. The choice of optimal cutting parameters is crucial in each machining procedure for increasing the quality of machined productions as well as to decrease the machining expenses. In [35], an adaptive neuro-fuzzy inference system is introduced to model and predict surface roughness in ball end milling of a die material. The algorithm developed in that paper presets the cutting parameters for a favorable level of surface roughness.

5 Conclusion

In this paper, the recent advances of the artificial neural network, fuzzy logic, and neuro-fuzzy applications in industrial engineering are provided. The artificial neural network, fuzzy logic, and neuro-fuzzy are the three major computational intelligence methods. Utilization of these models can be taken into account as a cheap, highly effective, and more reliable alternative devices. Therefore, these three methods can provide more ability to problem resolving than other techniques.

References

1. Razvarz S, Jafari R, Yu W, Khalili A (2017) PSO and NN modeling for photocatalytic removal of pollution in wastewater. In: 14th International Conference on Electrical Engineering, Computing Science and Automatic Control (CCE) Electrical Engineering, pp 1–6
2. Jafari R, Yu W (2015) Artificial neural network approach for solving strongly degenerate parabolic and burgers-fisher equations In: 12th International Conference on Electrical Engineering, Computing Science and Automatic Control. https://doi.org/10.1109/iceee.2015.7357914
3. Jafari R, Razvarz S, Gegov A (2018) A new computational method for solving fully fuzzy nonlinear systems. In: Computational Collective Intelligence, ICCCI 2018. Lecture Notes in Computer Science, vol 11055. Springer, Cham, pp 503–512
4. Razvarz S, Jafari R (2017) ICA and ANN modeling for photocatalytic removal of pollution in wastewater. Math Comput Appl 22:38–48
5. Razvarz S, Jafari R, Gegov A, Yu W, Paul S (2018) Neural network approach to solving fully fuzzy nonlinear systems. In: Fuzzy modeling and control Methods Application and Research, Nova science publisher Inc., New York, pp 45–68. ISBN: 978-1-53613- 415-5
6. Razvarz S, Jafari R (2017) Intelligent techniques for photocatalytic removal of pollution in wastewater. J Electr Eng 5:321–328. https://doi.org/10.17265/2328-2223/2017.06.004
7. Jafari R, Razvarz S (2017) solution of fuzzy differential equations using fuzzy Sumudu transforms. In: IEEE International Conference on Innovations in Intelligent Systems and Applications, pp 84–89
8. Jafari R, Razvarz S, Gegov A, Paul S (2018) Fuzzy modeling for uncertain nonlinear systems using fuzzy equations and Z-numbers. In: Advances in Computational Intelligence Systems: Contributions Presented at the 18th UK Workshop on Computational Intelligence, 5–7 September 2018, Nottingham, UK. Advances in Intelligent Systems and Computing, vol 840. Springer, Cham, pp 66-107
9. Jafari R, Razvarz S (2018) Solution of fuzzy differential equations using fuzzy sumudu transforms. Math Comput Appl 23(1):1–15
10. Jafari R, Razvarz S, Gegov A (2019) Solving differential equations with Z-Numbers by utilizing fuzzy Sumudu transform, intelligent systems and applications. IntelliSys 2018. Advances in Intelligent Systems and Computing, vol 869. Springer, Cham, pp 1125–1138
11. Yu W, Jafari R (2019) Fuzzy modeling and control with fuzzy equations and Z-number, IEEE Press Series on Systems Science and Engineering, Wiley-IEEE Press. ISBN-13: 978-1119491552
12. Razvarz S, Jafari R (2018) Experimental study of Al2O3 nanofluids on the thermal efficiency of curved heat pipe at different tilt angle. J Nanomater 2018:1–7
13. Razvarz S, Vargas-Jarillo C, Jafari R (2019) Pipeline monitoring architecture based on observability and controllability analysis. In: IEEE International Conference on Mechatronics (ICM), Ilmenau, Germany, vol 1, pp 420–423. https://doi.org/10.1109/icmech.2019.872287

14. Razvarz S, Vargas-jarillo C, Jafari R, Gegov A (2019) Flow control of fluid in pipelines using PID controller. IEEE Access 7:25673–25680. https://doi.org/10.1109/ACCESS.2019.2897992
15. Jafari R, Yu W, Razvarz S, Gegov A (2019) Numerical methods for solving fuzzy equations: a Survey. In: Fuzzy Sets and Systems. https://doi.org/10.1016/j.fss.2019.11.003. ISSN 0165–0114
16. Jafari R, Yu W, Li X (2016) Solving fuzzy differential equation with Bernstein neural networks. In: IEEE International Conference on Systems, Man, and Cybernetics, Budapest, Hungary, pp 1245–1250
17. Jafari R, Yu W (2017) Uncertain nonlinear system control with fuzzy differential equations and Z-numbers. In: 18th IEEE International Conference on Industrial Technology, Canada, pp 890–895. https://doi.org/10.1109/icit.2017.7915477
18. Jafarian A, Measoomy NS, Jafari R (2012) Solving fuzzy equations using neural nets with a new learning algorithm. J Adv Comput Res 3:33–45
19. Castellano G, Fanelli AM (2000) Variable selection using neural-network models. Neurocomputing 31:1–13
20. Babuska R, Verbruggen H (2003) Neuro-fuzzy methods for nonlinear system identification. Annu Rev Control 27:73–85
21. Murakami S, Maeda M (1985) Automobile speed control system using a fuzzy logic controller. In: Sugeno M (ed) Industrial Applications of Fuzzy Control. North-Holland, Amsterdam
22. Scharf EM, Mandix NJ (1985) The application of a fuzzy controller to the control of a multi-degree-of-freedom robot arm. In: Sugeno M (ed) Industrial Applications of Fuzzy Control. North-Holland, Amsterdam
23. Yagishita O, Itoh O, Sugeno M (1985) Application of fuzzy reasoning to the water purification process. In: Sugeno M (ed) Industrial Applications of Fuzzy Control. North-Holland, Amsterdam
24. Yasunobu S, Miyamoto S (1985) Automatic train operation system predictive fuzzy control. In: Sugeno M (ed) Industrial Applications of Fuzzy Control. North-Holland, Amsterdam
25. Wang LX (1994) Adaptive Fuzzy Systems and Control Design and Stability Analysis. Prentice Hall, NJ
26. Tarng YS, Tseng CM, Chung LK (1997) A fuzzy pulse discriminating system for electrical discharge machining. Int J Mach Tools Manuf 37:511–522
27. Taskin H, Cemalettin K, Uygun O, Arslankaya S (2006) Fuzzy logic control of a fluid catalytic cracking unit (FCCU) to improve dynamic performance. Comput Chem Eng 30:850–863
28. Rafiee J, Arvani F, Harifi A, Sadeghi MH (2006) Intelligent condition monitoring of a gearbox using artificial neural network. Mech Syst Signal Process 21(4):1746–1754
29. Arriagada J, Genrup M, Assadi M, Loberg A (2003) Fault Diagnosis System for an Industrial Gas Turbine by Means of Neural Networks. In: Proceedings of International Gas Turbine Congress, Tokyo, Japan
30. Bersini H, Nordvik J, Bonarini A (1993) A simple direct adaptive fuzzy controller derived from its neural equivalent. In: Proceedings 1993 IEEE International Conference on Fuzzy System, vol 1, pp 345–350
31. Wang LX, Mendel JM (1992) Fuzzy basis functions, universal approximation, and orthogonal least squares learning. IEEE Trans Neural Netw 3:807–814
32. Jang J (1993) ANFIS: adaptive network-based fuzzy inference systems. IEEE Trans Syst Man Cybern 23:665–685
33. Khajeh A, Modarress H, Rezaee B (2009) Application of adaptive neuro-fuzzy inference system for solubility prediction of carbon dioxide in polymers. Expert Syst Appl 36:5728–5732
34. Yazdi MRS, Khorram A (2010) Modeling and optimization of milling process by using RSM and ANN methods. Int J Eng Technol (IJET) IACSIT 2:474–480
35. Hossain MSJ, Ahmad N (2014) A neuro-fuzzy approach to select cutting parameters for commercial die manufacturing. Proc Eng 90:753–759

Foundation of Z-Numbers and Engineering Applications

Sina Razvarz[1(✉)], Felipe Hernández-Rodríguez[2], Raheleh Jafari[3], and Alexander Gegov[4]

[1] Departamento de Control Automático, CINVESTAV-IPN (National Polytechnic Institute), 07360 Mexico City, Mexico
srazvarz@yahoo.com

[2] School of Engineering and Science, Tecnológico de Monterrey, Prol. Juan de la Barrera 1241, 25270 Saltillo, Mexico
Felipe.hdz@tec.mx

[3] School of Design, University of Leeds, Leeds LS2 9JT, UK
r.jafari@leeds.ac.uk

[4] School of Computing, University of Portsmouth, Buckingham Building, Portsmouth PO1 3HE, UK
alexander.gegov@port.ac.uk

Abstract. Decision-making problems in economics, engineering, risk assessment, forecasting, manufacturing, design and process analysis are naturally determined by vagueness as well as partial reliability. Generating a formalism for dealing with Z-numbers is essential to handle these kinds of information. In this paper, the applications of Z-numbers in engineering are studied. The most current researches in the area of Z-numbers are covered in this work. Since some engineers have successfully applied Z-numbers, detailed discussions are supplied to stimulate future investigations.

Keywords: Z-number · Decision-making · Reliability · Engineering applications

1 Introduction

Mainly for the system with a mathematical design which is hard to obtain [1–13]. Fuzzy set theory is considered as an efficient technique to deal with uncertainty. It has been effectively implemented to resolve multi-criteria decision making problems. Nevertheless, the main issue in decision-making is the reliability of the information associated with the procedure.

Real-world information is often vague and also is normally represented in natural language. Moreover, this information is usually partially reliable and a degree of reliability is likewise expressed in natural language. To establish a basis to handle this vagueness and partial reliability, in [14] the idea of Z-number is developed. The Z-number is defined as an ordered pair $Z = (U, V)$ of fuzzy numbers utilized to explain a value of a random variable X, such that U is taken to be a fuzzy constraint on values of X, also V is taken

E. E. Hernandez et al. (Eds.): LASIRS 2019, MMS 86, pp. 15–24, 2020.
https://doi.org/10.1007/978-3-030-45402-9_3

to be a reliability which is a value of probability rate of U. Z-numbers are extensively utilized in a wide variety of implementations in different fields [15–22]. In [23] a novel analytic hierarchy process technique is proposed based on the Z-numbers for handling the linguistic decision-making problems. In [24] an arithmetic technique to discrete Z-numbers is suggested. In [25] an arithmetic technique to continuous Z-numbers is presented.

To demonstrate the usage of Z-numbers in engineering and also for providing a basis for future investigation, a literature review of the Z-numbers in engineering is conducted in this paper. This article remaining sections are organized into three Sections. Section 2 presents some basic definitions related to Z-numbers. In Sect. 3 the applications of Z-number in engineering are given. Conclusions are included in Sect. 4.

2 Mathematical Preliminaries

We consider the following definitions.

Definition 1. Suppose c is:

1) normal, there is $\varsigma_0 \in \Re$ where $c(\varsigma_0) = 1$,
2) convex, $c(\beta\varsigma + (1-\beta)\varsigma) \geq min\{c(\varsigma), c(\varrho)\}, \forall\varsigma, \varrho \in \Re, \forall\beta \in [0, 1]$,
3) upper semi-continuous on $\Re$, $c(\varsigma) \leq c(\varsigma_0) + \varepsilon, \forall\varsigma \in N(\varsigma_0), \forall\varsigma_0 \in \Re, \forall\varepsilon > 0$, $N(\varsigma_0)$ is a neighborhood,
4) $c^+ = \{\varsigma \in \Re, , c(\varsigma) > 0\}$ is compact, so c is a fuzzy variable, $c \in E : \Re \to [0, 1]$.

The fuzzy variable c is demonstrated as

$$c = \left(\underline{c}, \bar{c}\right) \tag{1}$$

in which c is the lower-bound variable also, c is the upper-bound variable.

Definition 2. The Z-number is made up of two components $Z = [c(\varsigma), p]$. The first component $c(\varsigma)$ is the restriction on a real-valued uncertain variable ς. The second component p is a measure of the reliability of c. p can be reliability, strength of belief, probability or possibility. The Z-number can be stated as Z^+-number, in a case $c(\varsigma)$ be a fuzzy number also p be the probability distribution of ς. If $c(\varsigma)$, as well as p, are fuzzy numbers, the Z-number can be stated as Z^--number.

The Z^+-number contains more information when compared with the Z^--number. In this paper, the definition of Z^+-number is utilized, i.e., $Z = [c, p]$, c is a fuzzy number also, p is a probability distribution.

The most common membership functions which define the fuzzy numbers are the triangular function

$$\mu_c = H(s, u, v) = \begin{cases} \frac{\varsigma - s}{u - s} s \leq \varsigma \leq u \\ \frac{v - \varsigma}{v - u} u \leq \varsigma \leq v \end{cases} \quad otherwise\, \mu_c = 0 \tag{2}$$

and trapezoidal function

$$\mu_c = H(s,u,v,w) = \begin{cases} \frac{\varsigma - s}{u - s} s \leq \varsigma \leq u \\ \frac{w-\varsigma}{w-v} v \leq \varsigma \leq w \ \ otherwise\, \mu_c = 0 \\ 1\ u \leq \varsigma \leq v \end{cases} \quad (3)$$

If c represents a fuzzy event in $\Re$, the real line, then the probability measure can be stated as

$$P(c) = \int_{\Re} \mu_c(\varsigma)p(\varsigma)d\varsigma \quad (4)$$

where p is the probability density of ς. For discrete Z-numbers we have

$$P(c) = \sum_{l=1}^{n} \mu_c(\varsigma_l)p(\varsigma_l) \quad (5)$$

Definition 4. The α-level of the Z-number $Z = (c, p)$ is defined as follows

$$[Z]^\alpha = ([c]^\alpha, [p]^\alpha) \quad (6)$$

where $0 < \alpha \leq 1$. $[p]^\alpha$ is computed by the Nguyen's theorem

$$[p]^\alpha = p([c]^\alpha) = p([\underline{c}^\alpha, \bar{c}^\alpha]) = [\underline{P}^\alpha, \bar{P}^\alpha] \quad (7)$$

where $p([c]^\alpha) = \{p(\varsigma)|\varsigma \in [c]^\alpha\}$. Therefore, $[Z]^\alpha$ is stated as

$$[Z]^\alpha = (\underline{Z}^\alpha, \bar{Z}^\alpha) = ((\underline{c}^\alpha, \underline{P}^\alpha), (\bar{c}^\alpha, \bar{P}^\alpha)) \quad (8)$$

where $\underline{P}^\alpha = \underline{c}^\alpha p(\underline{\varsigma}^\alpha)$, $\bar{P}^\alpha = \bar{c}^\alpha p(\bar{\varsigma}^\alpha)$, $[\varsigma_l]^\alpha = (\underline{\varsigma}^\alpha, \overline{\varsigma_l}^\alpha)$.

The same as with the fuzzy numbers [24], three main operations are defined for the Z-numbers: $\oplus$, $\ominus$ and $\odot$, which indicate sum, subtract and multiply respectively.

Suppose $Z_1 = (c_1, p_1)$ as well as $Z_2 = (c_2, p_2)$ are two discrete Z-numbers expressing the uncertain variables ς_1 and ς_2, so, $\sum_{l=1}^{n} p_1(\varsigma_{1l}) = 1$, $\sum_{l=1}^{n} p_2(\varsigma_{2l}) = 1$.

The following operation is defined

$$Z_{12} = Z_1 * Z_2 = (c_1 * c_2, p_1 * p_2) \quad (9)$$

Where $* \in \{\oplus, \ominus, \odot\}$.

The operations used for the fuzzy numbers $[c_1]^\alpha = [c_{11}^\alpha, c_{12}^\alpha]$ and $[c_2]^\alpha = [c_{21}^\alpha, c_{22}^\alpha]$ are stated as [26],

$$[c_1 \oplus c_2]^\alpha = [c_1]^\alpha + [c_2]^\alpha = [c_{11}^\alpha + c_{21}^\alpha, c_{12}^\alpha + c_{22}^\alpha]$$
$$[c_1 \ominus c_2]^\alpha = [c_1]^\alpha - [c_2]^\alpha = [c_{11}^\alpha - c_{21}^\alpha, c_{12}^\alpha - c_{22}^\alpha]$$

$$[c_1 \odot c_2]^\alpha = \begin{pmatrix} \min\{c_{11}^\alpha c_{21}^\alpha, c_{11}^\alpha c_{22}^\alpha, c_{12}^\alpha c_{21}^\alpha, c_{12}^\alpha c_{22}^\alpha\} \\ \max\{c_{11}^\alpha c_{21}^\alpha, c_{11}^\alpha c_{22}^\alpha, c_{12}^\alpha c_{21}^\alpha, c_{12}^\alpha c_{22}^\alpha\} \end{pmatrix} \tag{10}$$

For the discrete probability distributions, the following relation is defined for all $p_1 * p_2$ operations

$$p_1 * p_2 = \sum_{l} p_1(\varsigma_1, \iota) p_2(\varsigma_{2,(n-\iota)}) = p_{12}(\varsigma) \tag{11}$$

The Hukuhara difference is defined as [17],

$$\begin{aligned} Z_1 \ominus_H Z_2 &= Z_{12} \\ Z_1 &= Z_2 \oplus Z_{12} \end{aligned} \tag{12}$$

In a case that $Z_1 \ominus_H Z_2$ exists, the α-level can be defined as

$$[Z_1 \ominus_H Z_2]^\alpha = [\underline{Z}_1^\alpha - \underline{Z}_2^\alpha, \overline{Z}_1^\alpha - \overline{Z}_2^\alpha] \tag{13}$$

α Clearly, $Z_1 \ominus_H Z_1 = 0$, $Z_1 \ominus Z_1 \neq 0$.

Moreover, the generalized Hukuhara difference is defined as [22],

$$Z_1 \ominus_{gH} Z_2 = Z_{12} \Leftrightarrow \begin{cases} 1)\ Z_1 = Z_2 \oplus Z_{12} \\ 2)\ Z_2 = Z_1 \oplus (-1) Z_{12} \end{cases} \tag{14}$$

By taking into consideration the α-level, we have $[Z_1 \ominus_H Z_2]^\alpha = [\min\{\underline{Z}_1^\alpha - \underline{Z}_2^\alpha, \overline{Z}_1^\alpha - \overline{Z}_2^\alpha\}, \max\{\underline{Z}_1^\alpha - \underline{Z}_2^\alpha, \overline{Z}_1^\alpha - \overline{Z}_2^\alpha\}]$ and If also, if $Z_1 \ominus_H Z_2$ exists, $Z_1 \ominus_H Z_2 = Z_1 \ominus_{gH} Z_2$. The conditions for the existence of $Z_{12} = Z_1 \ominus_{gH} Z_2 \in E$ are

$$\begin{aligned} &1)\begin{cases} \underline{Z}_{12}^\alpha = \underline{Z}_1^\alpha - \underline{Z}_2^\alpha \text{ and } \overline{Z}_{12}^\alpha = \overline{Z}_1^\alpha - \overline{Z}_2^\alpha \\ \text{with } \underline{Z}_{12}^\alpha \text{ increasing, } \overline{Z}_{12}^\alpha \text{ decreasing, } \underline{Z}_{12}^\alpha \le \overline{Z}_{12}^\alpha \end{cases} \\ &2)\begin{cases} \underline{Z}_{12}^\alpha = \overline{Z}_1^\alpha - \overline{Z}_2^\alpha \text{ and } \overline{Z}_{12}^\alpha = \underline{Z}_1^\alpha - \underline{Z}_2^\alpha \\ \text{with } \underline{Z}_{12}^\alpha \text{ increasing, } \overline{Z}_{12}^\alpha \text{ decreasing, } \underline{Z}_{12}^\alpha \le \overline{Z}_{12}^\alpha \end{cases} \end{aligned} \tag{15}$$

where $\forall \alpha \in [0, 1]$.

Suppose c is a triangular function, the absolute value of the Z-number $Z = (c, p)$ is defined as

$$|Z(\varsigma)| = (|s_1| + |u_1| + |v_1|, p(|s_2| + |u_2| + |v_2|)) \tag{16}$$

Let c_1 as well as c_2 are triangular functions, the supremum metric for Z-numbers $Z_1 = (c_1, p_1)$ and $Z_2 = (c_2, p_2)$ is expressed as

$$D(Z_1, Z_2) = d(c_1, c_2) + d(p_1, p_2) \tag{17}$$

where $d(\cdot, \cdot)$ is the supremum metrics for fuzzy sets [22]. $D(Z_1, Z_2)$ has the below mentioned properties,

$$\begin{aligned} &D(Z_1 + Z, Z_2 + Z) = D(Z_1, Z_2) \\ &D(Z_2, Z_1) = D(Z_1, Z_2) \\ &D(bZ_1, kZ_2) = |b|\, D(Z_1, Z_2) \\ &D(Z_1, Z_2) \le D(Z_1, Z) + D(Z, Z_2) \end{aligned} \tag{18}$$

where $b \in \Re$, $Z = (c, p)$ is Z-number, also c is triangle function.

Definition 5. Suppose $\widehat{Z}$ is the space of Z-numbers. The $\alpha-$level of the Z-number valued function $H : [0, s] \rightarrow \widehat{Z}$ is defined as

$$H(c, \alpha) = [\underline{H}(c, \alpha), \overline{H}(c, \alpha)] \tag{19}$$

where $c \in \widehat{Z}$, for each $\alpha \in [0, 1]$.

Using the definition of Generalized Hukuhara difference, the gH-derivative of H at c_0 is defined as

$$\frac{d}{dt}H(c_0) = \lim_{\zeta \to 0} \frac{1}{\zeta}[H(c_0 + \zeta) \ominus_{gH} H(c_0)] \tag{20}$$

In (27), $H(c_0 + \zeta)$ as well as $H(c_0)$ represent symmetric pattern with Z_1 and Z_2 respectively given in (19).

In this book chapter, the aim of the modeling is to find the Z-number coefficient a_ι for the below fuzzy equation

$$w_r^* = \sum_{\iota=1}^{\gamma} a_\iota f_\iota(\vartheta_r) \tag{21}$$

in such a way that the output of the plant w_r approaches to the desired output w_r^*.

Definition 6. The distance between two Z-numbers, $\psi, \phi \in \widehat{Z}$, is described as the Hausdorff metric $D_H[\psi, \phi]$,

$$\begin{aligned} D_H[\psi, \phi] = \max\{&\sup_{(a_1, c_1)\in\psi} \inf_{(a_2, c_2)\in\phi}(d(a_1, a_2) \\ &+ d(c_1, c_2)), \sup_{(a_1, c_1)\in\phi} \inf_{(a_2, c_2)\in\psi}(d(a_1, a_2) + d(c_1, c_2)\} \end{aligned} \tag{22}$$

$d(a, c)$ is the supremum metrics defined for fuzzy sets.

Lemma 1. Let $\varpi \subset \widehat{Z}$ is a compact set, in this case, ϖ is uniformly support-bounded, i.e. there exists a compact set $\Psi \subset \Re$, in such a way that $\forall\psi \in \varpi$,

$$\text{Supp}(\psi) \subset \Psi. \tag{23}$$

Lemma 2. Suppose $\psi, \phi \in \widehat{Z}$, also $\alpha \in (0, 1]$, in this case the following is concluded:

(i) if $f, g : \Re \rightarrow \Re$ are continuous, $[f(\psi)]^\alpha = f([\psi^\alpha])$ holds;
(ii) if $f : \Re \rightarrow \Re$ are continuous, so $f(\text{Supp}(\psi)) = \text{Supp}(f(\psi))$.

Proof. As (i) concludes from (Yang and Gao 1993), so just (ii) will be proved. Initially, $\overline{f(D_1)} = f(\overline{D}_1)$ is shown for $D_1 \subset \Re$. As $f(D_1) \subset f(\overline{D}_1)$, also $f(\overline{D}_1)$ is closed by the continuity of f, therefore, $\overline{f(D_1)} \subset f(\overline{D}_1)$. Also, for randomly given $\theta \in f(\overline{D}_1)$, there exists a sequence $\{a_n | n \in N\} \subset \Re$, $a \in \Re$, in such a way that $a_n \rightarrow a$ $(n \rightarrow +\infty)$, $\theta = f(a)$. Since f is continues so, $\lim_{n\to+\infty} f(a_n) = f(a) = \theta$.

However, $f(a_n) \in f(D_1)$, therefore, $\theta \in \overline{f(D_1)}$. So $f(\overline{D}_1) \subset \overline{f(D_1)}$. Hence $\overline{f(D_1)} = f(\overline{D}_1)$.

Since,

$$\begin{aligned} \mathrm{Supp}(f(\psi)) &= \overline{\{\theta \in \Re \mid (f(\psi))(\theta) > 0\}} \\ f(\mathrm{Supp}(\psi)) &= f\left(\overline{\{a \in \Re \mid \psi(a) > 0\}}\right) \end{aligned} \tag{24}$$

the following is concluded

$$\begin{aligned} f(\mathrm{Supp}(\psi)) &= \overline{f(\{a \in \Re \mid \psi(a) > 0\})} \\ &= \overline{\{f(a) \in \Re \mid \psi(a) > 0\}} \end{aligned} \tag{25}$$

As, $\{\theta \in \Re \mid (f(\psi))(\theta) > 0\} = \{f(a) \mid \psi(a) > 0\}$, so

$$\mathrm{Supp}(f(\psi)) = f(\mathrm{Supp}(\psi)) \tag{26}$$

Lemma 3. Suppose $G \subset \Re$ is a compact set, also f_1 as well as f_2 are continuous on G, $\zeta > 0$, in addition

$$|f_1(a) - f_2(a)| < \zeta, \quad \forall a \in G \tag{27}$$

Hence $|\sup_{a \in G_1} f_1(a) - \sup_{a \in G_1} f_2(a)| < \zeta$ is valid for each compact set $G_1 \subset G$.

Theorem 1. Suppose $f : \Re \rightarrow \Re$ is a continuous function, so for each compact set $\varpi \subset \widehat{Z}_0$ (the set of all the bounded Z-number set) also $\chi > 0$, there exists $a_\iota \in \widehat{Z}_0$, $\iota = 1, 2, \ldots, \gamma$, in such a manner that

$$d(f(\tilde{\vartheta}), \textstyle\sum_{\iota=1}^{\gamma} f_\iota(\vartheta) a_\iota) < \chi \quad \forall \vartheta \in \varpi, \quad \forall \tilde{\vartheta} \in \Re \tag{28}$$

in which χ is a finite number.

Theorem 2. If $f : \Re \rightarrow \Re$ is a continuous function, so for each compact set $\varpi \subset \widehat{Z}_0$, $\tau > 0$ also arbitrary $\varepsilon > 0$, there exists $a_\iota \in \widehat{Z}_0$, $\iota = 1, 2, \ldots, \gamma$, in such a way that

$$d(f(\vartheta), \sum_{\iota=1}^{\gamma} f_\iota(\vartheta) a_\iota) < \tau, \quad \forall \vartheta \in \varpi \tag{29}$$

in which, τ is termed as a finite number. The lower as well as the upper limits of the α-level set of Z-number function approach to τ, however the center approaches to ε.

Theorem 3. Suppose $\varpi \subset \widehat{Z}_0$ is compact, also Ψ is the corresponding compact set of ϖ and f, $\hat{f} : \Re \rightarrow \Re$ are the continuous functions that,

$$|f(\vartheta) - \hat{f}(\vartheta)| < \zeta, \quad \forall \vartheta \in \Psi \tag{30}$$

where $\zeta > 0$. So, $\forall \psi_1 \in \varpi$, $d(f(\psi_1) - \hat{f}(\psi_1)) \leq \zeta$.

3 Applications of Z-Number in Engineering

Multi-sensor data fusion, a functional simulation of the process related to the decision-making executed by the human brain, has obtained a high growth level over the past few years across industries. In [27] a new technique with the combination of the Z-number and Dempster Shafer evidence theory is suggested for modeling as well as processing the uncertainties in a sensor data fusion system. In that technique, the Z-number is utilized for modeling the fuzziness as well as the reliability of the sensor data also the Dempster-Shafer evidence theory is utilized for fusing the uncertain information of Z-numbers. The major benefits of the suggested technique in [27] are that it produces a perfect robust degree of reliability to the sensor data, also the supplementary information of multi-sensors decreases the uncertainty of the fault detection, hence increasing the reliability of fault identification. The process of fault detection with the technique proposed in [27] is shown in Fig. 1.

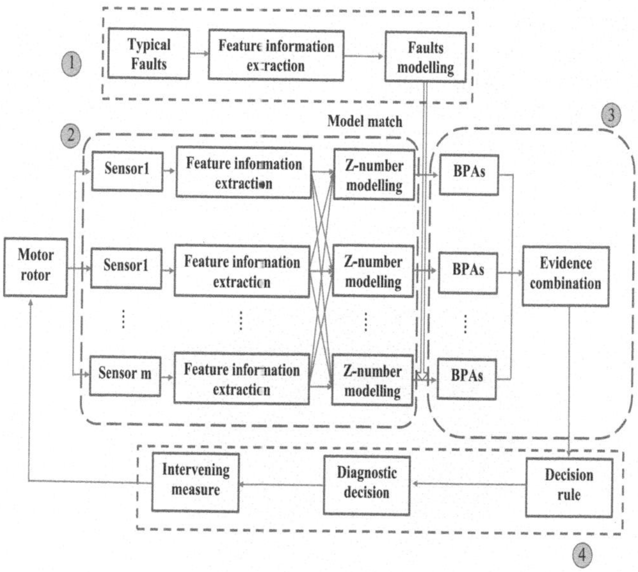

Fig. 1. The process of fault detection.

In [28] Bernstein neural network is used for finding the Z-number solutions of fuzzy differential equations. In [28] a generalized model of a tank system is proposed which is in the form of the fuzzy differential equation, see Fig. 2. The tank model is described as

$$\frac{d}{dt}y(t) = -\frac{1}{CJ}y(t) + \frac{k}{C} \tag{31}$$

where K is inflow disturbances of the tank that generates vibration in liquid level y, J is the flow obstruction which can be curbed utilizing the valve, also C is the cross-section of the tank.

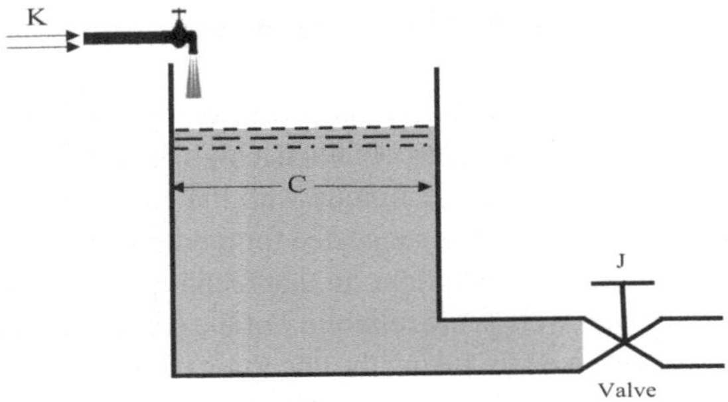

Fig. 2. Liquid tank system.

The project management has become a significant approach nearly in all fields of human activity. Every project is temporary by nature and also it has the beginning and finish dates. In [29], the capabilities of the Z-numbers in the improvement of the state of risk evaluation is studied.

In [30], fuzzy equations with Z-numbers are used to design the fuzzy controller. In [30], a heat source by insulating materials is proposed which is in the form of the fuzzy equation, see Fig. 3. The insulating materials center is considered to be the source of heat. The model is described as

$$\frac{A}{K_1} \oplus \frac{B}{K_2} = \frac{C}{K_3} \oplus \frac{D}{K_4} \oplus R \tag{32}$$

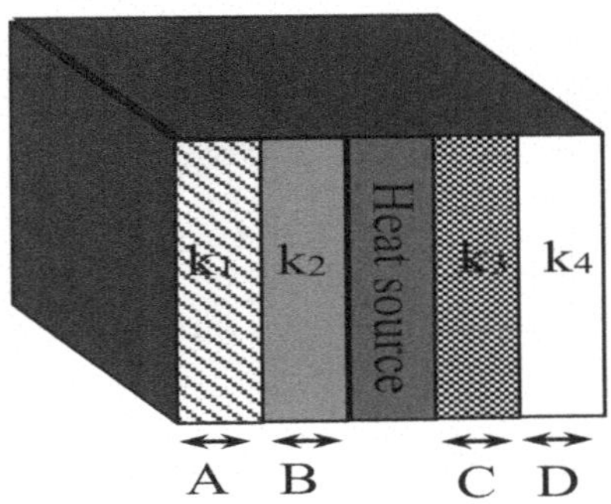

Fig. 3. Heat source by insulating materials.

The material widths are not precise and they are in the form of Z-numbers.

4 Conclusion

Engineers tend to deal with various types of measures as well as evaluations. However, all evaluation of information may not be presented as a clear number.

Z-number has considerable ability in explaining the vagueness of human knowledge. In this paper, the applications of Z-numbers in engineering are studied. The most current researches in the area of Z-numbers are covered in this work. Since some engineers have successfully applied Z-numbers, detailed discussions are supplied to stimulate future investigations.

References

1. Jafari R, Razvarz S, Gegov A (2018) A new computational method for solving fully fuzzy nonlinear systems. In: Computational collective intelligence, ICCCI 2018. Lecture notes in computer science, vol 11055. Springer, Cham, pp 503–512
2. Razvarz S, Jafari R, Gegov A, Yu W, Paul S (2018) Neural network approach to solving fully fuzzy nonlinear systems. In: Fuzzy modeling and control methods application and research. Nova Science Publisher, Inc., New York, pp 45–68. ISBN 978-1-53613-415-5
3. Jafari R, Razvarz S (2017) Solution of fuzzy differential equations using fuzzy Sumudu transforms. In: IEEE international conference on innovations in intelligent systems and applications, pp 84–89
4. Jafari R, Razvarz S (2018) Solution of fuzzy differential equations using fuzzy Sumudu transforms. Math Comput Appl 1–15
5. Negoita CV, Ralescu DA (1975) Applications of fuzzy sets to systems analysis. Wiley, New York
6. Zadeh LA (1968) Probability measures of fuzzy events. J Math Anal Appl 23:421–427
7. Zadeh LA (1975) Calculus of fuzzy restrictions. In: Zadeh LA, Fu KS, Tanaka K, Shimura M (eds) Fuzzy sets and their applications to cognitive and decision processes. Academic Press, New York, pp 1–39
8. Zadeh LA (1996) Fuzzy logic and the calculi of fuzzy rules and fuzzy graphs. Multi Val Logic 1:1–38
9. Jafari R, Yu W, Li X (2016) Solving fuzzy differential equation with Bernstein neural networks. In: IEEE international conference on systems, man, and cybernetics, Budapest, Hungary, pp 1245–1250
10. Jafari R, Yu W (2017) Uncertain nonlinear system control with fuzzy differential equations and Z-numbers. In: 18th IEEE international conference on industrial technology, Canada, pp 890–895. https://doi.org/10.1109/icit.2017.7915477
11. Jafarian A, Measoomy Nia S, Jafari R (2012) Solving fuzzy equations using neural nets with a new learning algorithm. J Adv Comput Res 3:33–45
12. Jafarian A, Jafari R (2019) A new computational method for solving fully fuzzy nonlinear matrix equations. Int J Fuzzy Comput Model 2:275–285
13. Jafari, R., Razvarz, S., Gegov, A (2018) A novel technique to solve fully fuzzy nonlinear matrix equations. In: 13th international conference on applications of fuzzy systems and soft computing, Warsaw, Poland. Advances in intelligent systems and computing. Springer
14. Zadeh LA (2011) A note on Z-numbers. Inform Sci 181:2923–2932
15. Jafari R, Razvarz S, Gegov A, Paul S (2018) Fuzzy modeling for uncertain nonlinear systems using fuzzy equations and Z-numbers. In: Advances in computational intelligence systems: contributions presented at the 18th UK workshop on computational intelligence, Nottingham, UK, 5–7 September 2018. Advances in intelligent systems and computing, vol 840. Springer, pp 66–107
16. Jafari R, Razvarz S, Gegov A (2019) Solving differential equations with Z-numbers by utilizing fuzzy Sumudu transform. In: Intelligent systems and applications, IntelliSys 2018. Advances in Intelligent Systems and Computing, vol 869. Springer, Cham, pp 1125–1138
17. Yu W, Jafari R (2019) Fuzzy modeling and control with fuzzy equations and Z-number. IEEE Press series on systems science and engineering. Wiley-IEEE Press, Piscataway. ISBN-13: 978–1119491552

18. Jiang W, Xie C, Luo Y, Tang Y (2017) Ranking Z-numbers with an improved ranking method for generalized fuzzy numbers. J Intell Fuzzy Syst 32:1931–1943
19. Yaakob AM, Gegov A (2015) Fuzzy rule based approach with Z-numbers for selection of alternatives using TOPSIS. In: Proceedings of the IEEE international conference on fuzzy systems, pp 1–8
20. Zamri N, Ahmad F, Rose ANM, Makhtar M (2016) A fuzzy TOPSIS with Z-numbers approach for evaluation on accident at the construction site. In: Proceedings of the international conference on soft computing and data mining, pp 41–50
21. Razvarz S, Jafari R, Yu W (2018) Numerical solution of fuzzy differential equations with Z-numbers using Fuzzy Sumudu transforms. Adv Sci Technol Eng Syst J (ASTESJ) 3:66–75. https://doi.org/10.25046/aj030108
22. Jafari R, Razvarz S, Gegov A, Paul S, Keshtkar S (2018) Fuzzy Sumudu transform approach to solving fuzzy differential equations with Z-numbers. In: Advanced fuzzy logic approaches in engineering science. IGI Global. https://doi.org/10.4018/978-1-5225-5709-8.ch002
23. Azadeh A, Saberi M, Atashbar NZ, Chang E, Pazhoheshfar P (2013) Z-AHP: a Z-number extension of fuzzy analytical hierarchy process. In: Proceedings of the 7th IEEE international conference on digital ecosystems and technologies, pp 141–147. https://doi.org/10.1109/DEST.2013.6611344
24. Aliev RA, Alizadeh AV, Huseynov OH (2015) The arithmetic of discrete Z-numbers. Inform Sci 290:134–155. https://doi.org/10.1016/j.ins.2014.08.024
25. Aliev RA, Huseynov OH, Zeinalova LM (2016) The arithmetic of continuous Z-numbers. Inform Sci 373:441–460. https://doi.org/10.1016/j.ins.2016.08.078
26. De Barros LC, Bassanezi RC, Lodwick WA (2017) The extension principle of Zadeh and fuzzy numbers. In: A first course in fuzzy logic, fuzzy dynamical systems, and biomathematics. Studies in fuzziness and soft computing, vol 347. Springer, Heidelberg, pp 23–41
27. Jiang W, Xie Ch, Zhuang M, Shou Y, Tang Y (2016) Sensor data fusion with Z-numbers and its application in fault diagnosis. Sensors 16:1509. https://doi.org/10.3390/s16091509
28. Jafari R, Yu W, Li X, Razvarz S (2017) Numerical solution of fuzzy differential equations with Z-numbers using Bernstein neural networks. Int J Comput Intell Syst 10:1226–1237
29. Nuriyev AM (2019) application of Z-numbers based approach to project risks assessment. Eur J Interdiscip Stud 5:67–73
30. Jafari R, Yu W, Li X (2017) Numerical solution of fuzzy equations with Z-numbers using neural networks. Intell Autom Soft Comput 1–7

Force Balancing of the 2RRR Planar Parallel Manipulator via Center of Mass Acceleration Control Using Fully Cartesian Coordinates

Mario Acevedo[1(✉)] and María T. Orvañanos-Guerrero[2]

[1] Facultad de Ingeniería, Universidad Panamericana,
Álvaro del Portillo 49, 45010 Zapopan, Jalisco, Mexico
mario.acevedo@up.edu.mx

[2] Facultad de Ingeniería, Universidad Panamericana,
Josemaría Escrivá de Balaguer 101, 20290 Aguascalientes, Aguascalientes, Mexico

Abstract. Shaking force balancing of mechanisms usually is achieved by an optimal redistribution of the moving masses, which allows the cancellation or the reduction of the variable dynamic loads on the manipulator's frame. This procedure usually leads to an increase in the mass of the original links, motivating the increment on the driving torques and on the shaking moment. In this article the concept of center of mass acceleration control, presented previously elsewhere, is resumed this time with the use of fully Cartesian coordinates. The trajectory of the general center of mass of the system, calculated out from the motion of the moving links, is defined as straight line and characterized with a "bang-bang" profile. In this way it is possible to reduce the maximum acceleration of the center of mass thus, reducing the shaking force. The suggested balancing technique is illustrated through computer simulations.

Keywords: Dynamic balancing · Shaking force · Parallel Manipulator · Fully Cartesian coordinates · Center of mass control

1 Introduction

Robotic industrial manipulators deal with the problem of vibrations transmitted to the fixed frame during high-speed motion. These vibrations are due to different factors, one of them is the unbalanced inertia forces which finally produce the increment of the shaking force and shaking moment. Thus, a primary objective in the dynamic balancing of manipulators is to eliminate or reduce the variable dynamic loads transmitted to the base and surrounding structures.

A traditional method used for the dynamic balancing of manipulators has been the mass redistribution. In this case the objective is to maintain the total center of mass of the moving links stationary, see [2,5,8]. In practice this is done adding a set of counterweights and counter-inertias increasing considerably the

E. E. Hernández et al. (Eds.): LASIRS 2019, MMS 86, pp. 25–35, 2020.
https://doi.org/10.1007/978-3-030-45402-9_4

mass. This causes an increment in the overall size of the robot-manipulator and in the efforts at the joints.

Another alternative also traditionally explored is the use of auxiliary structures. In the same way the aim is to keep the total center of mass of the mechanism static. The structures proposed and generally used are the parallelogram and the pantograph, [6,7,12]. Although this solution leads to relatively complex mechanical system increasing manipulator's sizes.

Recently a new approach has been proposed. It is based on the optimal control of the robot-manipulator center of mass, [4,11]. The method is based on planning the displacements of the total center of mass of the moving links. Its trajectory is defined as a straight line with "bang-bang" motion profile. This approach allows the reduction of the maximum value of the center of mass acceleration. Thus leads to a reduction in the shaking force. Using this method it is possible to ensure the initial and final positions of the end-effector. In this work this approach is presented using fully Cartesian coordinates, also known as Natural Coordinates, [10].

An advantage using fully Cartesian coordinates is that the expressions necessary to calculate the general center of mass of the complete system, the shaking force, and the shaking moment, can be calculated in a direct way. In the authors opinion this contributes to make the method based on the optimal control of the robot-manipulator center of mass easy to be programmed and fastest. A complete explanation of their application to find the balancing conditions of a 2RRR planar parallel manipulator (five-bar mechanism) along with different practical solution alternatives have been presented in detail in [1].

In this article fully Cartesian coordinates are used to obtain the expressions to calculate the global center of mass and the shaking force of a 2RRR Planar Parallel Manipulator (2RRR PPM). These expressions are used for its force balancing applying the method based on the optimal control of the robot-manipulator center of mass. Comparative results from computer simulations are presented.

2 Modeling of a 2RRR-PPM Using Fully Cartesian Coordinates

In Fig. 1 a generic 2RRR-PPM (five-bar mechanism) is presented. All links considered rigid. Kinematic revolute joints are identified with points A through E, and points A and E are fixed to the frame. Link lengths are indicated as l_i, where $i = 1, 2, 3, 4$. In this case the centers of mass of the links can be identified with the letter g_i, respectively.

The idea of modeling using fully Cartesian coordinates is to distribute the mass, ideally concentrated in the center of masses, to the "basic points". In this case this is to distribute the masses to the points A through E, the joints.

Thus the system can be seen as a cloud of points, with their corresponding equivalent mass, moving. Thanks to this change, it is possible to work exclusively with Cartesian coordinates. Angular coordinates traditionally used in modeling with Reference-Point coordinates can be avoided, as well as their associated moments of inertia. This facilitates the dynamic analysis greatly.

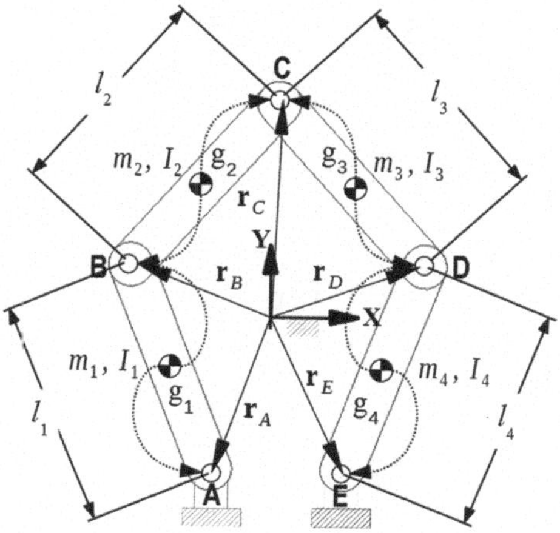

Fig. 1. Generic 2RRR Planar Parallel Manipulator.

In this way the manipulator has been modeled using the Cartesian coordinates of the "basic points", that can be organized in a positions vector $\mathbf{q}$:

$$\mathbf{q} = \left[\mathbf{r}_A \; \mathbf{r}_B \; \mathbf{r}_C \; \mathbf{r}_D \; \mathbf{r}_E\right]^T \tag{1}$$

where $\mathbf{r}_j = [x_j \; y_j]^T$ are the position vectors that locate the basic points A through E with respect to the origin of the reference frame.

The dependent coordinates are related by a set of four kinematic constraints, $\boldsymbol{\Phi}(\mathbf{q})$, obtained from the rigid body condition of the bodies:

$$\boldsymbol{\Phi}(\mathbf{q}) = \begin{bmatrix} (x_B - x_A)^2 + (y_B - y_A)^2 - l_1^2 \\ (x_C - x_B)^2 + (y_C - y_B)^2 - l_2^2 \\ (x_C - x_D)^2 + (y_C - y_D)^2 - l_3^2 \\ (x_D - x_E)^2 + (y_D - y_E)^2 - l_4^2 \end{bmatrix} = 0 \tag{2}$$

The coordinates are absolute and can be used to calculate the position of the center of mass of the system, as well as the shaking force, directly with respect to the global reference frame (inertial frame). The key point is the assembly of the mass matrix.

2.1 Calculation of the Mass Matrix

For the assemblage of the mass matrix of the system it is necessary to calculate first the mass matrix of the elements.

The way to calculate the (4×4) mass matrix of a single body using fully Cartesian coordinates can be reviewed in [10] and more recently in [1]. The general form of the mass matrix of the link i is shown in Eq. (3):

$$\mathbf{M}_i = \begin{bmatrix} \mathbf{M}_{1i} & \mathbf{M}_{2i} \\ \mathbf{M}_{3i} & \mathbf{M}_{4i} \end{bmatrix} \tag{3}$$

where:

$$\mathbf{M}_{1i} = \begin{bmatrix} m_i - \frac{2m_i x}{l} + \frac{J_i}{l^2} & 0 \\ 0 & m_i - \frac{2m_i x}{l} + \frac{J_i}{l^2} \end{bmatrix} \tag{4}$$

$$\mathbf{M}_{2i} = \begin{bmatrix} \frac{m_i x}{l} - \frac{J_i}{l^2} & -\frac{m_i y}{l} \\ \frac{m_i y}{l} & \frac{m_i x}{l} - \frac{J_i}{l^2} \end{bmatrix} \tag{5}$$

$$\mathbf{M}_{3i} = \begin{bmatrix} \frac{m_i x}{l} - \frac{J_i}{l^2} & \frac{m_i y}{l} \\ -\frac{m_i y}{l} & \frac{m_i x}{l} - \frac{J_i}{l^2} \end{bmatrix} \tag{6}$$

$$\mathbf{M}_{4i} = \begin{bmatrix} \frac{J_i}{l^2} & 0 \\ 0 & \frac{J_i}{l^2} \end{bmatrix} \tag{7}$$

and m_i is the mass of link i, and I_i is the moment of inertia of the link i with respect to its local reference frame.

All the mass matrices of the four mobile bodies of the manipulator can be calculated in the same way and their terms reorganized to form the (10×10) mass matrix of the system. Its form is presented in Eq. (8).

$$\mathbf{M} = \begin{bmatrix} \mathbf{M}_{11} & \mathbf{M}_{21} & 0 & 0 & 0 \\ \mathbf{M}_{31} & \mathbf{M}_{41} + \mathbf{M}_{12} & \mathbf{M}_{21} & 0 & 0 \\ 0 & \mathbf{M}_{32} & \mathbf{M}_{42} + \mathbf{M}_{43} & \mathbf{M}_{33} & 0 \\ 0 & 0 & \mathbf{M}_{32} & \mathbf{M}_{31} + \mathbf{M}_{44} & \mathbf{M}_{34} \\ 0 & 0 & 0 & \mathbf{M}_{24} & \mathbf{M}_{41} \end{bmatrix} \tag{8}$$

This matrix is symmetric and positive definite.

2.2 Calculation of the Global Center of Mass

Once changing to the use of fully Cartesian coordinates and assembling the global mass matrix of the system, the global center of mass of the system can be calculated directly as:

$$m_t \mathbf{g}_t = \mathbf{BMq} \tag{9}$$

where $m_t = \sum_{i=1}^{4} m_i$, is the total mass or the system, $\mathbf{g}_t$ is the position vector of the location of the global center of mass, $\mathbf{q}$ is vector of dependent Cartesian coordinates, and $\mathbf{M}$ is mass matrix, already defined by Eqs. (1) and (8) respectively. Matrix $\mathbf{B}$ is introduced here to sum x and y components in an automatic way and is composed by (2×2) identity matrices, $\mathbf{I}_{2\times2}$, its form is:

$$\mathbf{B} = \begin{bmatrix} \mathbf{I}_{2\times2} & \mathbf{I}_{2\times2} & \mathbf{I}_{2\times2} & \mathbf{I}_{2\times2} \end{bmatrix} \tag{10}$$

Substituting terms in Eq. (9), the global center of mass of the manipulator can be calculated as:

$$m_t \mathbf{g}_t = \begin{bmatrix} a & b & c & d & e & f & g & h \\ -b & a & -d & c & -f & e & -h & g \end{bmatrix} \begin{bmatrix} x_B \\ y_B \\ x_C \\ y_C \\ x_D \\ y_D \\ x_E \\ y_E \end{bmatrix} \tag{11}$$

where:

$$a = \left(-\frac{m_2\, x_2}{l_2} + \frac{m_1\, x_1}{l_1} + m_2 \right) \tag{12}$$

$$b = \left(\frac{m_2\, y_2}{l_2} - \frac{m_1\, y_1}{l_1} \right) \tag{13}$$

$$c = \left(\frac{m_3\, x_3}{l_3} + \frac{m_2\, x_2}{l_2} \right) \tag{14}$$

$$d = \left(-\frac{m_3\, y_3}{l_3} - \frac{m_2\, y_2}{l_2} \right) \tag{15}$$

$$e = \left(\frac{m_4\, x_4}{l_4} - \frac{m_3\, x_3}{l_3} + m_3 \right) \tag{16}$$

$$f = \left(\frac{m_4\, y_4}{l_4} - \frac{m_3\, y_3}{l_3} \right) \tag{17}$$

$$g = \left(m_4 - \frac{m_4\, x_4}{l_4} \right) \tag{18}$$

$$h = \left(\frac{m_4\, y_4}{l_4} \right) \tag{19}$$

where (x_i, y_i) with $i = 1 \dots 4$ are the coordinates of the corresponding center of mass of body i expressed in the local reference frame, see Fig. 2. Note that the origin of the reference frame coincides with point A, thus $(x_A, y_A) = (0, 0)$. Additionally note the orientation of the local reference frames presented in the same Fig. 2.

2.3 Calculation of the Shaking Force

The shaking force is the time derivative of the linear momentum. The linear momentum can be calculated directly deriving equation (9) with respect to time:

$$m_t \dot{\mathbf{g}}_t = \mathbf{B}\mathbf{M}\dot{\mathbf{q}} \tag{20}$$

where $\dot{\mathbf{q}}$ is the vector of velocities, the time derivative of vector $\mathbf{q}$.

Taking into account that the mass matrix (8) is constant, and that the velocities of point A and of point E are zero, the extended expression for the linear momentum is:

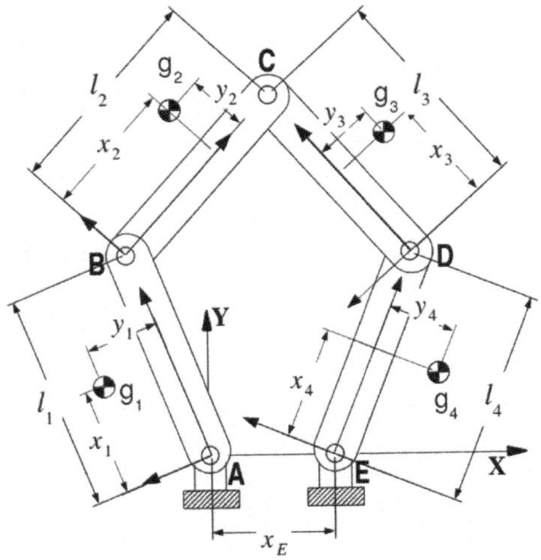

Fig. 2. Model of the 2RRR Planar Parallel Manipulator, showing the local reference frames orientation. The centers of mass are considered in a general position.

$$m_t \dot{\mathbf{g}}_t = \begin{bmatrix} a & b & c & d & e & f \\ -b & a & -d & c & -f & e \end{bmatrix} \begin{bmatrix} \dot{x}_B \\ \dot{y}_B \\ \dot{x}_C \\ \dot{y}_C \\ \dot{x}_D \\ \dot{y}_D \end{bmatrix} \tag{21}$$

The shaking force, $\mathbf{f}_{sh}$, can be calculated as the time derivative of the linear momentum. Thus the shaking force of the manipulator can be finally expressed in extended form as:

$$\mathbf{f}_{sh} = m_t \ddot{\mathbf{g}}_t = \begin{bmatrix} a & b & c & d & e & f \\ -b & a & -d & c & -f & e \end{bmatrix} \begin{bmatrix} \ddot{x}_B \\ \ddot{y}_B \\ \ddot{x}_C \\ \ddot{y}_C \\ \ddot{x}_D \\ \dot{y}_D \end{bmatrix} \tag{22}$$

Note that the position of the global center of mass, the linear momentum, and the shaking force are calculated in a very straightforward way once the mass matrix has been calculated. The mass matrix of the system is very easy to formulate once the corresponding mass matrix of a single body is known. The laborious task is the mass assembly, but this can be automated easily.

3 Illustrative Example

To show the application of the equations developed in the previous sections, a specific 2RRR PPM has been selected. A model with the following geometric parameters have been used: $l_1 = l_4 = 0.36$ m, $l_2 = l_3 = 0.3$ m, and $x_E = 0.48$ m. To locate the centers of mass the following values where used: $x_1 = x_4 = 0.18$ m, $x_2 = x_3 = 0.15$ m, and $y_1 = y_2 = y_3 = y_4 = 0$. The masses are: $m_1 = m_4 = 0.18$ kg and $m_2 = m_3 = 0.15$ kg.

3.1 Given Motion to the End-Effector

The trajectory of the end-effector C is given by a straight line limited between the initial position C_i, with the coordinates $x_i = 0.04$ m, and $y_i = 0.45$ m, and the final position C_f, with the coordinates $x_f = 0.44$ m, and $y_f = 0.45$ m. The straight line is traveled following a cycloidal profile with parameters $\Delta s = 0.4$ m in a period of time $T = 0.05$ s. This can be expressed as in Eq. (23).

$$s(t) = \Delta s \left[\frac{t}{T} - \frac{1}{2\pi} \sin \left(\frac{2\pi t}{T} \right) \right] \tag{23}$$

The coordinates of points B and D, and the corresponding position of the global center of mass of the system were obtained solving the inverse kinematics problem, using Eqs. (2) and (11) simultaneously, by a standard Newton-Raphson algorithm. Note that these equations are non-linear. The trajectory followed by the tool, point C, and by the global center of mass (CoM) can be seen in see Fig. 3.

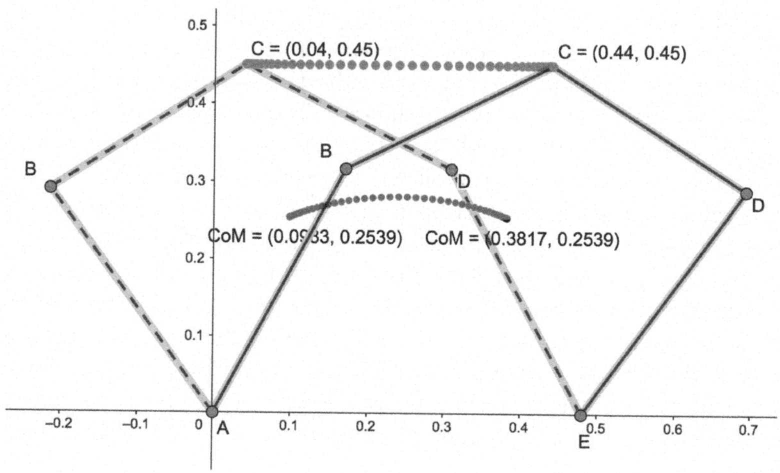

Fig. 3. Tool trajectory, showing initial and final points of the end-effector (point C) and of the global center of mass (CoM). Dashed lines correspond to the initial position of the manipulator. Solid lines correspond to the final position.

The global center of mass describes an arc. Its initial position is at (0.0983, 0.2539), while its final position is (0.3817, 0.2539).

To solve the velocities of point B, D, and the velocity of the CoM, the linear momentum equations (21) and the kinematic equations for velocities are solved simultaneously. This last equations are obtained deriving the kinematic constraints (2) with respect to time as:

$$\frac{d\boldsymbol{\Phi}(\mathbf{q})}{dt} = \frac{\partial \boldsymbol{\Phi}}{\partial \mathbf{q}}\dot{\mathbf{q}} \equiv \boldsymbol{\Phi}_{\mathbf{q}}\dot{\mathbf{q}} = 0 \tag{24}$$

where $\frac{\partial \boldsymbol{\Phi}}{\partial \mathbf{q}}$ is the Jacobian matrix. Note that these are a set of simultaneous linear equations.

In a similar way, to solve the accelerations of point B, D, and the acceleration of the CoM, the shaking force equations (22) and the kinematic equations for accelerations are solved simultaneously. This last equations can be obtained deriving the equations for velocities (24) with respect to time as:

$$\boldsymbol{\Phi}_{\mathbf{q}}\ddot{\mathbf{q}} + \dot{\boldsymbol{\Phi}}_{\mathbf{q}}\dot{\mathbf{q}} = 0 \tag{25}$$

where $\dot{\boldsymbol{\Phi}}_{\mathbf{q}}$ is the time derivative of the Jacobian matrix. Again these are a set of simultaneous linear equations.

3.2 Center of Mass Acceleration Control

As proposed, a second motion profile is given to the manipulator. A trajectory to the global center of mass of the moving links is defined as straight line and expressed with a "bang-bang" motion profile. See Fig. 4.

On one hand using this profile it is possible to ensure only the initial and the final positions of the end-effector but not a straight line trajectory between them. On the other hand such a control approach allows the reduction of the maximal value of the center of mass acceleration, leading to the elimination of the variation in the shaking force. Additionally the shaking force is reduced.

The "bang-bang" law is based on theoretical considerations. Actuators are unable to achieve discontinuous efforts. Thus this law is proposed to be modified by trapezoidal profiles to account for the actuators properties in terms of admissible effort variations. It has been observed that for reasonable actuator parameters, the results obtained in the cases of the "bang-bang" and trapezoidal profiles are very close (less than 1%). See [3,9] and [4] for further information.

The corresponding values for points B, D, and C were obtained solving the inverse kinematics problem, with Eqs. (2) and (11) simultaneously, by the standard Newton-Raphson algorithm. Note that these equations are non linear. The position coordinates of the global CoM are used as the degrees of freedom. The trajectories followed by the tool, point C, and by the global center of mass (CoM) can be seen in see Fig. 5.

To solve the velocities of points B, C, and D, the linear momentum equations (21) and the kinematic equations for velocities (24) are solved simultaneously. The velocity coordinates of the global CoM were taken as degrees of freedom.

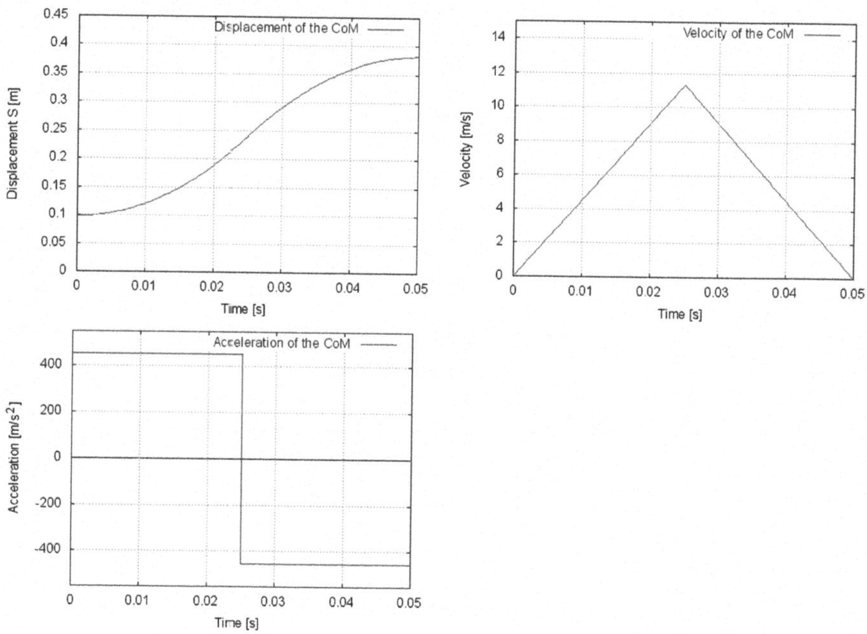

Fig. 4. Bang-bang profile applied to the motion of the global center of mass of the manipulator.

In a similar way, to solve the accelerations of points B, C, and D, the shaking force equations (22) and the kinematic equations for accelerations (25) are solved simultaneously. The acceleration coordinates of the global CoM were taken as degrees of freedom.

3.3 Shaking Force Comparison

Once solving the kinematics problems: positions, velocities and accelerations, simultaneously with the equations for the center of mass location, the linear momentum and the shaking force, it is possible to calculate in straightforward way the corresponding shaking force.

The procedure, followed for the case of imposing a straight line trajectory to the end-effector and for a bang-banng trajectory imposed to the global CoM, gave the corresponding shaking forces compared in Fig. 6.

Note that the variation in shaking force is eliminated after imposing the "bang-bang" profile to the motion of the global CoM. Additionally the maximum shaking force is reduced.

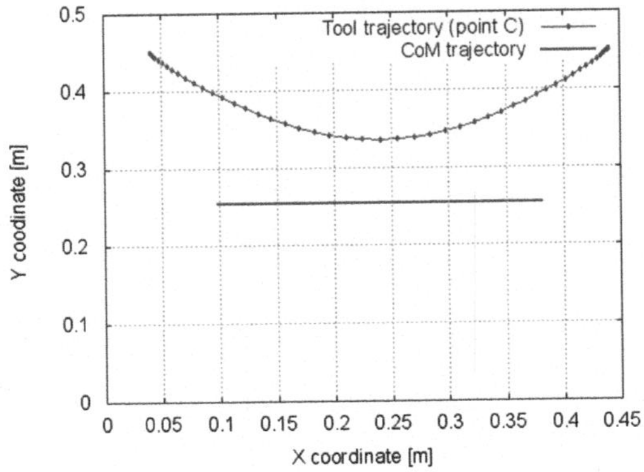

Fig. 5. End-effector and global CoM trajectories.

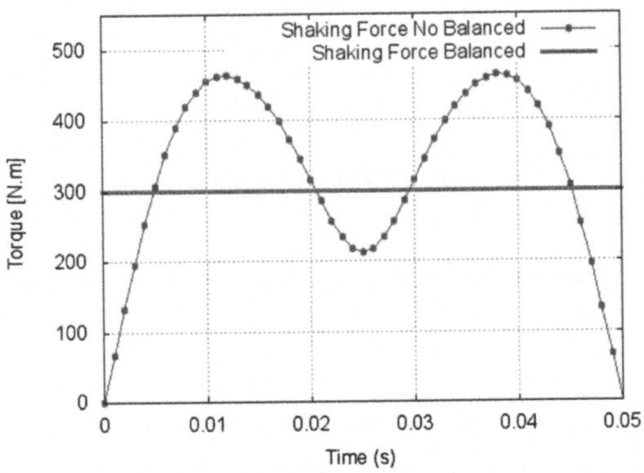

Fig. 6. Shaking force compared. In both cases magnitude is plotted.

4 Conclusions

A method, originally presented in [3] and [4], based on the center of mass acceleration control of the manipulator to eliminate the variation and reduce the shaking force balanced is resumed here using fully Cartesian coordinates. The procedure to apply this kind of coordinates to obtain the expressions of the global center of mass of the system, as well as the shaking force, is presented for a 2RRR PPM. The expressions are obtained in a direct explicit form. These expressions are used in computer simulations for the shaking force reduction of the manipulator.

References

1. Acevedo M (2015) Conditions for dynamic balancing of planar parallel manipulators using natural coordinates and their application. In: 14th world congress in mechanism and machine science, Taipei, Taiwan, 25–30 October 2015

2. Agrawal S, Fattah A (2004) Reactionless space and ground robots: novel design and concept studies. Mech Mach Theory 39(1):25–40. https://doi.org/10.1016/S0094-114X(03)00102-2
3. Arakelian V (2016) Design of partially balanced 5R planar manipulators with reduced center of mass acceleration (RCMA). In: Parenti-Castelli V, Schiehlen W (eds) ROMANSY21 - Robot design, dynamics and control. ROMANSY21 2016. CISM international centre for mechanical sciences (courses and lectures), vol 569. Springer, Cham
4. Briot S, Arakelian V, Le Baron JP (2012) Shaking force minimization of high-speed robots via center of mass acceleration control. Mech Mach Theory 57:1–12. https://doi.org/10.1016/j.mechmachtheory.2012.06.006
5. Buganza A, Acevedo M (2011) Dynamic balancing of a 2-DOF 2RR planar parallel manipulator by optimization. In: 13th world congress in mechanism and machine science, Guanajuato, Mexico, 19–25 June 2011
6. Fattah A, Agrawal S (2003) Design and modeling of classes of spatial reactionless manipulators. In: Proceedings of the IEEE international conference on robotics and automation, ICRA 2003, Taipei, Taiwan, pp. 3225–3230
7. Fattah A, Agrawal S (2005) Design arm simulation of a class of spatial reactionless manipulators. Robotica 3(1):75–81. https://doi.org/10.1017/S0263574704000670
8. Filaretov V, Vukobratovic M (1993) Static balancing and dynamic decoupling of the motion of manipulation robots. Mech Mach Theory 3(6):767–783. https://doi.org/10.1016/0957-4158(93)90062-7
9. Geng J, Arakelian V (2019) Design of partially balanced planar 5R symmetrical parallel manipulators via an optimal motion planning. In: Uhl T (ed) Advances in mechanism and machine science. IFToMM WC 2019. Mechanisms and machine science, vol 73. Springer, Cham
10. de Jalón JG, Bayo E (1994) Kinematic and dynamic simulation of multibody systems: the real-time challenge. Springer, New York. https://doi.org/10.1007/978-1-4612-2600-0
11. Nenchev DN (2012) Reaction null space of a multibody system with applications in robotics. Mech Sci 4:97–112. https://doi.org/10.5194/ms-4-97-2013
12. van der Wijk V, Herder JL (2012) Synthesis method for linkages with center of mass at invariant link point – pantograph based mechanisms. Mech Mach Theory 48:15–28. https://doi.org/10.1016/j.mechmachtheory.2011.09.007

Approximation of the Workspace of a Cable-Driven Parallel Robot with a Movable Gripper

Larisa Rybak, Elena Gaponenko, and Dmitry Malyshev(✉)

Belgorod State Technological University named after V.G. Shukhov, Belgorod, Russian Federation
rlbgtu@gmail.com

Abstract. The article discusses the non-uniform covering method application to solve the problem of the workspace determination of a cable-driven parallel robot with a movable gripper for cargo handling in special conditions. The robot is a rigidly fixed frame connected by tensioned ropes to a platform containing an axial movement device. The non-uniform covering method is one of the most well-known deterministic methods for solving global optimization problems. An algorithm for approximating the workspace of a cable-driven parallel robot using interval analysis methods has been synthesized. Visualization of simulation results is carried out by converting a set of three-dimensional boxes into an STL file.

Keywords: Cable-driven parallel robot · Workspace · Non-uniform covering method · Interval analysis

1 Introduction

Currently, there is a wide variety of parallel structure mechanisms [1–3]. Of particular interest are cable-driven parallel robots - a special type of kinematic structure, consisting mainly of a working body connected to a fixed base platform using cables. Cable-driven parallel robots use cables instead of extension rods to control the position of the output link. In these mechanisms, the position of the output link is controlled by changing the length of the cables. Cables are usually wound around coils attached to the base and driven by a rotating motor. Cable-driven parallel robots have special advantages, such as a large workspace, ease of assembly and disassembly, high mobility, high load capacity and ease of reconfiguration. By controlling cable lengths over a wide range, you can access a very large workspace from a few tens of centimeters to several tens of meters or more. The use of cables instead of rigid links further reduces weight because the drives do not change position and are attached to a fixed base so that the only moving parts are cables and the output link. As a result, we get a robot with higher speed and maneuverability and increased payload. The cost of a cable-driven parallel robots manufacturing is significantly lower than conventional manipulators. Cable-driven parallel robots are easy to install. Such a manipulator can be assembled using several inexpensive winches

E. E. Hernandez et al. (Eds.): LASIRS 2019, MMS 86, pp. 36–43, 2020.
https://doi.org/10.1007/978-3-030-45402-9_5

and cables. Also, since motors do not need to be installed near a moving platform, they are suitable for use in hazardous environments. Also, their carrying capacity is relatively high, even comparable to construction cranes.

Particular attention is paid to the topic of cable-driven parallel robots by J.P. Merlet, A. Pott, C. Gosselin, J.B. Izard, M. Michelin, S. Baradat, and others. So, Pott and others developed a family of cable-driven parallel robots called IPAnema that are designed for industrial processes, in [4] methods for planning and modeling a cable robot are described in detail. J.P. Merlet and D. Daney developed the light and mobile cable parallel robot Marionet [5], which is designed for rescue operations in hard-to-reach places. In [6], dynamic planning of the trajectory of 3-DOF CDPRS is considered. The proposed approach to planning trajectories can be used to plan dynamic trajectories that go beyond the static working space of the mechanism, opening up new applications and opportunities for cable robots. In [7], dynamic analysis and classification of the workspace were considered based on the general equation of motion of a cable robot and the one-way property of cables. A qualitative comparison of various types of workspaces. In [8], based on a structural analysis, the classification of cable robots was compiled and considered in detail: by the method of placing the drums, by the number of degrees of freedom of the working body and the number of flexible links, by the method of application, by the type of base, by the method of cables fixing.

An important characteristic of parallel structure mechanisms is their workspace, within which a working tool should be located when performing technological operations. To determine the workspace, it is known to use various methods: geometric, numerical, and discretization methods. In [9], the method of non-uniform coverings for approximating the set of solutions of a system of nonlinear inequalities was considered, and in [10, 11], the application of this method to determine the workspace of some types of planar robots was considered.

2 Problem Statement

Consider the application of the non-uniform covering method to determine the workspace of a Cable-driven parallel robots designed to work in special conditions. For this purpose, we will use the methods of interval analysis described in detail in [12, 13]. The circuit technical solution of such a robot is shown in Fig. 1. The manipulator consists of four columns, four cables, which are connected at one end to a moving platform at points B_1, B_2, B_3, B_4. A gripping device with the possibility of axial movement is fixed on the platform. The second ends of the cables passed through the pulleys, indicated by points A_1, A_2, A_3, A_4, mounted on columns, are fixed on the drums D_1, D_2, D_3, D_4, respectively. Under the action of the weight of the load mg, fixed at point C, tensile forces T_1, T_2, T_3, T_4 arise in the cables. A change in the position of the load securing point occurs due to a change in the cable lengths during rotation of the drums by gear motors M_1, M_2, M_3, M_4. The gripping device is a movable platform connected with cables. An output link C is located on the mobile platform, which has the possibility of axial movement of h along the Z axis. This mechanism allows you to move objects in hard-to-reach places with special conditions without human intervention, for example, working with radioactive elements, radiation sources, working with elements that can irradiate cells lead to their mutation. They can cause irreparable harm to health. This category also includes work at nuclear and thermal power plants.

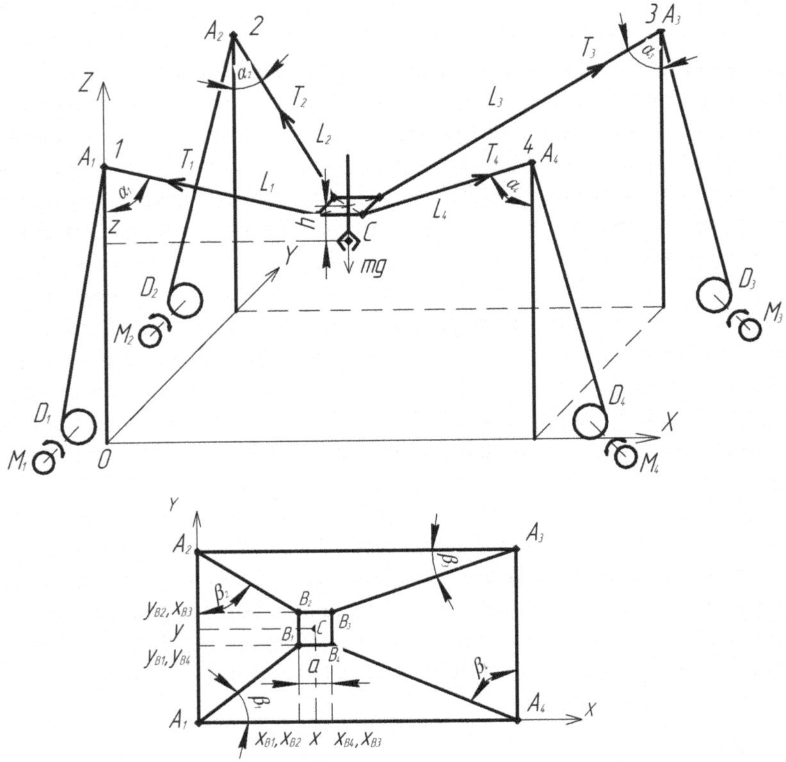

Fig. 1. Scheme of a cable-driven parallel robot for cargo moving

Despite their many advantages, there are several problems associated with controlling the movement of robot cables. One of the drawbacks is that the cables can pull, but not push, this leads to a one-way restriction, in which the cables must always be kept in a tight state. The provisions in which at least one of the cables is loosened are special and are not considered in this article.

We introduce restrictions on the cable tension forces $\boldsymbol{T}_i$

$$\boldsymbol{T}_i := \left[\underline{\boldsymbol{T}_i}, \overline{\boldsymbol{T}_i}\right] = \left\{0 \le \underline{\boldsymbol{T}_i} \le T_i \le \overline{\boldsymbol{T}_i}\right\} \quad (1)$$

We write down the intervals that describe the ranges of variation of the cosines and sines of the angles α_i and β_i:

$$\boldsymbol{S_{Ai}} := \left[\underline{\boldsymbol{S_{Ai}}}, \overline{\boldsymbol{S_{Ai}}}\right] = \left\{\underline{\boldsymbol{S_{Ai}}} \le \sin\alpha_i \le \overline{\boldsymbol{S_{Ai}}}\right\}, i \in 1, \ldots, 4.$$

$$\boldsymbol{S_{Bi}} := \left[\underline{\boldsymbol{S_{Bi}}}, \overline{\boldsymbol{S_{Bi}}}\right] = \left\{\underline{\boldsymbol{S_{Bi}}} \le \sin b_i \le \overline{\boldsymbol{S_{Bi}}}\right\}, i \in 1, \ldots, 4.$$

$$\boldsymbol{C_{Ai}} := \left[\underline{\boldsymbol{C_{Ai}}}, \overline{\boldsymbol{C_{Ai}}}\right] = \left\{\underline{\boldsymbol{C_{Ai}}} \leq \cos\alpha_i \leq \overline{\boldsymbol{C_{Ai}}}\right\}, i \in 1, \ldots, 4.$$

$$\boldsymbol{C_{Bi}} := \left[\underline{\boldsymbol{C_{Bi}}}, \overline{\boldsymbol{C_{Bi}}}\right] = \left\{\underline{\boldsymbol{C_{Bi}}} \leq \cos b_i \leq \overline{\boldsymbol{C_{Bi}}}\right\}, i \in 1, \ldots, 4.$$

The tension forces in the flexible links can be calculated by solving systems of nonlinear equations having the form:

$$\begin{cases} \boldsymbol{T}_1\boldsymbol{S}_{A1}\boldsymbol{C}_{B1} + \boldsymbol{T}_2\boldsymbol{S}_{A2}\boldsymbol{S}_{B2} - \boldsymbol{T}_3\boldsymbol{S}_{A3}\boldsymbol{C}_{B3} - \boldsymbol{T}_4\boldsymbol{S}_{A4}\boldsymbol{S}_{B4} = 0 \\ \boldsymbol{T}_1\boldsymbol{S}_{A1}\boldsymbol{S}_{B1} - \boldsymbol{T}_2\boldsymbol{S}_{A2}\boldsymbol{C}_{B2} - \boldsymbol{T}_3\boldsymbol{S}_{A3}\boldsymbol{S}_{B3} + \boldsymbol{T}_4\boldsymbol{S}_{A4}\boldsymbol{C}_{B4} = 0 \\ \boldsymbol{T}_1\boldsymbol{C}_{A1} + \boldsymbol{T}_2\boldsymbol{C}_{A2} + \boldsymbol{T}_3\boldsymbol{C}_{A3} + \boldsymbol{T}_4\boldsymbol{C}_{A4} - mg = 0 \end{cases} \tag{2}$$

where

$$\boldsymbol{C_{Ai}} = \frac{z_{Ai} - \boldsymbol{Z_{Bi}}}{\sqrt{(\boldsymbol{X_{Bi}} - x_{Ai})^2 + (\boldsymbol{Y_{Bi}} - y_{Ai})^2 + (z_{Ai} - \boldsymbol{Z_{Bi}})^2}};$$

$$\boldsymbol{S_{Ai}} = \frac{\sqrt{(\boldsymbol{X_{Bi}} - x_{Ai})^2 + (\boldsymbol{Y_{Bi}} - y_{Ai})^2}}{\sqrt{(\boldsymbol{X_{Bi}} - x_{Ai})^2 + (\boldsymbol{Y_{Bi}} - y_{Ai})^2 + (z_{Ai} - \boldsymbol{Z_{Bi}})^2}};$$

$$\boldsymbol{S_{B1}} = \frac{\boldsymbol{Y_{B1}} - y_{A1}}{\sqrt{(\boldsymbol{X_{B1}} - x_{A1})^2 + (\boldsymbol{Y_{B1}} - y_{A1})^2}};$$

$$\boldsymbol{S_{B2}} = \frac{\boldsymbol{X_{B2}} - x_{A2}}{\sqrt{(\boldsymbol{X_{B2}} - x_{A2})^2 + (y_{A2} - \boldsymbol{Y_{B2}})^2}};$$

$$\boldsymbol{S_{B3}} = \frac{y_{A3} - \boldsymbol{Y_{B3}}}{\sqrt{(x_{A3} - \boldsymbol{X_{B3}})^2 + (y_{A3} - \boldsymbol{Y_{B3}})^2}};$$

$$\boldsymbol{S_{B4}} = \frac{x_{A4} - \boldsymbol{X_{B4}}}{\sqrt{(x_{A4} - \boldsymbol{X_{B4}})^2 + (\boldsymbol{Y_{B4}} - y_{A4})^2}};$$

$$\boldsymbol{C_{B1}} = \frac{\boldsymbol{X_{B1}} - x_{A1}}{\sqrt{(\boldsymbol{X_{B1}} - x_{A1})^2 + (\boldsymbol{Y_{B1}} - y_{A1})^2}};$$

$$\boldsymbol{C_{B2}} = \frac{y_{A2} - \boldsymbol{Y_{B2}}}{\sqrt{(\boldsymbol{X_{B2}} - x_{A2})^2 + (y_{A2} - \boldsymbol{Y_{B2}})^2}};$$

$$\boldsymbol{C_{B3}} = \frac{x_{A3} - \boldsymbol{X_{B3}}}{\sqrt{(x_{A3} - \boldsymbol{X_{B3}})^2 + (y_{A3} - \boldsymbol{Y_{B3}})^2}};$$

$$\boldsymbol{C_{B4}} = \frac{\boldsymbol{Y_{B4}} - y_{A4}}{\sqrt{(x_{A4} - \boldsymbol{X_{B4}})^2 + (\boldsymbol{Y_{B4}} - y_{A4})^2}}.$$

Wherein

$$\boldsymbol{X_{B1}} = \boldsymbol{X_{B2}} = \boldsymbol{X_C} - \frac{a}{2},$$

$$X_{B3} = X_{B4} = X_C + \frac{a}{2},$$

$$Y_{B1} = Y_{B4} = Y_C - \frac{a}{2},$$

$$Y_{B2} = Y_{B3} = Y_C + \frac{a}{2},$$

$$Z_{Bi} = Z_C + H,$$

where

$$X_C := [\underline{X_C}, \overline{X_C}] = \{\underline{X_P} \le x_C \le \overline{X_P}\}$$

$$Y_C := [\underline{Y_C}, \overline{Y_C}] = \{\underline{Y_C} \le y_C \le \overline{Y_C}\}$$

$$Z_C := [\underline{Z_C}, \overline{Z_C}] = \{\underline{Z_C} \le z_C \le \overline{Z_C}\}$$

$$H := [\underline{H}, \bar{H}] = \{\underline{H} \le h \le \bar{H}\}$$

The left sides of the system of Eq. (2) are functions in the form $g_j,\ j \in 1, \ldots, 3$: $g_1 = T_1 S_{A1} C_{B1} + T_2 S_{A2} S_{B2} - T_3 S_{A3} C_{B3} - T_4 S_{A4} S_{B4}$, ..., $g_3 = T_1 C_{A1} + T_2 C_{A2} + T_3 C_{A3} + T_4 C_{A4} - mg$. The above equations allow us to find solutions to the system of Eq. (2) that determine the limitations of the workspace.

3 Algorithm Synthesis

Taking into account formulas (1–2), an algorithm for approximating the workspace of a cable-driven parallel robot was synthesized, which is shown in Fig. 2. It works with a system of equations written in a general form:

$$\begin{cases} g_1(x) = 0, \\ \ldots \\ g_m(x) = 0 \\ a_k \le x_k \le b_k,\ k = 1, .., n. \end{cases} \tag{3}$$

The initial box Q, which covers the entire set of solutions X, is determined by the interval constraints $a_k \le x_k \le b_k, k = 1, .., n$. Consider an arbitrary box B. Let $m(B) = \max_{j=1,\ldots,m} \min_{x \in B} g_j(x)$ and $M(B) = \min_{j=1,\ldots,m} \max_{x \in B} g_j(x)$. If $m(B) > 0$ or $M(B) < 0$, then B does not contain possible points for the system (3). If the box can not be excluded, it is divided into two smaller boxes, if its diameter is greater than the specified approximation accuracy δ. The algorithm works with three lists of n-dimensional boxes $\mathbb{P}$ (current list), $\mathbb{P}_A$ (coverage) and $\mathbb{P}_E$ (excluded boxes). Each of the dimensions of the boxes corresponds to the intervals X_C, Y_C, and Z_C.

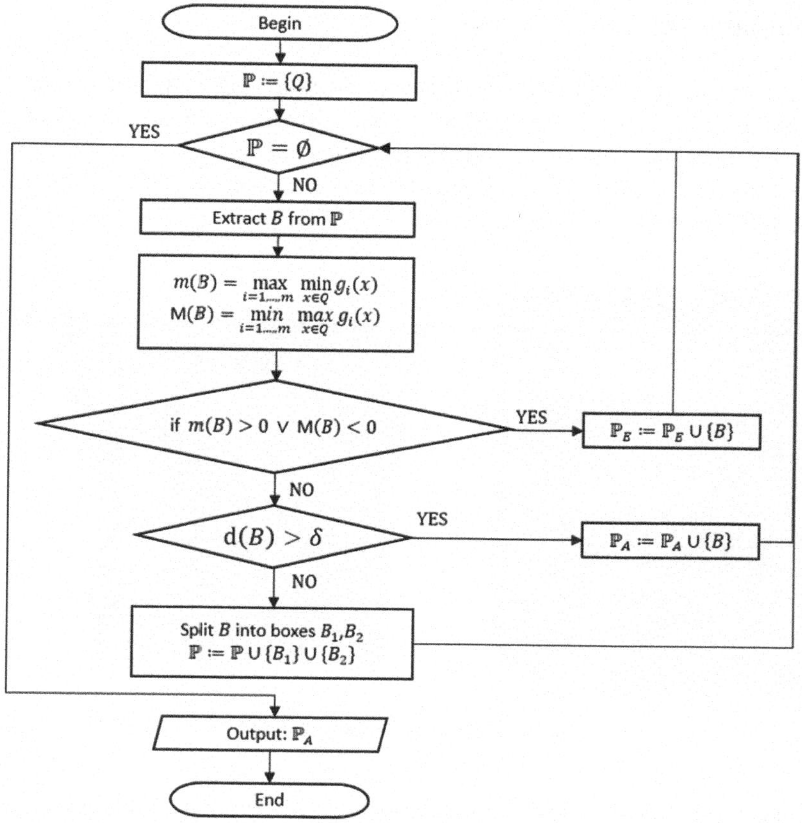

Fig. 2. The block diagram of the algorithm for determining the workspace and optimization of geometric parameters.

The algorithm works as follows:

1. Set the geometric parameters of the robot, the intervals **H**, $T_i, i \in 1, \ldots, 4$ and approximation accuracy δ.

2. At the first step of the algorithm, lists $\mathbb{P}_A$ and $\mathbb{P}_E$ are empty, list $\mathbb{P}$ consists of only one box Q, including intervals $\boldsymbol{X_C}$, $\boldsymbol{Y_C}$ and $\boldsymbol{Z_C}$: $\boldsymbol{X_C} = [0, x_{A3}]$, $\boldsymbol{Y_C} = [0, y_{A3}]$, $\boldsymbol{Z_C} = [0, z_{A3}]$.

3. Remove the box B from the list $\mathbb{P}$.

4. Calculate $m(B) > 0$ and $M(B)$.

5. If $m(B) > 0$ or $M(B) < 0$, then add B to the list $\mathbb{P}_E$ and go to step 8.

6. If $d(B) < \delta$, then add B to the list $\mathbb{P}_A$ and go to step 8.

7. In either cases, B is divided into two equal boxes along the edge with the longest length. Insert these boxes at the end of the list P, that is, $\mathbb{P} := \mathbb{P} \cup \{B_1\} \cup \{B_2\}$.

8. If $\mathbb{P} \neq \emptyset$, then go to step 3.

The algorithm is implemented in the C++ programming language using the Snowgoose interval analysis library [14]. Visualization of the simulation results is carried out by converting the list of boxes $\mathbb{P}_I$ to the universal format of 3D models - the STL file.

4 Simulation Results

A computational experiment was conducted for the following geometric parameters of the mechanism: $\mathbf{H} = [0\,\text{mm}, 200\,\text{mm}]$, $\boldsymbol{T} = [20\,\text{N}, 100\,\text{N}]$, $x_{A3} = x_{A4} = y_{A2} = x_{A2} = z_{Ai} = 1000$ mm. The simulation results are presented in Fig. 3. The calculation time for the approximation accuracy $\delta = 10$ mm on a personal computer was 5 min 36 s. The volume of the workspace is 0.474 m^3.

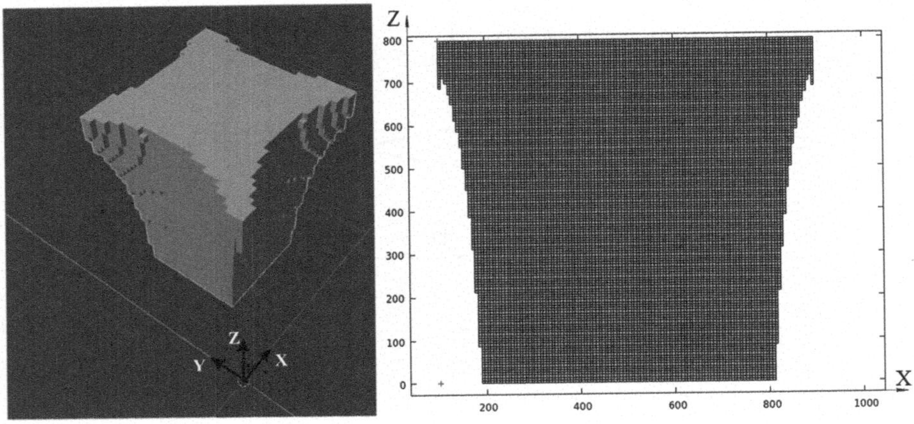

Fig. 3. The results of the simulation.

Figure 4 shows the dependence of the workspace volume of the cable-driven parallel robot on the range of the output link C movement along the Z axis.

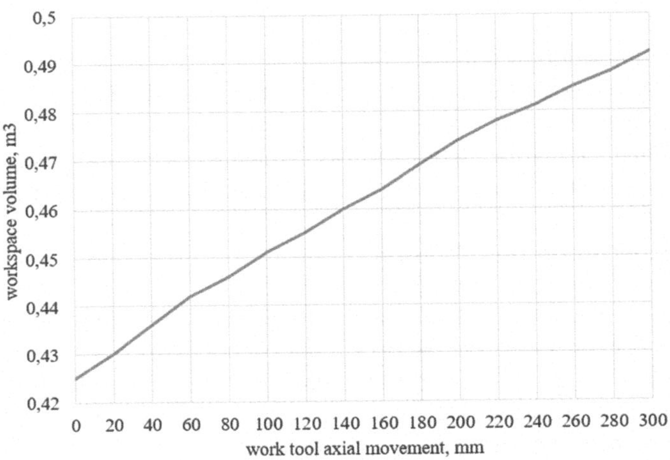

Fig. 4. Dependence of the workspace volume on the work tool axial movement.

5 Conclusion

Thus, for the proposed cable-driven parallel robot, effective numerical methods and algorithms were developed and tested that allowed to determine the workspace. The calculation time for the approximation accuracy $\delta = 10$ mm on a personal computer was 5 min 36 s. A further area of research is the optimization of the geometric parameters of the robot to ensure the required workspace.

Acknowledgments. This research has been supported by the RFBR according to the research project No. 18-31-20060.

References

1. Merlet J-P (2007) Parallel robots, 2nd edn. Springer, Heidelberg
2. Kong H, Gosselin CM (2007) Type synthesis of parallel mechanisms. Springer, Heidelberg
3. Aleshin AK, Glazunov VA, Rashoyan GV, Shai O (2016) Analysis of kinematic screws that determine the topology of singular zones of parallel-structure robots. J Mach Manuf Reliab 45(4):291–296
4. Pott A, Mütherich H, Kraus W, Schmidt V, Miermeister P, Verl A (2013) IPAnema. A family of cable-driven parallel robots for industrial applications. In: Bruckmann T, Pott A (eds) Cable-driven parallel robots. Mechanisms and machine science, vol 12. Springer, Heidelberg, pp 119–134
5. Merlet JP, Daney DA (2010) A portable, modular parallel wire crane for rescue operations. In: 2010 IEEE international conference on robotics and automation (ICRA), pp 2834–2839
6. Gosselin C (2013) Global planning of dynamically feasible trajectories for three-DOF spatial cable-suspended parallel robots. In: Bruckmann T, Pott A (eds) Cable-driven parallel robots. Springer, Heidelberg, pp 3–22
7. Duan QJ, Xuechao D (2014) Workspace classification and quantification calculations of cable-driven parallel robots. In: Advances in mechanical engineering. Hindawi Publishing Corporation, pp 1–9
8. Gaponenko EV, Rybak LA, Kholoshevskaya LR (2019) Structural analysis and classification of robotic systems with driving mechanisms on the basis of cable elements. Bulletin BSTU named after V.G. Shukhov 9:126–136 (2019)
9. Evtushenko Yu: Numerical methods for finding global extreme (case of a non-uniform mesh). USSR Comput Maths Math Phys 11: 1390–1403 (1971)
10. Evtushenko Y, Posypkin M, Rybak L, Turkin A (2018) Approximating a solution set of non-linear inequalities. J Global Optimization 7:129–145
11. Gorchakov AYu, Ignatov AD, Malyshev DI, Posypkin MA (2019) Parallel algorithm for approximating the working area of the robot. Int J Open Inf Technol 7(1):1–7
12. Jaulin L (2001) Applied interval analysis: with examples in parameter and state estimation, robust control and robotics. Springer, New York
13. Moore RE (1966) Interval analysis. Prentice-Hall, Englewood Cliffs
14. Posypkin M, Usov A. Basic numerical routines. https://github.com/mposypkin/snowgoose. Accessed 18 Nov 2019

Design of a New Game for Teaching Assembly Process

Tianxiong Wang[1] and Rafiq Ahmad[2](✉)

[1] Beijing Institute of Technology, Beijing, China
[2] University of Alberta, Edmonton, Canada
rafiq.ahmad@ualberta.ca

Abstract. Assembly practice is regarded as a great way to deal with the training needs of the increasing number of mechanical companies as students can be equipped with hands-on skills and working experience from assembly training. However, the current assembly-related courses with hands-on activities have a series of problems including cost and safety. This paper analyzes the possibility of applying serious games to assembly training and introduces a novel serious game to help mechanical engineering students participate the assembly process in a fun environment. The game is developed by Unity3D and based on BCN3D MOVEO robotic arm. Learners can get familiar with the assembly process from the game and instructors can customize the assembly process for their students to learn. The benefits this game can have on mechanical engineering education are examined

Keywords: Assembly · Serious game · Gamification

1 Introduction

Learning mechanical engineering requires students to be actively involved in the learning process. However, a majority of courses which are related to mechanical engineering are given by lecture-based approach [1]. This traditional teaching method has obvious disadvantages because it is not an effective motivator as students can only receive information from textbooks and instructors without taking active part in education [2]. Furthermore, the traditional approach leaves students with poor practical abilities as students cannot get enough opportunities to deal with workplace problems which are complex, ambiguous and ill-structured without a clear goal and a single solution [3], compared with the clear and direct principles taught by traditional lectures. In order to help students to get love for learning and prepare them for the real world of practice, project-based and training-based teaching approaches are supposed to be carried out [4].

Our game is designed to respond to this demand and help more mechanical engineering learners take part in the assembly process conveniently and economically. Assembly training is in the core of integrated courses which are related to mechanical engineering. Through assembly practice, students are able to develop operative skills in real manufacturing environment [5].

E. E. Hernandez et al. (Eds.): LASIRS 2019, MMS 86, pp. 44–52, 2020.
https://doi.org/10.1007/978-3-030-45402-9_6

Our game is designed in Unity, which means that the game is three dimensional, portable and machine independent. These characteristics make the game a good supporting tool for mechanical engineering instructors. Furthermore, the scoring system and the game-based design principle enable students to study not only independently but also cooperatively while also having fun.

Although our design focuses on the assembly process of one machine, our research can be extended to other assembly simulator design projects in the field of engineering.

The paper is organized as follows. Section 2 introduces related work about education of assembly in mechanical engineering. The concept of gamification and other related theoretical background information are explained in Sect. 3. Section 4 briefly introduces BCN3D MOVEO, which is the machine chosen for our game. The development of the game will be introduced in Sect. 5. Section 6 provides conclusions.

2 Related Work

Nowadays, many companies and laboratories have already designed games or software-based simulators to help students learn practical skills in the field of mechanical engineering. NIU-Torcs is a car racing video game designed to help mechanical engineering undergraduates learn computational techniques and concepts. In the game, a player is required to write a C++ program to get his or her car to move on the track in the way prescribed by the game [6]. Although the game can help students learn engineering calculations strategies, it is not able to improve students' hands-on skills because the game concentrates on programming rather than assembly.

Mecagenius® is a game designed to teach students mechanical manufacturing and machining process [7]. But the game is developed by Adobe Flash technology, which implies that all the components in the game is two dimensional so students cannot have a complete perspective of all the assemblies in the game. Moreover, the correct operational procedures in each activity of the game cannot be changed, which means that teachers are not able to design their own operational procedures for their students.

On the contrary, our game is three dimensional so users can observe every component of the assembly clearly in our game and the simple user interface makes the game easy-to-use and easy-to-learn for new learners. In addition, teachers can change the correct assembly procedure according to their own teaching plans and project assessments.

3 Theoretical Background

3.1 Gamification

The term "gamification" was first documented used in 2008 but there are many parallel or similar concepts which have already existed before 2008 or continue being introduced after the appearance of gamification. Although the term originated in 2008, it was not widely adopted at the beginning. Gamification was a contested term before 2011 with many researchers using other terms for their similar practice [8].

Deterding [8] proposed a definition of "gamification" as the use of game design elements in non-game context in 2011, which is regarded as the most acknowledged definition so far. In order to reach this definition, Deterding make a distinction between playing and gaming. Gaming is characterized by explicit rule systems and goals. In contrast, the main goal of playing is to entertain the users so playing denotes a free-form and open recombination of behaviors and meanings [9]. The goal of our project is to apply gamification in the educational context, so the following discussion will focus on gaming rather than playing.

3.2 Serious Games

The term "serious game" refers to games which are designed to help users develop new knowledge and skills rather than mere entertainment [7]. Through the application of computer, users of serious games can experience situations which are impossible for them to meet due to safety and cost. Nowadays, serious games have already been widely applied in a large number of fields, e.g. health care, education, government, military.

Serious games can be classified according to their fields of application: America's Army, which was released in 2002, is one of the most well-known serious games in military. The U.S. military invested the game and the game successfully helped U.S. military to recruit new volunteer soldiers with 15% of the cost of other recruiting problems. Dance Revolution is a healthcare serious game, which requires associated hardware. The game successfully combined game elements with physical activities to make players be more engaged in physical exercises [10]. E and Eve's Electrical Endeavors is an example of educational serious games. The game was developed by the Eindhoven University of Technology and aims to help players to learn knowledge about electrical engineering. According to the investigation of the players of the game, students who used the game to study performed remarkably better than those who received information via textbooks [11].

3.3 Advantages of Serious Games in Engineering Education

It has been proved that adding game elements in educational context can significantly improve students' attention, participation and interests [12, 13]. Moreover, games can provide students with opportunities to improve communication skills, cooperate with other students and experiment dangerous conditions in a safe manner. By integrating game elements in curriculum or offering game-based courses, schools can increase their students' learning effects and enthusiasm towards study. [14] provides a study in which students' engagement in a mechanical engineering course built around a video/computer game was measured. The authors get the conclusion that students are eager to learn and take on different challenges in a game-based learning environment.

Furthermore, the evolution of industries has already increased the demand for employees with hands-on experience. Although assembly is a good way to improve students' hands-on skills and computer-based assembly exercise is able to satisfy this demand, there are no assembly-oriented serious games available on the market. A serious game about assembly with the characteristics of efficiency, convenience and fun can fill the gap of gamified experiences for assembly process and help to cope with the training needs of industries.

4 Open Source MOVEO Robot

We chose BCN3D MOVEO [15] as the assembly of our game. The picture of MOVEO can be seen in Fig. 1. MOVEO is a fully open source 3D printed robotic arm developed by BCN3D Technologies. The structure of MOVEO is simple and it is therefore suitable for education. This robotic arm consists of five major parts and is controlled by the software Arduino. There are total 232 components in MOVEO, including screws, locknuts, printed parts, etc. As BCN3D Technologies puts it, MOVEO was developed with an open and education wish.

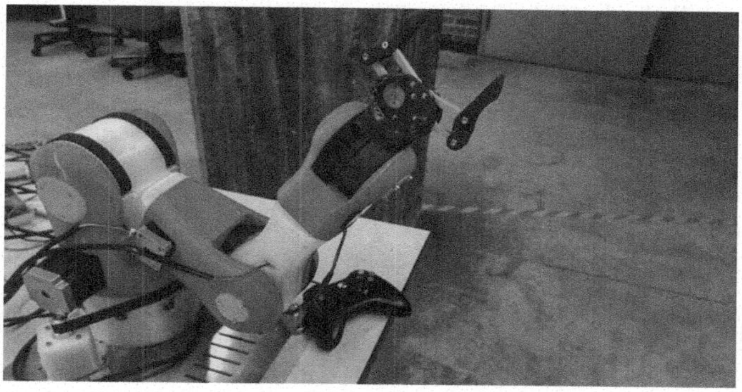

Fig. 1. MOVEO robotic arm (BCN3D Technologies)

All the necessary information about assembling a MOVEO robotic arm is available on GitHub [16]. Thanks to BCN3D Technologies and GitHub, one can get the CAD and STL files of all the components of MOVEO. These files were converted to OBJ files in SolidWorks and used in Unity3D, the development platform of our game. Furthermore, BCN3D Technologies also provides the user manual, which gives the instruction of the assembly process on GitHub. The official user manual is used in the instruction module of the designed game.

5 Proposed Game

De Paz [17] divided the general guidelines to gamification into three phases, which will be briefly introduced afterwards.

Phase 1: **Preparation.** In this phase, designers of are supposed to set up the goals of their game, be clear about what problems they want to gamify and gain some knowledge about different game elements which can be involved in their game.

Phase 2: **Design.** At this stage, the main objective is to determine the basic design and the use of game elements. As new ideas may come up throughout the project, the ideas that designers have in this stage can be changed from time to time.

Phase 3: **Implementation.** The target of this stage is to build the physical system and put it into play.

Then, we will introduce how we design and implement our game according to the design principle proposed by De Paz.

5.1 Preparation

The goal of the design is to provide students with a computer-based system on which they can assemble a machine, learn some knowledge about assembly process and get to know parameters of different basic components (screws, motors, bearing and so on). In addition, instructors are supposed to be authorized to change the assembly sequence or even the final machine according to their own teaching plans. In order to motivate students to play the game, some elements of competition are expected to be added into the game design [18]. A good scoring system can not only help teachers know the performance of different students but also make students compete with their friends and improve their skills further.

5.2 Design

Level Design
According to the user manual of MOVEO, the whole assembly can be divided into five sub-assemblies, so there are five levels in the game. The order of these five levels corresponds to the complexity of the five sub-assemblies. Level one is the easiest, while level five is the hardest. Before entering a new level, students should finish all the easier levels first.

Selection and Sequencing of 3D Printed Components
The assignment of students in these section is to choose the right 3D printed components and put them in a correct order according to the instruction of their current level. Figure 2 shows the user interface of this section.

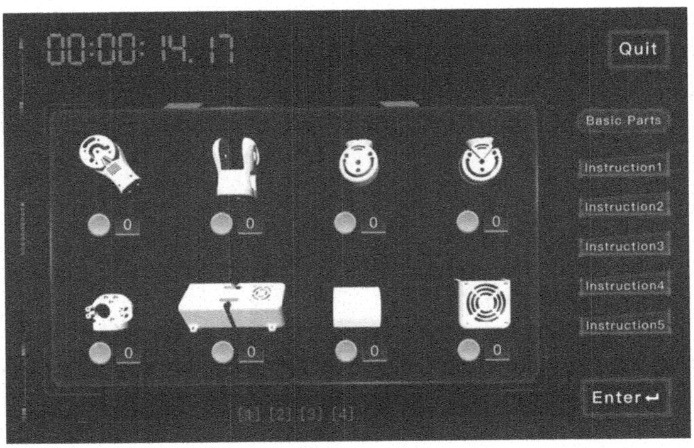

Fig. 2. User interface of the section of 3D printed components

Different from basic components, 3D printed components are complex, asymmetric and irregular, so traditional 2D game development platform, such as Adobe Flash, cannot be used. That is the reason why we choose Unity3D to develop our assembly game.

Once students put the mouse on one component, it will rotate so students can observe the component from all perspectives. This design is to make the game get close to real assembly environment and make students see the components more clearly in a computer game.

In order to make novice players play the game easily, the user interface was designed as concisely as possible. First, the player selects the "Instruction" button to read the instruction of the sub-assembly in his/her level. Then, he/she can click on the white square to select the 3D printed parts he/she wants to use in the sub-assembly. After selection, the player need to input the sequence numbers of the 3D printed parts he/she has selected in the input field next to the white square. "P1", "P2", "P3" and "P4" buttons at the lower right corner of the interface control the page transition as there are total thirty-two 3D printed parts in MOVEO but one page can only display eight of them. The clock at the top left corner displays the time the player has already spent in his level. The "Enter" button enables players to check whether their selection and sequencing are correct or not. If it is right, the system will transfer to next module which will be explained.

Selection of Basic Components

In this section, players are required to choose the right components and input the quantity according to the instruction of their current level. Figure 3 shows the user interface of this section.

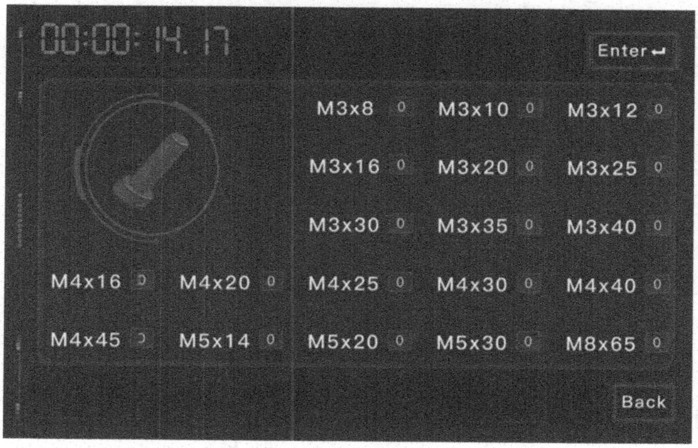

Fig. 3. User interface of the section of basic components

Players can click on the "Enter" button to check whether their input values are correct or click on the "Back" button to go back to the welcome window of this section to choose other basic components. Moreover, players can transfer between the section of basic components and 3D printed components at any time with all the data saved automatically.

After these two sections, the scoring system of the game will give out the final score of the player.

Scoring System

The final score depends on the total time a player spend and the number of clicks he/she makes in an assembly process. The score is calculated by the following equation:

$$\text{Final score} = 5000 - \text{Total Time(mins)} * 100 - \text{Number of Clicks} * 100$$

Time is a very important parameter to judge a player's level of proficiency in assembly. The number of clicks is also very important as some players may make many mistakes and make many changes in a short time or even make random attempts in the game. Therefore, the scoring system is designed to make players who can finish the assembly both quickly and accurately to get high scores.

Custom Settings

Instructors who have correct usernames and passwords can get into the custom setting section in which they can define the required 3D printed components from the thirty-two components and the assembly sequence according to their preference. Figure 4 shows the user interface of custom setting.

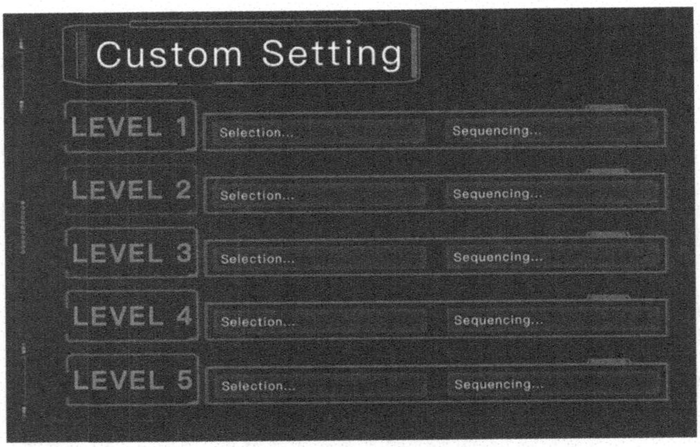

Fig. 4. User interface of custom setting

For each level, instructors need to input two strings with the length of 32 characters. The first string represents selection, which means whether a certain component should be selected in this level. 1 stands for yes, while 0 stands for no. The second string represents sequencing. Instructors need to set the sequence number for each of the components that should be selected.

5.3 Implementation

The game can be played in the class as a way to train students' assembly skills. Instructors can use the official assembly methods of MOVEO or define their own assembly strategies and provide students with custom instructions. Schools can offer game-based curriculum with this game or offer real-assembly courses and use this game as a teaching aid. For

Students, they can play the game after class to preview or review. The game can be also used in tests due to the scoring system. For an initial evaluation, the game was tested by several group members. They expressed their interest and judged that the game is a good learning tool for assembly practice. We intend to evaluate the performance of the game in future works.

6 Conclusions

This paper presented a brief research on gamification and serious games and discussed the value of applying serious games to the education of mechanical engineering. Assembly serious games can greatly improve the academic performance and hands-on skills of students but there is a lack of assembly-oriented serious games. Then, the paper introduced the design of an assembly game, which is based on BCN3D MOVEO and developed by Unity3D. The purpose of this design was to help students learn basic assembly skills and help schools set up assembly courses. We believe that this game can be improved in several ways, which will be considered in future work. For example, the graphics will be upgraded and the custom setting is better not to require users to set all the five levels at one time. The future work will also focus on validating this game in engineering courses. In future, the game evaluation will also be extended to school students.

References

1. Yadav A, Shaver GM, Meckl P (2010) Lessons learned: implementing the case teaching method in a mechanical engineering course. J Eng Educ 99(1):55–69
2. Hamada M, Sato S (2011) A game-based learning system for theory of computation using Lego NXT Robot. Procedia Comput Sci 4:1944–1952
3. Jonassen D, Strobel J, Lee CB (2006) Everyday problem solving in engineering: lessons for engineering educators. J Eng Educ 95(2):139–151
4. Prince MJ, Felder RM (2006) Inductive teaching and learning methods: definitions, comparisons, and research bases. J Eng Educ 95(2):123–138
5. Ahmad R et al (2018) Alberta learning factory for training reconfigurable assembly process value stream mapping. Procedia Manuf 23:237–242
6. Coller B (2007) Implementing a video game to teach principles of mechanical engineering. In: 114th annual ASEE conference and exposition
7. Pons-Lelardeux C et al (2015) Didactic study of a learning game to teach mechanical engineering. Procedia Eng 132:242–250
8. Deterding S, et al (2011) From game design elements to gamefulness: defining gamification. In: Proceedings of the 15th international academic MindTrek conference: envisioning future media environments. ACM
9. Müller BC, Reise C, Seliger G (2015) Gamification in factory management education–a case study with Lego Mindstorms. Procedia CIRP 26:121–126
10. Michael DR, Chen SL (2005) Serious games: games that educate, train, and inform. Muska & Lipman/Premier-Trade, Boston (2005)
11. Stege L, Van Lankveld G, Spronck P (2011) Serious games in education. Int J Comput Sci Sport 10(1):1–9

12. Barata G et al (2013) Improving participation and learning with gamification. In: Proceedings of the first international conference on gameful design, research, and applications. ACM
13. Barata G et al (2013) Engaging engineering students with gamification. In: 2013 5th international conference on games and virtual worlds for serious applications (VS-GAMES). IEEE
14. Coller BD, Shernoff DJ (2009) Video game-based education in mechanical engineering: a look at student engagement. Int J Eng Educ 25(2):308
15. BCN3D. BCN3D MOVEO – A fully Open Source 3D printed robot arm. https://www.bcn3dtechnologies.com/en/bcn3d-moveo-the-future-of-learning/
16. BCN3D. BCN3D-MOVEO: Open Source 3D Printed Robotic Arm for educational purposes. https://github.com/BCN3D/BCN3D-Moveo
17. Paz BM (2013) Gamification: a tool to improve sustainability efforts. A dissertation Submitted to the University of Manchester for the Degree of Master of Science in the Faculty of Engineering and Physical Sciences
18. Ahmad NL, Omhover J (2014) Game methodology for design methods and tools selection. J Learn Des 7(1):1–9

An Improved Robot Path Planning Algorithm for a Novel Self-adapting Intelligent Machine Tending Robotic System

Feiyu Jia, Juan Tzintzun, and Rafiq Ahmad(✉)

University of Alberta, Edmonton, AB T6G 2R3, Canada
rafiq.ahmad@ualberta.ca

Abstract. Autonomous industrial mobile manipulation systems (AIMMS) are widely used in manufacturing processes. AIMMS can help with part handling and delivering, part insertion and extraction, loading and unloading, and some other auxiliary tasks in a machine tending workshop environment. However, nowadays most AIMMS cannot truly realize the complete automation because the path for the mobile platform needs setting in advance in terms of different environment and inspection need to be executed to ensure safety due to lack of proper path planning algorithms. Therefore, this paper proposes an improved path-planning algorithm based on Rapidly-exploring Random Tree (RRT) and the quintic B-spline curve technique to generate a collision-free and smoother path for our designed Novel Self-adapting Intelligent Machine Tending Robotic System in the workspace. In the end, the proposed algorithm is demonstrated to generate paths for five different scenarios to test its performance and reliability.

Keywords: Autonomous industrial mobile manipulation systems · Machine tending · Path planning · Quintic B-spline curve

1 Introduction

Recently, robots are widely used in industry to perform repetitive, dumb, dangerous, dull, and dirty tasks. Therefore, industry robotics forms an essential part of the manufacturing backbone [1]. Task flexibility and robotic mobility are two main advantages that mobile manipulators bring to industrial applications. Furthermore, a mobile robot shall adapt to changing environments and perform a variety of tasks. It is essential to combine locomotion capabilities with manipulation abilities to extend the capabilities of future robotics. The concept of mobile manipulation refers to a robotic system that integrates a mobile platform, a robot arm, supported by a vision system. "Mobile manipulation" is introduced by Schuler in 1984 [2], where a robot arm is mounted on a mobile platform to execute tool handling and delivering duty in a workshop environment.

This paper focuses exclusively on the machine-tending domain by developing an improved path-planning algorithm for our designed robotic system. Path planning is a purely geometric process that is only concerned with finding a collision-free path

E. E. Hernandez et al. (Eds.): LASIRS 2019, MMS 86, pp. 53–64, 2020.
https://doi.org/10.1007/978-3-030-45402-9_7

regardless of the feasibility of the path [3–5]. Early efforts to develop deterministic path planning techniques showed that it is computationally demanding even for simple systems [6]. Recently, sampling-based path planning (SBP) algorithms have received considerable attention due to their capability in complex and time-critical real-world planning problems [7]. SBP is unique in the fact that planning occurs by sampling the configuration space (C-space). In a sense, SBP attempt to capture the connectivity of the C-space by sampling it. This randomized approach has its advantages in terms of providing fast solutions for severe problems. Furthermore, the computational capacity of the robot controller unit is limited in my applications; SBP is a better choice because of its lower computational cost.

Sampling-based planning is by no means a novel concept in robotics [8]. It was proposed to overcome the complexity of deterministic robot planning algorithms for a robot with six degrees of freedom. The use of random computations to solve otherwise rather tricky problems have been immensely successful [9, 10]. Both sampling-based planners and the success of random computations inspired the development of the Randomized Potential Planner (RPP) [11]. RPP used random walks to escape local minima of the potential field planner. Later on, a planner based entirely on random walks, with adaptive parameters, was proposed [12].

The work of Barraquand and Latombe [11] paved the way for a new generation of path planning algorithms that employ randomization. Arguably, the most commonly used SBP algorithms are Probabilistic Roadmap Method (PRM) [13–15] and Rapidly-exploring Random Trees (RRT) [16]. Several other algorithms developed at the same time that outperformed RPP. The intuitive implementation of both RRT and PRM and the quality of the solutions lead to their widespread adoption in robotics and many other fields. Although the idea of connecting points sampled randomly is fundamental in both approaches, these two methods are different in the manner that they construct a graph connecting the points [17]. A comprehensive survey of work in SBP is presented [18]. The PRM algorithm has been recorded to implement well in high-dimensional state spaces. The PRM is created by curves or straight lines that enable the robot to go anywhere in its free space. The two well-known methods, named visibility graph (VG) and Voronoi diagram (VD) have achieved very good results for very different types of roads. However, the path generated by VG touches obstacles at the vertices or edges and thus is dangerous for the robot. Contrary, the solution paths based on VD are not optimal because it tends to maximize the distance between the robot and the obstacles.

RRT has received a considerable amount of attention, because of its computational efficiency and effectiveness and its ability to find a feasible motion plan relatively quickly, even in high-dimensional space [19]. In [20], RRT is implemented and combined with a slightly modified dynamic window method. RRT explores a robot working area by incrementally building a tree, creating new branches by generating points randomly and linking them to the closest location for which an obstacle-free path is obtained. A problem in RRT is that it produces paths with many branches in the workspace by using the randomized technique. To overcome this problem, an improved robot path-planning algorithm based on the RRT and quintic B-spline curves is proposed to generate collision-free and smooth paths for a novel self-adapting intelligent machine tending robotic system.

The paper is organized as follows: In Sect. 2, a novel self-adapting intelligent machine tending robotic system is introduced. The principle of the algorithm is demonstrated in Sect. 3. In Sect. 4, the results and discussion are presented. Finally, Sect. 5 concludes the paper.

2 Self-adapting Intelligent Machine Tending Robotic System

A mobile manipulation system, named self-adapting intelligent robotic system, is designed for machine tending in the manufacturing field, as shown in Fig. 1.

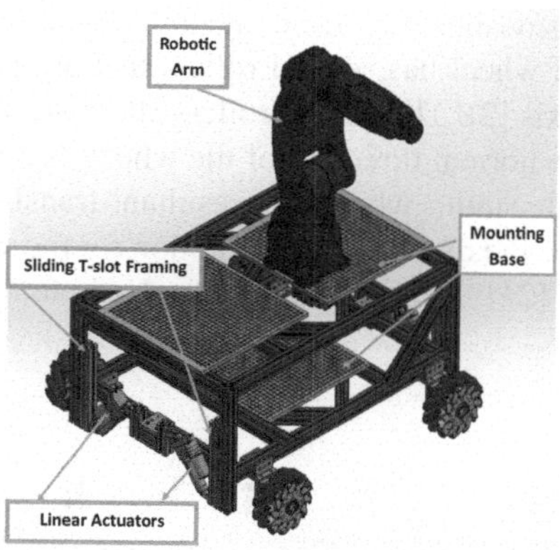

Fig. 1. The CAD model of the self-adapting intelligent robotic system

The linear actuator concept is designed to optimize the operation stability for robotic arm movements. This mobile platform has a wide-body design to prevent any tipping during the loading operation. By eliminating the need for lifting for arm operation, the vehicle stabilization is executed by four sliding t-slot frames that are driven by four short traveling linear actuators.

By using short traveling distance linear actuators, the vehicle can uphold to 225 kg extra payloads such as batteries, sensors, and laboratory (workshop) tools. Because these heavy duty linear actuators, the design could easily be expanded for larger vehicle body and more sophisticated robotic arms. The stabilizing mechanism of the design is shown in Fig. 2.

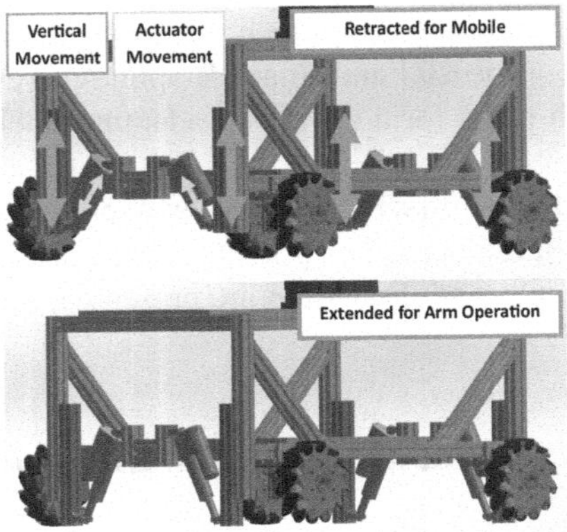

Fig. 2. Stabilizing mechanism

The mobile platform is designed by using a combination of four Mecanum wheels. A conventional Mecanum wheel has several rollers that are positioned at an angle of 45° to the central wheel axis [2]. The angled rollers allow the vehicle to translate a part of rotational force into the normal direction of the wheel. By controlling the rotational speed of the individual Mecanum wheel, the resultant translated force of four-wheel combination can determine the direction of the motion for the vehicle without changing the direction of the wheel [21]. A demonstration for basic four-wheel Mecanum wheel motions is shown in Fig. 3.

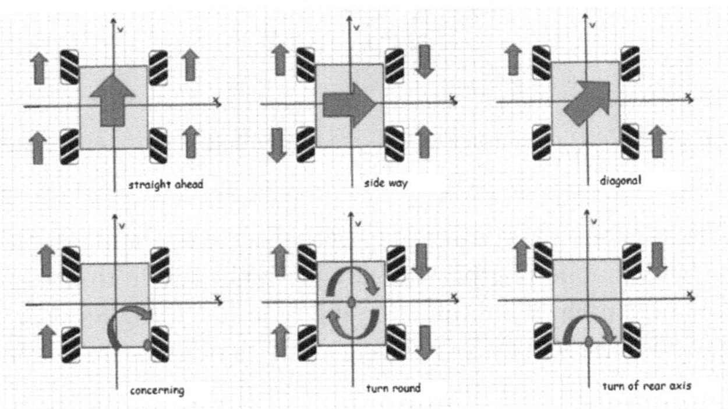

Fig. 3. Basic motions in v-x Axis [21]

By using Mecanum wheels, the vehicle can travel in any direction under any orientation. This allows the vehicle to break the non-holonomic constraints [2]. The purpose of this design is to develop an autonomous mobile platform for workshop and laboratory, which is often an environment with dynamic obstacles and confined spaces. To accomplish vehicle movement without complicated maneuvers and path-planning in such a narrow space, the omnidirectional character of Mecanum wheel has become the best solution for the design. The bottom view of the mobile platform is shown in Fig. 4.

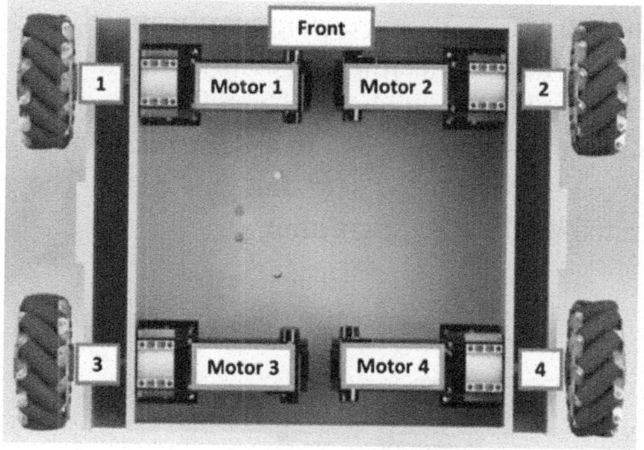

Fig. 4. Bottom view of mobile platform

In order to control individual motor precisely, two dual-channel controllers are used. By using a controller area network (CAN bus), one controller will operate as the master, and the other will serve as the slave [22]. A systematic drawing for the Mecanum wheel motion control is shown in Fig. 5.

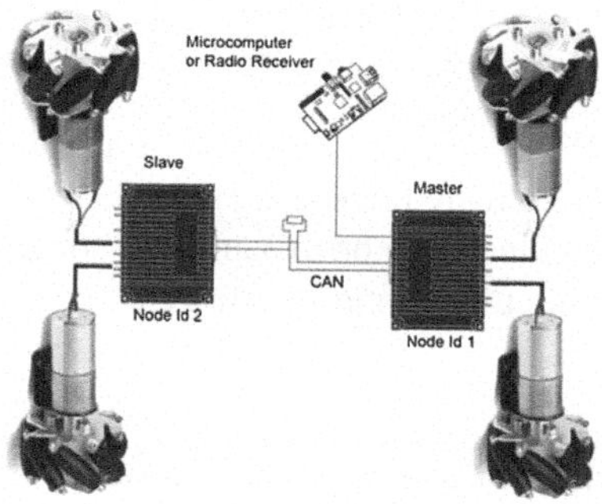

Fig. 5. Systematic illustration of CAN bus [22]

3 Outline of the Proposed Algorithm

3.1 Principle of the RRT Algorithm

To introduce the proposed algorithm, the necessary procedures of RRT must be defined. They are demonstrated as follows: (1) sampling: this procedure is used to select a configuration randomly and add it to the tree. The samples can be either in the free, or obstacle configuration space. This is the core of the RRT; (2) metric function: this procedure returns a value, or cost that indicates the effort required to reach C1 from C2, given two configuration C1 and C2 in workspace; (3) nearest neighbor: this procedure uses a search algorithm which returns that closest node to the new sample. The value is based on the predefined metric function; (4) select parent: this procedure selects a nearest existing node to connect to the newly sampled node. That current node is considered parent; (5) local planning: given two configurations C1 and C2, this procedure attempts to establish a connection between them; (6) collision checking: it is mostly a Boolean function that returns success or failure when connecting two configurations C1 and C2. A connection is successful if it does not intersect an obstacle area [18].

The ability of RRT to explore free space in a workspace with obstacles is illustrated in Fig. 6. Two Red Cross represents the start and goal point. The green straight-line is the tree or path generated by the RRT algorithm.

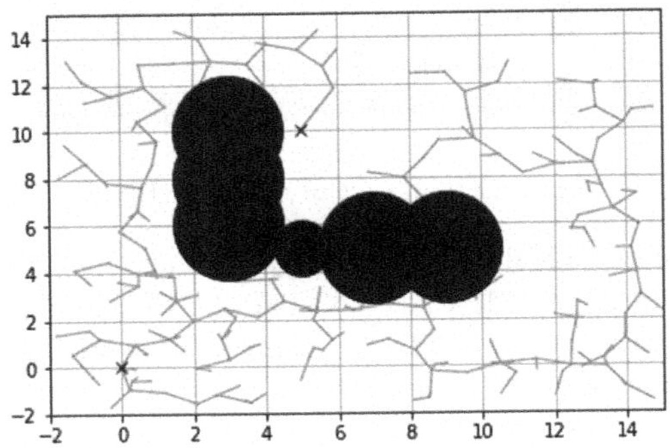

Fig. 6. RRT

It is known that the RRT algorithm is intuitive to employ straight-line paths, and the tree produced contains too many branches. However, for most robotic systems this is not a feasible plan due to kinematic or dynamic constraints. Thus, we integrate the quintic B-spline curves to help the robotic system generate a curvature continuous path in a practical workspace.

3.2 Path Generation Using Quintic B-spline Curves

In this part, we use a quintic B-spline curve to interpolate points Q_k, $k = 0 \ldots n$. The quintic B-spline curve is presented as follows:

$$\mathbf{S}(t) = \sum_{i=0}^{m} \boldsymbol{P}_i b_{i,5}(t) \tag{1}$$

Where $\boldsymbol{P}_i$ is the control point, $m = n + 4$, $b_{i,5}(t)$, $i = 0, \ldots, m$, is a quintic B-spline basis function. According to the chord length method, if an interpolating curve follows very closely to the data polygon, the length of the curve segment between two adjacent data points would be very close to the length of the chord of these two data points. The length of the interpolating curve would also be very close to the total length of the data polygon. Then, we assign parameter value τ_i to each data point $\boldsymbol{Q}_i$. It is shown as follows:

$$\tau_0 = 0 \tag{2}$$

$$\tau_i = \tau_{i-1} + \frac{|\boldsymbol{Q}_i - \boldsymbol{Q}_{i-1}|}{L}, i = 1, \ldots, n \tag{3}$$

Where L is the sum of the lengths of these chords:

$$L = \sum_{i=1}^{n} |\boldsymbol{Q}_i - \boldsymbol{Q}_{i-1}| \tag{4}$$

Therefore, the ratio of the chord length from the data point $\boldsymbol{Q}_0$ to data point $\boldsymbol{Q}_k$, denoted as L_k, over the length of the data polygon is

$$L_k = \frac{\sum_{i=1}^{k} |\boldsymbol{Q}_i - \boldsymbol{Q}_{i-1}|}{L} \tag{5}$$

For a clamped quintic B-spline curve, we compute a knot vector for interpolation without end conditions [23]:

$$t_0 = \ldots = t_5 = 0, t_{n+1} = \ldots = t_{n+6} \tag{6}$$

$$t_{j+5} = \frac{1}{5} \sum_{i=j}^{j+6} \tau_i, j = 1, \ldots, n-5 \tag{7}$$

The performance of the quintic B-spline curve to smooth a straight-line path is shown in Fig. 7.

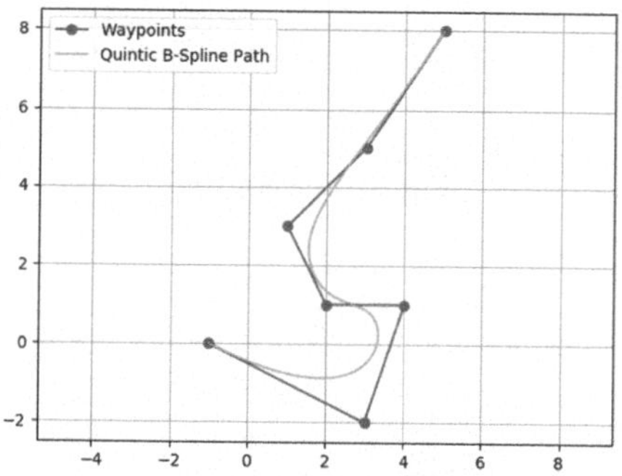

Fig. 7. Quintic B-spline curve

4 Results and Discussion

In this paper, five different scenarios (the goal point in each scenario in a different position) are executed to evaluate the developed path-planning algorithm. In each figure, the black circle areas represent the obstacles in the workspace. The green line represents the tree, the yellow line is the original path generated by RRT algorithm, the blue line is trimmed path, and the red is the smooth path produced by our path-planning algorithm. The results are presented as belows (Figs. 8, 9, 10, 11 and 12):

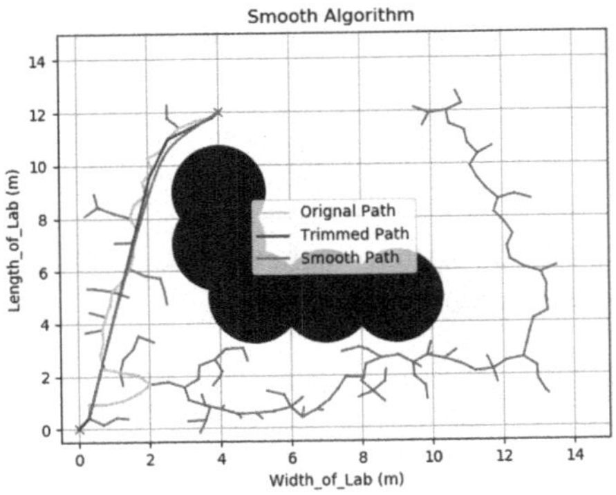

Fig. 8. Scenario 1

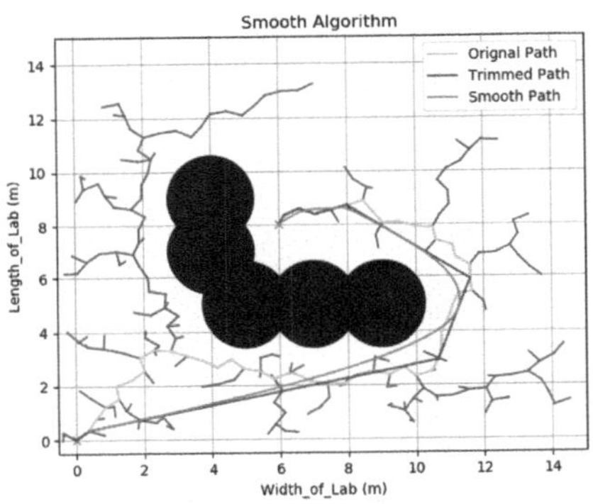

Fig. 9. Scenario 2

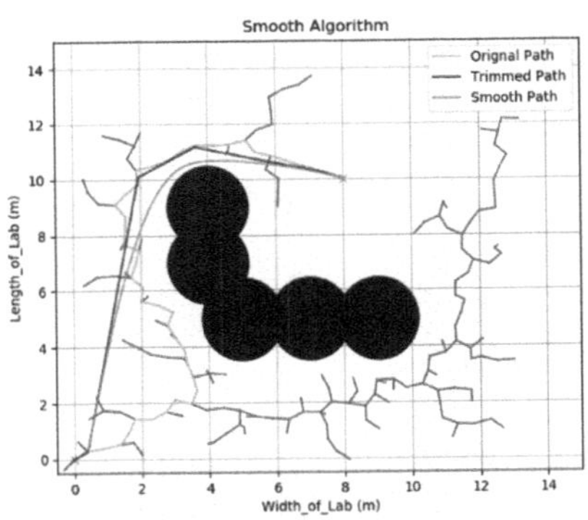

Fig. 10. Scenario 3

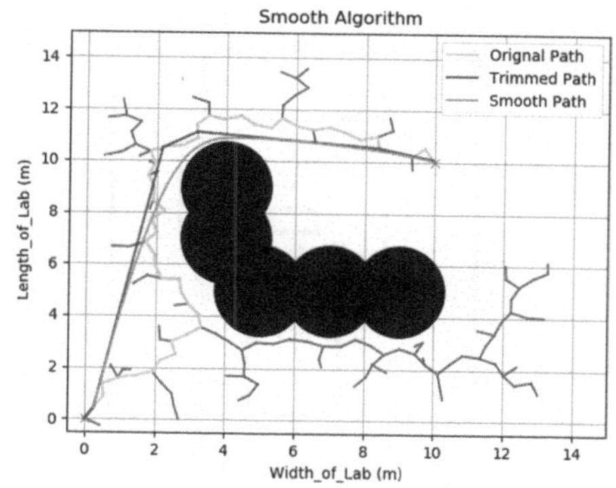

Fig. 11. Scenario 4

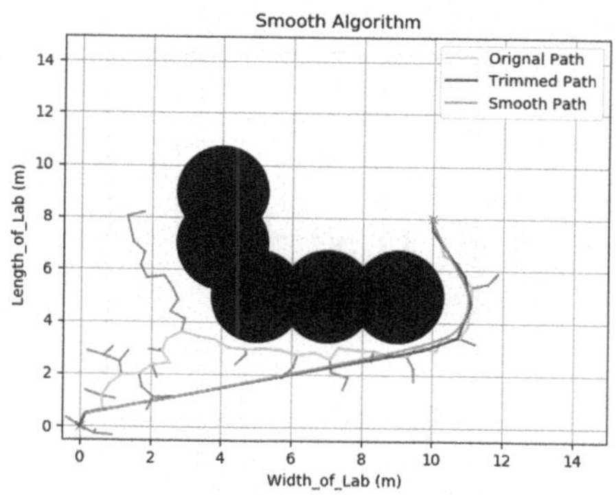

Fig. 12. Scenario 5

Though by combining the quantic B-spline curve with RRT can find the shorter and smoother path, and make it feasible for vehicle motion according to the kinematic and dynamic constraints, there still exists a chance that the vehicle might collide with obstacles. Therefore, continuous collision checking must be executed. A rectangular is placed in the simulation environment as a mobile robotic system, and detection about the collision between the four edges and the boundary of the obstacles is performed. When a path cannot be generated, the error message will be reported.

If there is no collision, the path generated. After the path is generated, the robotic system moves along the path at a constant rate, and its orientation coincides with the tangent of the curve. The algorithm is applied to three cases to evaluate the performance. The simulation is shown as follows (Figs. 13, 14 and 15):

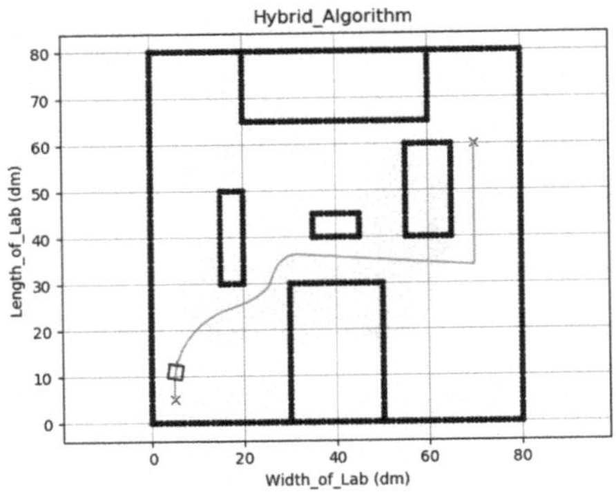

Fig. 13. Case 1

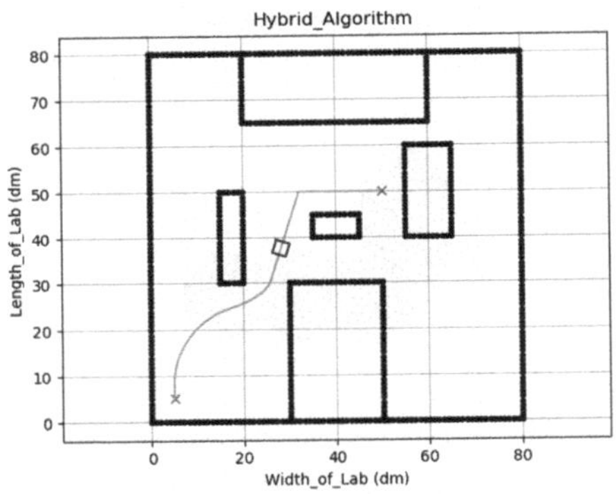

Fig. 14. Case 2

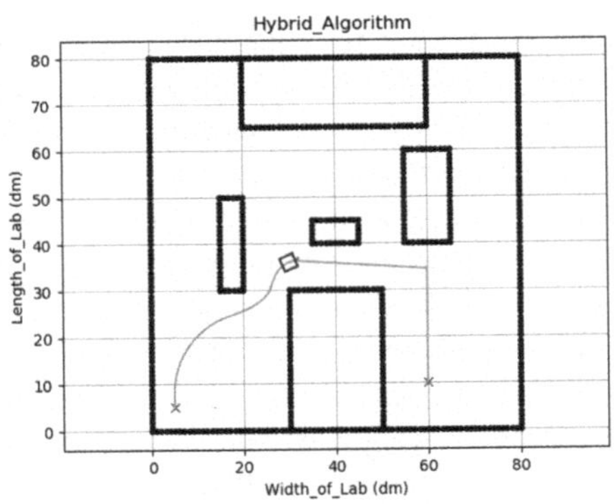

Fig. 15. Case 3

The results show that the proposed path-planning algorithm has excellent performance for the mobile robotic system in a 2-D virtual environment from the above three cases.

5 Conclusions

In this paper, we propose an improved path-planning algorithm based on quintic B-spline curve to effectively generate smooth and collision-free paths for our designed Novel Self-adapting Intelligent Machine Tending Robotic System. The advantage of our algorithm is that it makes the original path trimmed and smooth, so the mobile robotic system can continuously move following the kinematic and dynamic rules, and slightly rotate its wheel, which extends the life of tires. However, this algorithm is not suitable for a very narrow path between obstacles.

The next stage, our laboratory model will be created in V-REP and the novel self-adapting intelligent machine tending robotic system be placed in this 3-D virtual environment to test the performance of the proposed path planning algorithm.

References

1. Hvilshøj M, Bøgh S, Nielsen OS, Madsen O (2012) Autonomous industrial mobile manipulation (AIMM): past, present and future. Ind Robot Int J 39(2): 120–135
2. Adascalitei F, Doroftei I (2011) Practical applications for mobile robots based on mecanum wheels - a systematic survey. Technical University of Iasi
3. Wallace R, Stentz A, Thorpe C, Moravec H, Whittaker W, Kanade T (1985) First results in robot road following. In: Proceedings of the IJCAI, pp 381–387
4. Antonelli G, Chiaverini S, Fusco G (2007) A fuzzy-logic-based approach for mobile robot path tracking. IEEE Trans Fuzzy Syst 15(2):211–221
5. Jazar RN (2010) Mathematical theory of auto-driver for autonomous vehicles. J Vib Control 16(2):253–279
6. Reif JH (1979) Complexity of the mover's problem and generalizations. In: Proceedings of the 20th annual symposium foundation computer science, October 1979, pp 421–427
7. Lee J, Kwon O, Zhang L, Yoon SE (2014) A selective retraction-based RRT planner for various environments. IEEE Trans Robot 30(4):1002–1011
8. Donald BR (1987) A search algorithm for motion planning with six degrees of freedom. Artif Intell 31(3):295–353
9. Metropolis N, Ulam S (1949) The Monte Carlo method. J Amer Statist Assoc 44(247):335–341
10. Metropolis N, Rosenbluth AW, Rosenbluth MN, Teller AH, Teller E (1953) Equation of state calculations by fast computing machines. J Chem Phys 21(6):1087–1092
11. Barraquand J, Latombe J-C (1991) Robot motion planning: a distributed representation approach. Int J Robot Res 10(6):628–649
12. Carpin S, Pillonetto G (2005) Motion planning using adaptive random walks. IEEE Trans Robot 21(1):129–136
13. Kavraki L, Latombe JC (1994) Randomized preprocessing of configuration space for path planning: articulated robots. In: Proceedings of the IEEE/RSJ/GI international conference on IROS, vol 3, September 1994, pp 1764–1771

14. Amato NM, Wu Y (1996) A randomized roadmap method for path and manipulation planning. In: Proceedings of the IEEE international conference on robotics and automation, vol 1, April 1996, pp 113–120
15. Kavraki LE, Svestka P, Latombe JC, Overmars MH (1996) Probabilistic roadmaps for path planning in high-dimensional configuration spaces. IEEE Trans Robot Autom 12(4):566–580
16. LaValle SM (1998) Rapidly-exploring random trees: a new tool for path planning. Dept. Comput. Sci., Iowa State Univ, Ames, IA, USA, Technical report, TR 98-11
17. Karaman S, Frazzoli E (2011) Sampling-based algorithms for optimal motion planning. Int J Robot Res (IJRR) 30(7):846–894
18. Elbanhawi M, Simic M (2014) Sampling-based robot motion planning: a review. IEEE Access 2:56–77
19. Hsu D, Latombe JC, Kurniawati H (2006) On the probabilistic foundations of probabilistic roadmap planning. Int J Robot Res (IJRR) 25(7):627–643
20. Murtra AC, Trulls E, Sandoval O, Perez-Ibarz J, Vasquez D, Mirats-Tur JM, Ferrer M, Sanfeliu A (2010) Autonomous navigation for urban service mobile robots. In: IEEE/RSJ International Conference on Intelligent Robots and Systems, IROS, Taipei, Taiwan, pp 4141–4146
21. NEXUS Robot (2016) Mecanum wheel - NEXUS Robot, September 2016
22. RoboteQ (2015) AN1543-Driving Mecanum Wheels, 25 October 2015
23. Piegl L, Tiller W (2000) Curve interpolation with arbitrary end derivatives. Eng Comput 16:73–79

Stability Analysis of a Rotor Systems with Flow Forces

Jaime Cruz Cruz[1(✉)], Miguel Toledo Velázquez[1], and Oliver M. Huerta Chávez[2]

[1] SEPI ESIME Zacatenco IPN, 07580 Ciudad de México, Mexico
jcruzc0500@alumno.ipn.mx, mtv49@yahoo.com
[2] SEPI ESIME Ticoman IPN, 07580 Ciudad de México, Mexico
lecram_21@hotmail.com
https://www.sepi.esimez.ipn.mx/,
https://www.sepi.esimetic.ipn.mx/

Abstract. The vibration that occurs in Turbomachinery modifies parameters in its performance and depending on the application, such performance can compromise the entire system. The vibrations that occur in these cases can be divided depending on the source that exits, it: forced and self-excited, where the self-excited vibration extracts energy from the source, which generates instability and can cause catastrophic failures since it is very difficult to predict these effects during the design of the elements of the turbomachine. There are studies to predict some effects, from energy methods to coupled solutions Fluid-Structure and by using numerical solutions with neural networks that estimate the forces to which the rotor system is subjected, so it is It is important to investigate the effects of these forces on the structural response of the rotor system [5], therefore fluid-structure interaction studies (such as those applied in aircraft structures) are needed for rotodynamic instability. Therefore, this work presents a structural model of a rotor system considering loads due to the flow to obtain the aerodynamic forces and analyse the structural stability of the rotor system, numerical and analytical solutions have been chosen to see the behaviour of the system. The CFD solution is made to obtain the forces due to the flow in the rotor system and later these forces feed the structural model of the same to obtain its response (unbalance) using the finite element method.

Keywords: Stability · Rotor dynamics · CFD · Flow · Unbalance

1 Introduction

The problems associated with vibrations in compressors and turbines have become more frequent in recent years. This is partly due to the turbomachine's tendency towards higher speed and power units that operate well above the first critical speed. In particular, some recent high pressure centrifugal compressors

E. E. Hernandez et al. (Eds.): LASIRS 2019, MMS 86, pp. 65–75, 2020.
https://doi.org/10.1007/978-3-030-45402-9_8

have had serious non-synchronous vibrations. Recent analyzes on stability and destabilizing forces associated with couplings, bearings, seals, impellers, etc., attest to the importance of the problem. Within the current state of knowledge, a stability analysis of rotor bearing systems can be performed through the available software. The simplest analyzes are based on the energy dissipation associated with the dampened free vibrations of the system, while the more advanced analyzes consider real-time transient solutions in the rotor and can incorporate non-linear rolling characteristics, destabilization forces induced by the impeller, non-symmetrical supports, etc [7].

The purpose of this work is to develop a simplified stability analysis for a flexible rotor in rigid bearings based on the dampened free vibrations of the system due to the surface forces that are generated due to the interaction with the flow [6]. The analysis provides a quick method to establish guidelines during the design of the rotor and to compare existing rotor bearing systems, since the forces caused by the flow generate stresses whose resultant modifies the critical speeds when compared with a constant mass analysis. Although this procedure incorporates several simplifying assumptions and is obviously not intended to replace advanced analyzes (such as bearing instrumentation and aeroacoustic analysis), it is useful when considering system damping, a parameter that is an important factor with respect to stability [3].

A. Assumptions

The following simplifying assumptions are invoked in the present analysis.

- The stability criterion is based on the dampened free vibrations of the rotor bearing system (with cross coupling rotor forces ignored) at the first critical frequency.
- They are considered ideal bearings. (The effects of cross-coupling of the bearing are neglected depending on the inherent stability of the bearing of the tilt pad with respect to rotation).
- The rotor is supposed to be symmetric.
- In addition, vibration at the joints of the supports with the shaft is not considered.

2 Spring-Mass Model

The equations of movement and response of the rotor for the synchronous movement of an imbalanced Jeffcott rotor (massless axis, mass concentrated in half of the space) in linear isotropic bearings and flexible supports have been derived. In particular, it is shown that the response with rigid pedestals has an identical shape to that of a simple and one-dimensional spring-mass system [5].

$$\frac{F}{X_1} = K_E + i\omega C - M_S\omega^2 \tag{1}$$

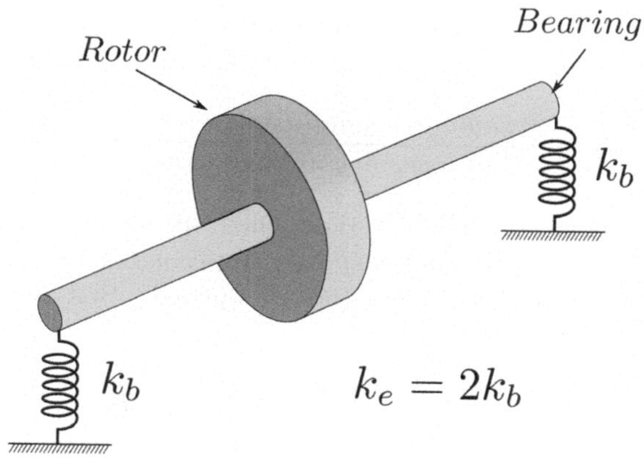

Fig. 1. Spring mass model

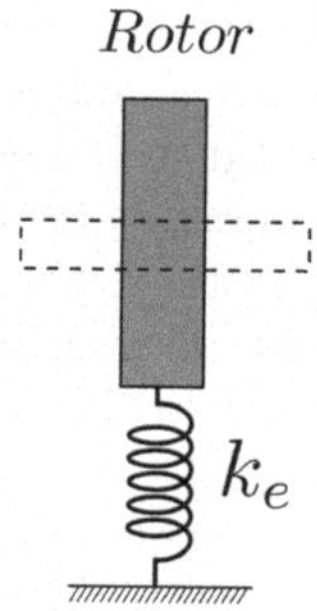

Fig. 2. Spring mass model 2D

Where

$$K_E = \frac{K_2\left(1+\frac{K_2}{K_1}\right)+\left(\frac{\omega}{K_1}\right)^2(C_2^2K_1+C_1^2K_2)}{\left(1+\frac{K_2}{K_1}\right)^2+\left(\frac{\omega}{K_1}\right)^2(C_1+C_2)^2} \tag{2}$$

$$C_E = \frac{C_1\left(\frac{K_2}{K_1}\right)+C_2+\left(\frac{\omega}{K_1}\right)^2C_1(C_2^2+C_1^2C_2)}{\left(1+\frac{K_2}{K_1}\right)^2+\left(\frac{\omega}{K_1}\right)^2(C_1+C_2)^2} \tag{3}$$

Since Eq. (1) is identical in shape to the analog equation for the simple spring-mass damping model shown in Fig. 2, then the spring and the damper associated with a simple equivalent model of the two-level system given by Eqs. (2) and (3), [4] develops the response of the simple system in Fig. 2 to a rotating unbalance of (Mse) as

$$X = \frac{M_S e\omega^2}{\sqrt{(K_E - M_S\omega^2)^2 + (C_E\omega)^2}} \tag{4}$$

Equations (2), (3) and (4) are identical to the dynamic results of the rotor with zero-axis damping, that is, with C $\backslash = 0$. Considering that the simple spring mass damping system of Fig. 2 is an equivalent model of a rotor bearing system, an important parameter is the classical damping factor f, defined as

$$Q = \frac{dynamic\ resonant\ displacement}{zero\ frequency\ displacement} \backsimeq \frac{1}{2\zeta} \quad (5)$$

As indicated above, the slightly dampened spring mass damping system (f < 0.2) shown in Fig. 2 will experience free vibrations or forced vibrations resonant at almost its natural frequency (not dampened), that is

$$Response\ Frequency \backsimeq \omega_n = \sqrt{\frac{K_E}{M_S}} \quad (6)$$

Since Ke is a frequency function, the determination of w, in Eq. (7) implies a fourth order expression. When solving the frequency, the parameters KE, CE and f can be evaluated from Eqs. (2), (3) and (5), respectively.

However, as the damping of the system decreases towards zero, the parameters KB, CE and, therefore, f become independent of the frequency and the analysis is considerably simplified. Since it has been stated that stability can become a factor of f < 0.055, it is appropriate to investigate a first order disturbance on this simple type solution to establish the application limits [3].

Considering a rotor bearing system with zero-axis damping (Ci = 0) and using Eqs. (5) and (7), the expressions for KE and CE given by Eqs. (2) and (3) become [1]

$$K_E = \frac{K_1 K_2}{K_1 + K_2} = \frac{1 + \left(1 + \frac{K_2}{K_1}\right)\left(\gamma \frac{K_E C_2}{C_E}\right)^2}{1 + \left(\gamma \frac{K_E C_2}{C_E}\right)^2} \quad (7)$$

$$C_E = \frac{C_2}{\left(1 + \frac{K_2}{K_1}\right)^2}\left[\frac{1}{1 + \left(\gamma \frac{K_E C_2}{C_E}\right)^2}\right] \quad (8)$$

where

$$\gamma = \frac{2\zeta}{(K_1 + K_2)} \quad (9)$$

As the damping factor tends to very small values, Eqs. (8) and (9) become

$$K_{E0} \equiv K_E \mid \zeta \to 0 = \frac{K_1 K_2}{K_1 + K_2} \quad (10)$$

$$C_{E0} \equiv C_E \mid \zeta \to 0 = \frac{C_2}{(1 + K_2/K_1)^2} \quad (11)$$

Substituting the above "zero" order solutions into the brackets of (8) and (9) results in the first order corrected solutions

$$K_{E1} = K_{E0}\left(\frac{A}{B}\right) \quad (12)$$

$$C_{E1} = \left(\frac{C_{E0}}{B}\right) \quad (13)$$

Where

$$A = 1 + 4\left(\frac{K_2}{K_1}\right)^2 (\zeta_0)^2 \left(1 + \frac{K_1}{K_2}\right) \quad (14)$$

$$B = 1 + 4\left(\frac{K_2}{K_1}\right)^2 (\zeta_0)^2 \tag{15}$$

$$\zeta_0 = \frac{C_{E0}}{2\sqrt{M_S K_{E0}}} \equiv \frac{C_{E0}}{2M_S \omega_{n0}} \tag{16}$$

From the given expressions, the ratios between the first and zero order approximations for ω_n and ζ are

$$\frac{\omega_{n1}}{\omega_{n0}} = \sqrt{\frac{A}{B}} \tag{17}$$

$$\frac{\zeta_1}{\zeta_0} = \frac{1}{\sqrt{AB}} \tag{18}$$

Thus, first order corrections to the "zero" order solution (which assumes very light dumping) may be obtained through Eqs. (18) and (19) if required.

3 Fluid Forces

The loads due to the flow that are presented in a rotating element depend on the angular velocity to which it is subjected since velocity components are generated that modify [2] the generated forces and the flow fields have a different behavior for each velocity value (Fig. 3).

Fig. 3. Blade fluid

For a three-dimensional model a rotor model like the one shown in the figure is used, with the obtained equations, the geometry of a rotor for finite element analysis and CFD respectively, where the initial model is as shown (Fig. 4):

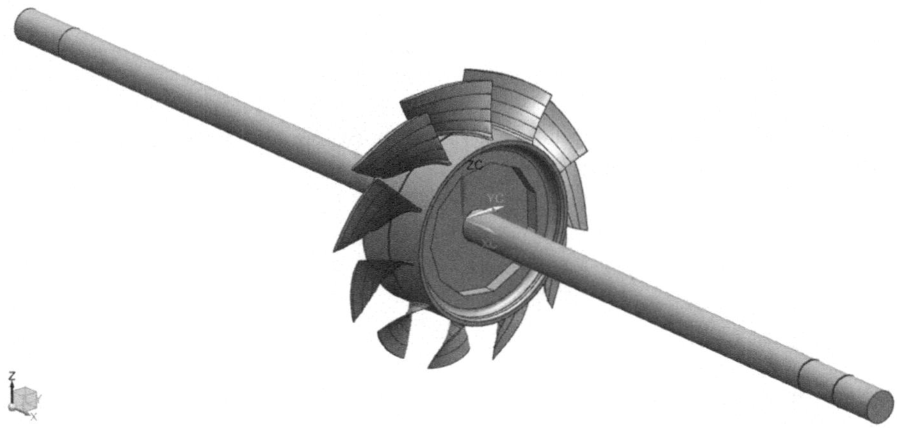

Fig. 4. Rotor model

With a discretization in mixed elements you get a mesh for structure and CFD analysis as shown in the figure (Fig. 5).

Fig. 5. Rotor mesh

Using the same configuration as in Fig. 1, and defining ideal supports and an angular speed of 3600 rpm, a behavior of forces was obtained as shown in the figure displacement in the mode 2. Where the displacements due to the flow produces a lateral displacement in the axis, which in turn modifies the flow field in the rotor (Figs. 6, 7, 8 and 9).

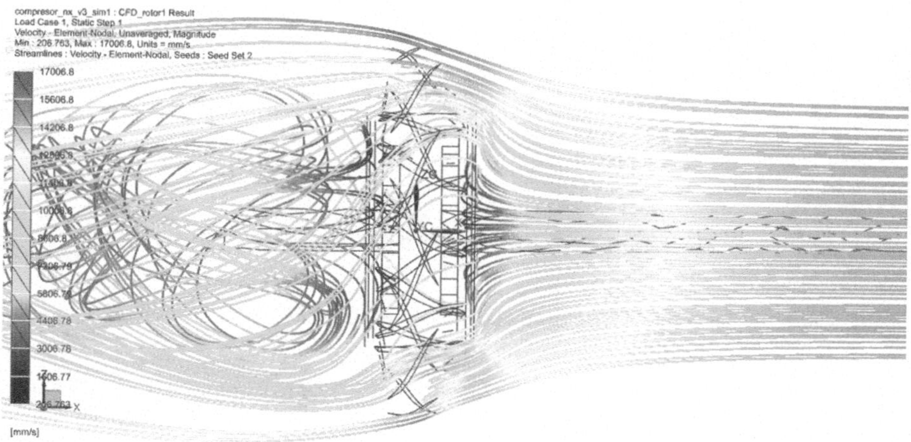

Fig. 6. Streamlines

Fig. 7. Deformation

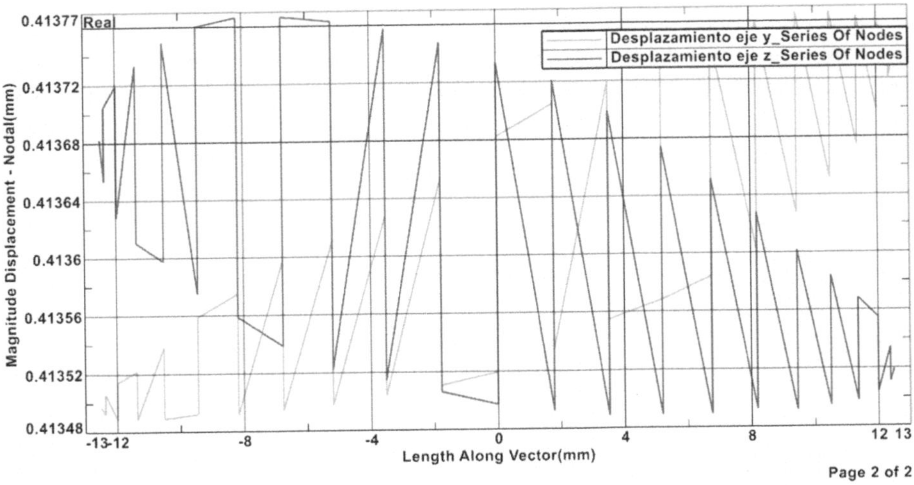

Fig. 8. Displacement x

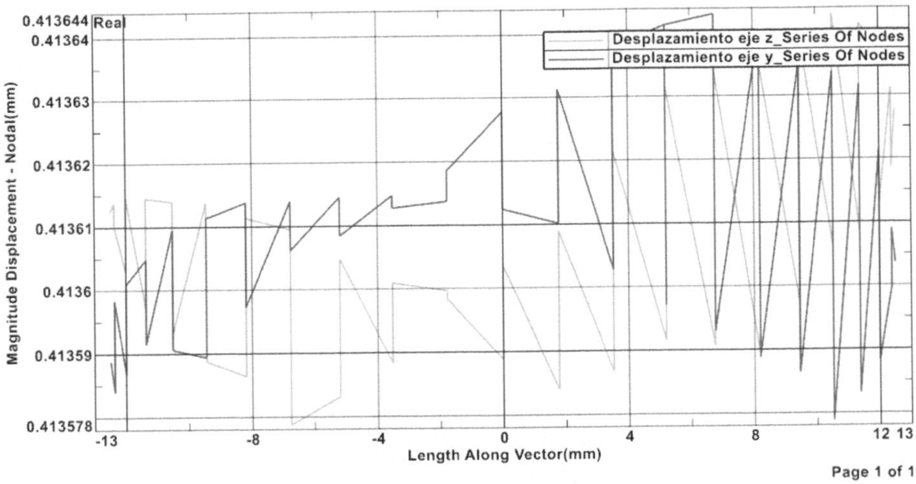

Fig. 9. Displacement y

4 Stability Criterion

Having established a means of determining the damping factor as a function of the rotor-bearing system parameters, it is appropnate to consider guidelines as to the values of the damping factor which can be associated with rotor stability. For this consideration, the log decrement "o" is a useful parameter. For a simple mass-spring-damper system (Fig. 2) with light damping

$$\delta \backsimeq 2\pi\zeta \tag{19}$$

$$\frac{U_n - U_{n+1}}{U_n} = 1 + e^{-2\delta} \tag{20}$$

$$UR_{1\,cycle} = (U - \epsilon U) + \epsilon U e^{-2\delta} \tag{21}$$

where U n + 1 is the energy associated with the (n + l)th cycle. Assume now that a constant amount of external energy "U" is supplied to the rotor-bearing system during each cycle of v1bration and that the dissipation function opera'tes on the total energy existing at the end of the previous cycle but only on a P?1·tion "E" of the energy supplied during the particular cycle. 'l hus the energy remaining in the system after one cycle is

$$UR_{ncycles} = U\left[1 + e^{-2\delta} + e^{-4\delta}..........e^{-2(n-1)\delta}\right] - \epsilon U(1 - e^{-2n\delta}) \tag{22}$$

and after cycles

$$UR_{ncycles} = U\left[1 + e^{-2\delta} + e^{-4\delta}..........e^{-2(n-1)\delta}\right] - \epsilon U(1 - e^{-2n\delta}) \tag{23}$$

in the limit, as the number of elapsed cycles become very large (i. e. $n \rightarrow \infty$)

$$UR_{n\rightarrow\infty} = U\left(\frac{1}{1 - e^{-2\delta}} - \epsilon\right) \tag{24}$$

In order to satisfy the boundary condition of $UR_{n\rightarrow\infty,\delta\rightarrow\infty} = 0$, the parameter ϵ must equal on (1.0), and hence

$$UR_{n\rightarrow\infty} = U\left(\frac{e^{-2\delta}}{1 - e^{-2\delta}} - \epsilon\right), \ \textit{where } \delta = 2\pi\zeta \tag{25}$$

Considering that detrimental nonsynchronous vibrations occur due to an accumulation of energy by the system as opposed to the specific quantity of input energy U associated with each cycle. For instance, if the destabilizing forces are small (small energy input per cycle). The usefulness or need of a more advanced analysis where required.

5 Sample Calculations

Sample calculations for a high-speed compressor rotors are shown using the loads due to the flow obtained in the CFD analysis. A numerical approximation was made and the behavior shown in the figures was obtained (Figs. 10 and 11).

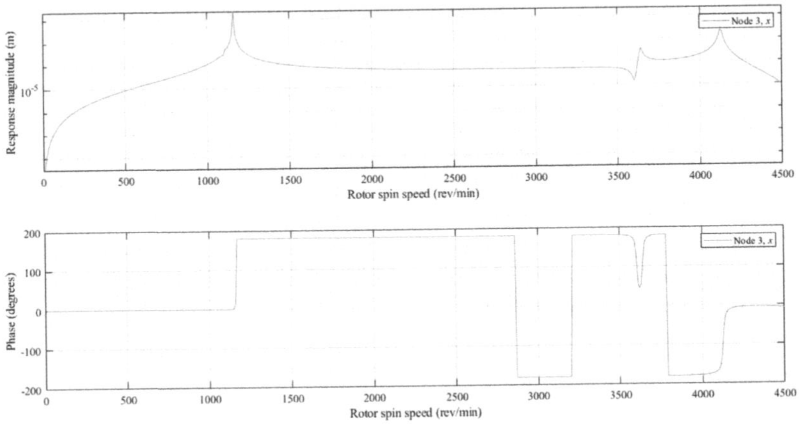

Fig. 10. Unbalance by fluid force

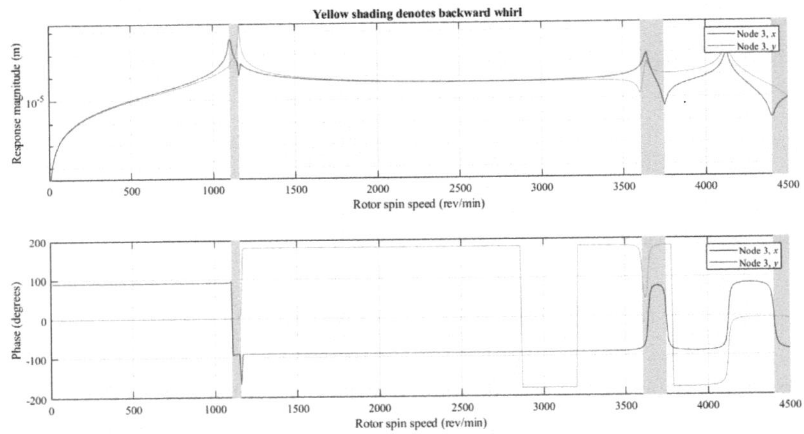

Fig. 11. Unbalance by fluid force

6 Conclusions

The paper presents a procedure for the analysis of stability due to forces generated by the flow in turbomachinery elements. The pressure excitations calculated by the harmonic methods of CFD are assigned to the structural model where a frequency response analysis is performed. This way, the influence of pressure fluctuations is estimated in the flow in the rotor towards the lateral displacements in the axis. In a post-processing stage an evaluation of for life that takes into account the vibrations induced by the flow. The procedure is considered for the use of models newspapers that are often applied in simulations of turbomachinery The vibration responses in the blades to the pressure fluctuations are calculated and the points are located of greater deformation in the axis. The procedure presented can be easily transferred to other areas of application, such as electromagnetically induced vibrations in motors or generators. A result of the frequency response analyzes can also be the sound pressure level for acoustic evaluations.

References

1. Genta G (2005) Dynamics of rotating systems. Mechanical engineering series. Springer, New York
2. Friswell MI, Penny JET, Lees AW, Garvey SD (2010) Dynamics of rotating machines. Cambridge University Press, Cambridge
3. Rao SS (2016) Mechanical vibrations. Prentice Hall, Upper Saddle River
4. Dimarogonas AD, Paipetis SA, Chondros TG (2013) Analytical methods in rotor dynamics. Springer, New York
5. Krämer E (2013) Dynamics of rotors and foundations. Springer, Heidelberg
6. Khader N (1997) Stability analysis for the dynamic design of rotors. J Sound Vib 207(3):287–299
7. Sinha SK (2004) Dynamic characteristics of a flexible bladed-rotor with Coulomb damping due to tip-rub. J Sound Vib 273(4–5):875–919

Flutter Analysis of Rotor Based on a Fluid–Structure Method

Jaime Cruz Cruz[2(✉)], Miguel Toledo Velázquez[1], Oliver M. Huerta Chávez[2], Gibran Jalil Garnica Castro[2], and Rafael Sánchez López[1]

[1] SEPI ESIME Zacatenco IPN, 07580 Ciudad de México, Mexico
mtv49@yahoo.com
[2] Tecnológico de Estudios Superiores de Ecatepec, 55210 Estado de México, Mexico
jcruzc0500@alumno.lpn.mx, lecram_21@hotmail.com
http://www.ipn.mx, http://www.tese.edu.mx

Abstract. The vibration that occurs in every rotating element (specifically in turbomachinery elements) generates effects that modify its operation and therefore its performance, which causes the system to compromise and not operate in nominal conditions. Of all the types of vibration that can occur due to various sources of excitement: body or flow scattering, what is clear is the generation of a forced vibration: where a self-excited vibration is generated and whose source energy It generates dynamic instability, which can lead to catastrophic failures due to the complexity of predicting these effects during design. Recent studies focus on aeroelasticity in the rotating elements, when using this approach, the elasticity equations and the aerodynamic forces generated in the blades are used, thus obtaining the equations of aeroelasticity, thus these equations are discretized and it is solved numerically to obtain the corresponding approximations.

The present work presents a structural model of a rotor considering the shifts that are presented by the fluid, for this case the complete rotor is studied not before analyzing the behavior of the flow in a blade, thereby obtaining displacements in the structure due to the flow and with it the vibration modes in the blade. The analysis in the blade is performed by FEM and CFD to obtain the behavior of the vibration modes considering the flow.

Keywords: Fluid-structure · Rotor dynamics · CFD · FEM

1 Introduction

The tendency of increasing the efficiency of gas-turbine engines implies the necessity of designing wide-chord unshrouded fan blades of complex spatial aerodynamic shape. However, with such blades, the risk of aeroelastic instability (e.g., flutter) development increases. For reliable and safe operation of gas-turbine engines with such fans, it is necessary to be able to accurately predict

E. E. Hernandez et al. (Eds.): LASIRS 2019, MMS 86, pp. 76–84, 2020.
https://doi.org/10.1007/978-3-030-45402-9_9

the aeroelastic behavior of fan blades at the design stage. The conventional approach to the utter calculations of the bladed disks is based on the frequency analysis method assuming that blade motions are described by harmonic time-dependent functions with a constant phase shift between the adjacent blades, whereas the natural modes and frequencies of the rotor low system remain similar to those in vacuum. This method ignores the feedback effect of the gas ow on blade oscillations.

This paper presents the solution methodology of the unsteady multidimensional aerodynamic problem coupled with the elastic blade oscillation problem in the gas ow. Also presented are the results of calculations for aeroelastic characteristics of a fan bladerow (BR) of a modern bypass gas-turbine engine.

2 Governing Equations

On the one hand, when considering the gas ow BR with oscillating blades, one has to take into account that the gasdynamic parameters depend on the blade velocity and position determined by a specied oscillation law (kinematically forced oscillations). On the other hand, when considering the problem of forced oscillations, the blade velocity and positions depend on the aerodynamic loads forcing the blades.

The adequate accounting of energy exchange can be attained by solving a coupled aerodynamic flow aeroelastic oscillations problem, based on simultaneous integration, using the time-marching scheme, of the complete system of equations of unsteady aerodynamics and aeroelastic blade oscillations [3,5].

The three-dimensional transonic ow of inviscid gas with zero thermal conductivity through the axial turbomachine is considered in the physical domain, including a fan BR, rotating at a constant angular velocity. The ow is described by the complete system of unsteady Euler equations presented in the integral form [4].

$$\frac{\partial}{\partial t}\int_{\Omega} f d\Omega + \oint_{\sigma} \vec{F}\vec{n} d\sigma + \int_{\Omega} H d\Omega \tag{1}$$

where

$$f = \begin{bmatrix} \rho \\ \rho v_1 \\ \rho v_2 \\ \rho v_3 \\ E \end{bmatrix} \qquad \vec{F} = \begin{bmatrix} \rho\vec{v} \\ \rho v_1\vec{v} + \delta_{1i}p \\ \rho v_2\vec{v} + \delta_{2i}p \\ \rho v_3\vec{v} + \delta_{3i}p \\ (E+p)\vec{v} \end{bmatrix}$$

$$H = \begin{bmatrix} 0 \\ \rho a_{e1} - 2\rho\omega v_2 \\ \rho a_{e2} + 2\rho\omega\psi_2 \\ 0 \\ 0 \end{bmatrix} \qquad \delta_{ji} = \begin{cases} 1, & j = i \\ 0, & j \neq i \end{cases} \qquad E = \rho(\varepsilon + \frac{v_1^2 + v_2^2 + v_3^2 + r^2\omega^2}{2})$$

Ω is the volume of the finite-volume computational cell with moving boundaries σ; $\overrightarrow{n}$ is the unit exterior normal; p and ρ are the pressure and density, respectively; ω is the angular rotation velocity; v_1, v_2 and v_3 are the components of the velocity vector $\overrightarrow{v}$; a_{e1} and a_{e2} are the projections of traslational acceleration; E is the total energy per unit volume; ε is the internal energy per unit mass; and r is the distance to the rotation axis.

The system of Eqs. (1) is supplemented with the ideal gas equation of state:

$$\varepsilon = \frac{1}{k-1}\frac{p}{\rho}$$

where k is the specific heat ratio.

Equations (1) are integrated in the computational domain, including the complete circumference, i.e., the number of interblade channels is equal to the number of blades. The number of interblade channels, N, and interblade phase angle δ are related to each other as $N\delta = 2\pi j$ (where j is the integer number). [2].

The computational grid is divided into z segments, either of which includes one blade and possesses the circumferential length equal to a rotor pitch. Each segment is discretized using the hybrid $H - H$ grid for rotor channels. The external H-grid remains fixed, while the internal H-grid is reconstructed at every iteration in accordance with the position of the oscillating blade [7]

The discrete form of Eqs. (1), obtained for an arbitrary spatially deforming grid [1]:

$$\begin{gathered}
\tfrac{1}{\Delta t}\left[f^{i+1/2,j+1/2,k+1/2}\Omega^{i+1/2,j+1/2,k+1/2} - f_{i+1/2,j+1/2,k+1/2}\Omega_{i+1/2,j+1/2,k+1/2}\right] \\
+\left[-\left(f\sigma\omega_n\right)_{i+1} + \left(f\sigma\omega_n\right)_i - \left(f\sigma\omega_n\right)_{j+1} + \left(f\sigma\omega_n\right)_j - \left(f\sigma\omega_n\right)_{k+1} + \left(f\sigma\omega_n\right)_k\right] \\
+\left[\left(F_1\sigma\right)_{i+1} - \left(F_1\sigma\right)_i + \left(F_2\sigma\right)_{j+1} - \left(F_2\sigma\right)_i + \left(F_3\sigma\right)_{k+1} - \left(F_3\sigma\right)_k\right] \\
+H_{i+1/2,j+1/2,k+1/2}\Omega_{i+1/2,j+1/2,k+1/2} = 0
\end{gathered}$$

Where subscripts and superscripts stand for "new" and "old" cells; $f = \{\rho, \rho\overrightarrow{v}, E\}$ is the symbolic vector of unknown variables; F_1, F_2 and F_3 are the values of variables in the centers of cell faces; σ and ω_n are the face area and normal velocity in the face center, respectively. The gasdynamic parameters on the faces are determined from the solution of Riemann problem with the use of iterative process [6].

The dynamic model of the oscillating blade in linear formulation is described by the matrix equation [8]:

$$[M]\{\ddot{u}(x,t)\} + [C]\{\dot{u}(x,t)\} + [K]\{u(x,t)\} = 0$$

Where $[M]$, $[C]$, and $[K]$ are mass, damping, and stiffness matrices, respectively; $\{u(x,t)\}$ is the blade displacement; and $[F]$ is the vector of unsteady aerodynamic loads. Using the modal approach:

$$\{u(x,t)\} = [U(x)]\{q(t)\} = \overset{[}{i} = 1]N\sum\{U_i(x)\}q_i(t)$$

Where $U_i(x)$ is the blade displacement vector in the ith mode, and $q_i(t)$ is the modal coefficient. Based on the condition of orthogonality of natural modes, the Eq. (2) can be transformed to the system of independent differential equations with respect to the modal coefficient of natural modes:

$$\ddot{q}_i(t) + 2h_i\dot{q}_i(t) + \omega_i^2 q_i(t) = \lambda_i(t)$$

With h_i is the damping coefficient of the ith mode; ω_i is the natural frecuency of the ith mode; λ_i is the modal force corresponding to the displacement within These ith mode, calculated in the iteration based on the pressure distribution on the blade.

$$\lambda_i = \frac{\iint\limits_{\sigma} p\overline{U}_i\overline{n}d\sigma}{\iiint\limits_{v} \rho\overline{U}_i^2 dv}$$

Where ρ is the pressure on the blade surface.

3 Numerical Modeling

The numerical simulation of aeroelastic characteristics was performed for a fan, consisting of 24 blades (Fig. 1) for the regime with n = 0:7 at the operation line. The rst ve natural modes and frequencies of blade oscillation are shown in Fig. 1 Fan bladerow Table 1 (all natural frequencies are related to the rst natural frequency). The calculations were performed using 10 25 58 (radial tangential axial) computational grid for each interblade channel.

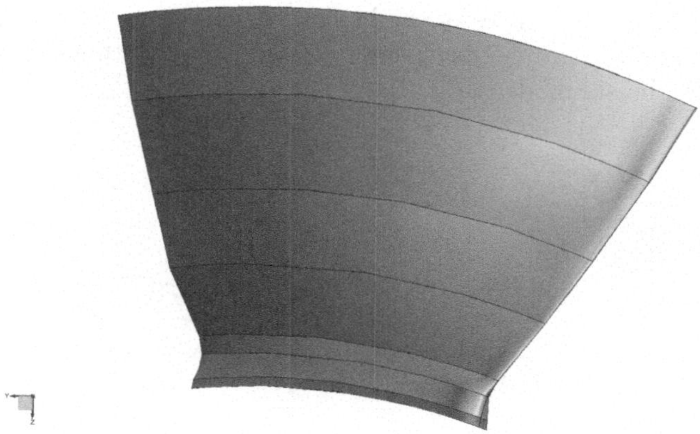

Fig. 1. Blade rotor CAD

At the first stage, the aerodynamic calculations for a rotating fan BR with a preset blade oscillation law were made. All blades were assumed to exhibit harmonic oscillations according to each of the natural modes following the same law with a constant interblade phase angle (IBPA) Æ:

$$q_{ij} = q_{i0}\sin\left[2\pi\nu_i t + (j-1)\delta\right]$$

where qij is the modal coefficient; i is the natural mode number; j is the blade number; qi0 is the oscillation amplitude of the ith mode; and i is the natural frequency.

The mesh created with tetrahedral elements adjusting to the volume of the blade (Fig. 2).

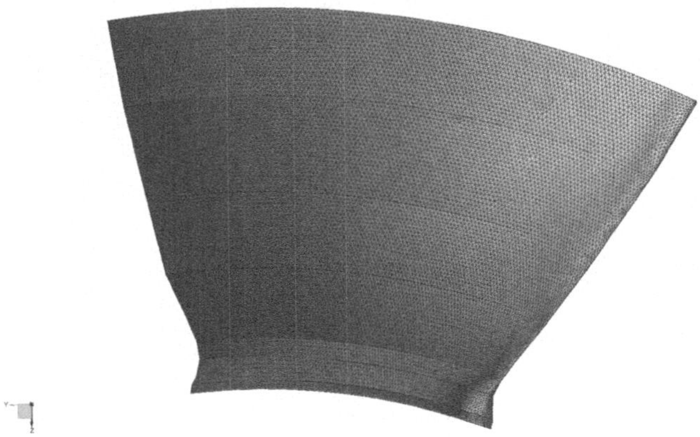

Fig. 2. Blade rotor Mesh

The calculation of the aerodynamic forces that occur in the wing in the influence due to the flow generate the oscillations in them and at the same time generate oscillation frequencies in the blades.

The calculations were performed for harmonic oscillations of blades on a 3600 rpm with the six modes (Fig. 3).

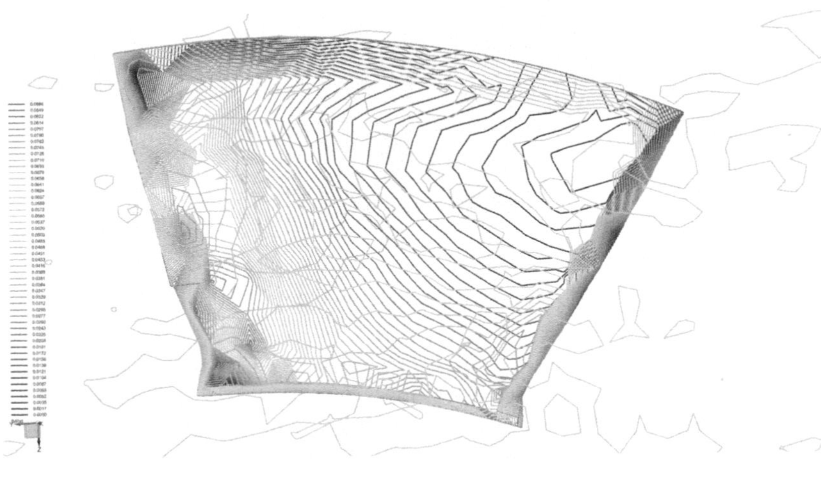

Fig. 3. Pressure distribution

Of the previous figure, the behavior of the pressure distribution is observed and the following figure, a graph of its behavior (Fig. 4).

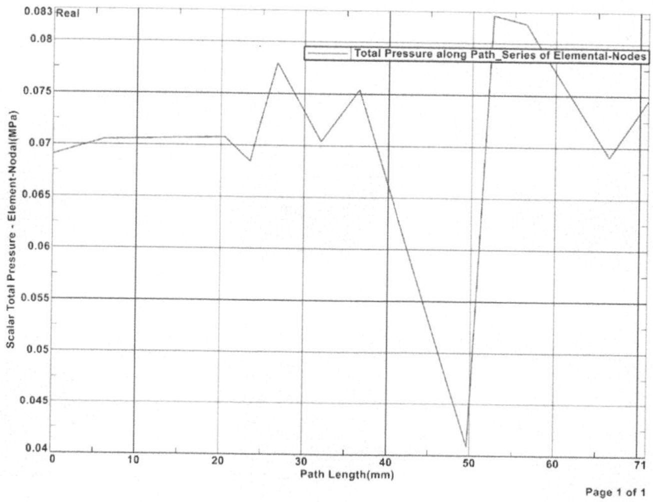

Fig. 4. Pressure

A cascade configuration shows the behavior of a blade to obtain its vibration modes due to flow loads. For a speed of 3600 rpm the distribution of pressures in the blade in Fig. 5 is observed.

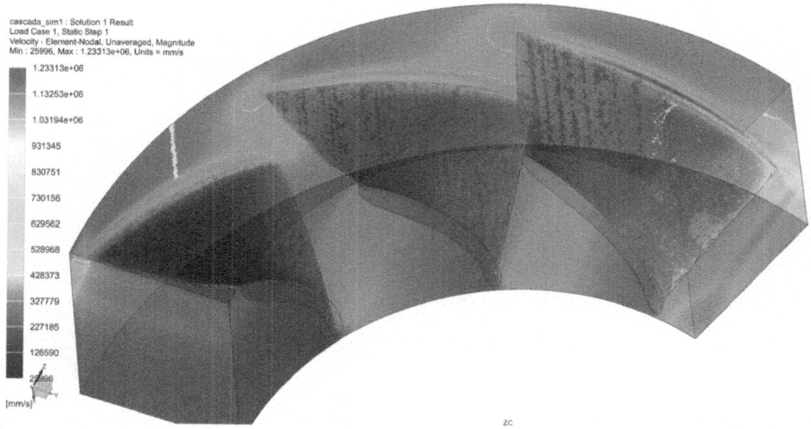

Fig. 5. Pressure distribution

Depending on the angular velocity, the variations change in each mode so that a whole system must be analyzed in order to obtain its natural frequencies and thus have data that help the prediction of failures in the rotating elements (Figs. 6, 7, 8 and 9).

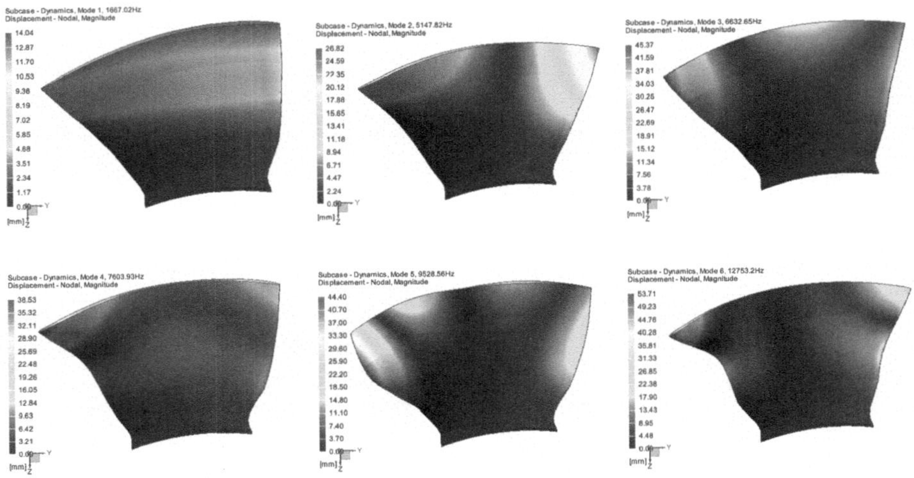

Fig. 6. Six modes of rotor

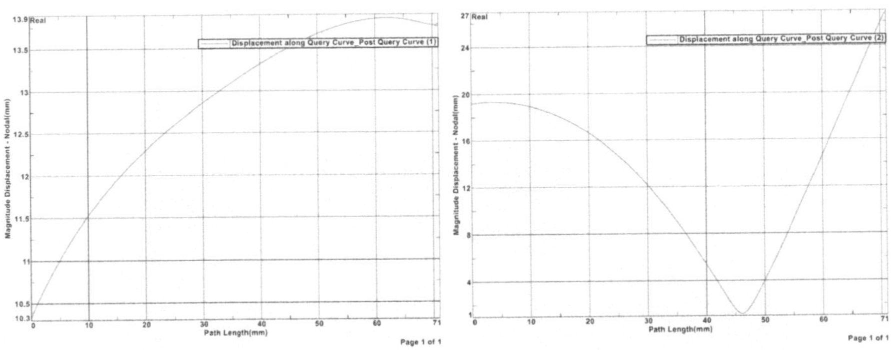

Fig. 7. Displacement mode 1 and 2

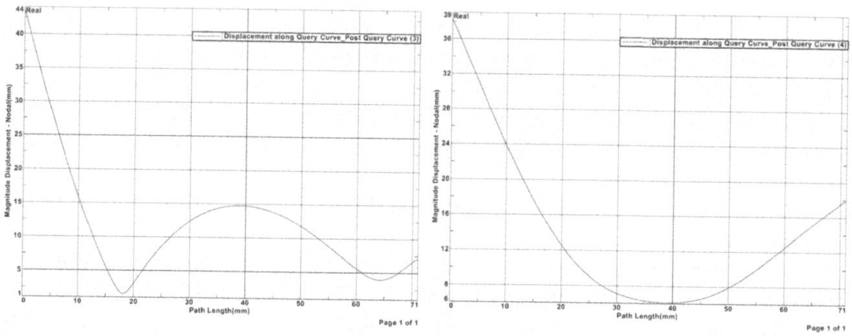

Fig. 8. Displacement mode 3 and 4

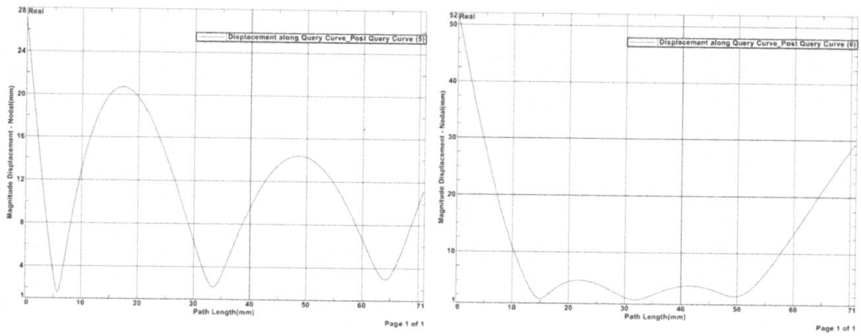

Fig. 9. Displacement mode 5 and 6

4 Conclusions

The present work presented the numerical simulation of the aerolastic characteristics of the Arabic using the Fluid structure model where flow properties were found that affect the modal behavior for each frequency.

It was found that there is a damping in the case of harmonic oscillations of the blades, for each of the natural modes with the interaction between the modes taken into account, that is, the energy of the oscillating blade is transferred to the main flow and vice versa as well as a damping due to the flow that can reduce the amplitudes and oscillation frequencies in the blade by means of the constant feeding of fluid during the operation of the rotor, that is to say in the gas-rotor coupling.

References

1. Hughes TJR (2012) The finite element method: linear static and dynamic finite element analysis. Dover Civil and Mechanical Engineering
2. Genta G (2005) Dynamics of rotating systems. Mechanical engineering series. Springer, New York

3. Blake WK (2017) Mechanics of flow-induced sound and vibration, general concepts and elementary sources, vol 1, 2nd edn. Academic Press, Cambridge
4. Friswell MI, Penny JET, Lees AW, Garvey SD (2010) Dynamics of rotating machines. Cambridge aerospace series. Cambridge University Press, Cambridge
5. Paidoussis MP (1998) Fluid-structure interactions: slender structures and axial flow. Elsevier Science, Oxford
6. Chakrabarti SK (2005) Numerical models in fluid-structure interaction. WIT Press, Southampton
7. Taylor NV, Allen CB, Gaitonde A, Jones DP (2004) A structure-coupled CFD method for time-marching flutter analysis. Aeronaut J 108:389–401
8. Sayma AI, Vahdati M, Imregun M (2000) An integrated nonlinar approach for turbo-machinery forced response prediction part 1: formulation. J Fluids Struct 14:87–101

Amphibious Propeller Optimization for Aerial and Underwater Operation

C. Genaro Carreon[1(✉)], C. Oliver M. Huerta[2], M. Alfredo Arias[2], C. Sergio Torres[2], and José Ángel L. Ortega H.[1]

[1] IPN-ESIME Unidad Zacatenco, SEPI, Av. Luis Enrique Erro S/N, Nueva Industrial Vallejo, 07738 Gustavo A. Madero, Ciudad de México, Mexico
gcarreonc1000@alumno.ipn.mx

[2] IPN-ESIME Unidad Ticoman, SEPI, Av. Ticoman 600, San José Ticoman, 07340 Gustavo A. Madero, Ciudad de México, Mexico
lecram_21@hotmail.com, aarias@ipn.mx, sgtorres.c@gmail.com, https://www.ipn.mx/

Abstract. In this paper a novel methodology for biphasic flow (air-water) is presented. It involves a CST-parametrization and the optimization by NSGA-II coupled to design and optimize amphibious propellers, through CL, CD and CP coefficients, which allow to operate in aerial and underwater environment. Finally, the results of CL, CD and CP, were evaluated by XFoil CFD software before and after of the optimization and then compared to each other, Evolving the optimization process for 140 generations, give a total of 70000 individuals and obtain a relative improvement for the new propeller design as nearly as 30%.

Keywords: Propeller · Multiobjetive optimization · CFD simulation · Genetic algorithm

1 Introducción

Unmanned vehicles are more present in daily activities. We can find drone projects from that can deliver mail packages to even the most sophisticated ones as used in unfriendly zones like hurricanes. However, always we need new systems that they are able to operate in two or more environment to maximize their respective operations, minimize response time and reduce production costs for each vehicle to specific condition.

Proof of it, are the investment did by [9], who made a functional diagram of different vehicles sittings, whit capabilities to work in two distinct media. The Fig. 1 shows Drones for two o more media. This paper allowed us to say that the actual hybrid vehicles are not only concepts, but they already exist and are developed for various applications.

Supported by LASIRS.

E. E. Hernandez et al. (Eds.): LASIRS 2019, MMS 86, pp. 85–96, 2020.
https://doi.org/10.1007/978-3-030-45402-9_10

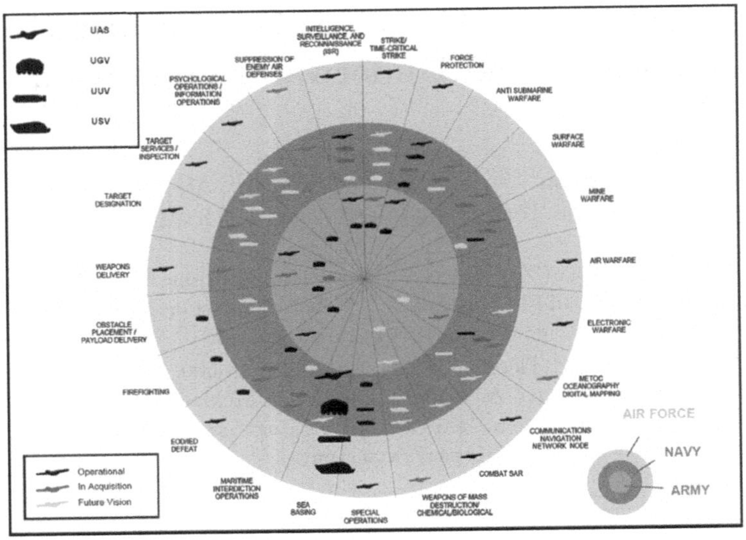

Fig. 1. Unmanned autonomous vehicle classification for operations in two media [9]

Research like this one has 4 principal aspects to develop: "Structure", "Electronics", "Control" and "Propulsion". In the propulsion section, this will adapt to the operation parameters in both media, specially to try to balance and if possible, maximize the operational performance. In our case, we set an operation rate as 50% aerial and 50% underwater.

The first problem we find is the different density between air and water which is nearly 1000 times different. This entails that the propulsion system has different operation objectives. While the aerial propeller tends to have high revolutions, the underwater propeller is quite the opposite.

The second problem is, consequence by the first, the cavitation effect of high revolutions of the propeller, phenomena that reduce the lifecycle and his performance in underwater operation. In the other hand, in the aerial operation phase, the working losses are present by compressible effects in the propeller blade tips.

The third and last problem lies in the optimization itself, since, by improving the underwater operation, aerial performance is decreased, therefore is necessary to find the best operation ratio, that would get us the maximum functioning for both media. To do this, is necessary to make use of a multi-objective optimization method.

The multi-objective optimization methods allow to find the best values for two or more variables. Following this idea, we use NSGA-II technique [5]. This code, which is a NSGA successor, present some advantages:

- Introduction of a non-dominance control.
- Relatively simple setting to adapting our problem.
- Scanning of the most solutions quantities possible.

To realize the optimizations we need a software with abilities to solve the coefficients with the lower time possible to have potentials to analyze the highest individuals number enough to ensure the population variety during the optimization.

The solution is to use XFoil [6]. XFoil software is based on a panel method for CFD, it is able to evaluate in 2–4 s, depending of variable number (i.e. Reynolds number, angle of attack, viscid or inviscid flow, etc). However, the precision that offer this package is suitable for the study to be realized.

The selected coefficients for this paper are related to lift, drag and pressure. The first two are applied to the aerial operation while the last is related to the cavitation phenomena through the Thoma coefficient (cavitation coefficient). This dimensionless value allows us to predict the existence of cavitation phenomena and its calculation is relatively simple, thus, is the perfect way to solve the problem mentioned before.

2 Theoretical Background

In this section, we approach the necessary process to set the geometric parametrization of the airfoil. The objective functions for the optimization and its relationship with de above point and finally de CFD analysis with XFoil to calculate the main coeffcients.

There are several types of techniques to use for the parametrization of the airfoil, which are listed below:

- Bsplines
- Parsec
- Bezier Curves

These are some of the most used procedures for the airfoil's parametrization process. However, these approaches have different troubles that make them nonfunctional for this paper. For example, Parsec [1,7] use airfoils characteristics having 12 equations with 12 control variables. Nevertheless, this produce a small design space, additianlly to the complexity to code the equations that define this method.

In other hand, the Kulfan CST method [2–4] use two functions to represent many different curves and volumes in 2D or 3D that are used most commonly in the aircrafts.

$$\zeta(\Psi) = C(\Psi)S(\Psi) + \Psi\zeta_{te} \tag{1}$$

The Eq. 1 is composed of one form and one class function. The form function indicate what kind of geometry is used; the class function delimits the curve in the plane (i,j). The strength of this method lies in the Bernstein polynomial, which one can be worked from second order (8 control variables by airfoil) to infinite order number.

Nevertheless, the higher the degree of the equation, the greater number of unknown quantities control and this turn more complex the optimization

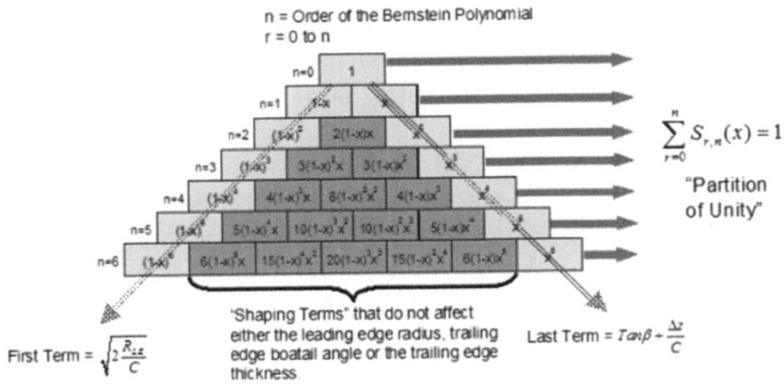

Fig. 2. Bernstein polynomial decomposition used in shape function 1 [4]

method, moreover, increasing the calculation time. For that, a suitable polynomial of grade 4 is used. This gives us 12 variables and obtain a great accuracy for the geometry parametrization (Fig. 2).

Applying the Bernstein fourth grade polynomial to the form function and combining in the CTS general equation we obtain the final version of the equations for the optimization, in which we calculate the Bernstein coefficients to find the expression that handle the airfoil used in the amphibious propeller. To Find the coefficients of the Bernstein polynomial fast and suitable, we use a stochastic search method based in the geometry comparation.

This search method is "fminsearch" MATLAB funcion, and the comparative geometry function is as follows.

$$\sum_{x=0}^{x=i} = abs(z_{par} - z_{ref}) \tag{2}$$

where z_{par} refers to the distance of the geometry by the parametrization method while z_{ref} indicates the original distance by the initial geometry. This function is applied both in extrados and intrados airfoil's surfaces.

For the selection of the optimization we apply the NSGA-II algorithm. This technique, in GPL license and code, operate under a non-dominated control to evaluate all individual generated. The use of this methodology is based on the work done by [8] where they shows all benefits of an analysis of a large number of individuals, which guarantees a solutions variety and thus be able to conclude in a optimal solution for the objective functions previously set.

It necessary keep in mind that the optimization techniques always result in a minimal value, but if the function is coded in specific form, we can obtain as a result a maximum value. Preserving this in mind, the objective functions must fulfill the following:

- C_L and C_P must be a maximum
- C_D must be a minimum

The aerodynamical coefficients evaluations must be done in the shortest time possible in order to analyze as many geometries as possible and, to achieve this, the XFoil [6] software is used. This software allows to have results oh the right precision to this problem with the advantage of being fast (in approximately-break 2–5 s). Another important advantage is its ease of integration with other softwares like MatLab.

3 Implementation and Results

The mathematical expression that represents the airfoil geometry, is a equation combining both CST methodology and Bernstein polynomial. This combination result as next shape function (3).

$$\begin{aligned} S(x) = \; & A_1(1-x)^5 + 5A_2(1-x)^4x + 10A_1(1-x)^3x^2 \\ & + 10A_4(1-x)^2x^3 + 5A_5(1-x)x^4 + A_6x^5 \end{aligned} \tag{3}$$

The class function 4, with the coefficients to NACA airfoil by Kulfan procedure are as follows:

$$C(\Psi) = (\Psi)^{0.5}[1-\Psi]^{1.0} \tag{4}$$

The Ψ value is equal to x/c where c represent the airfoil chord. To simplify, the class function is set a unitary chord remaining all the equation depending only of x.

Combining the Eqs. 3 and 4 into the Eq. 1 is obtained the function value to represent the position in the "y" axis; while the x values belongs to are taking by the x positional variable.

This process is inserted in the search method with fminsearch to establish the value of A_i coefficients since they are the ones that determine the behavior profile. Initially these quantities are assigned a unitary value, nevertheless, the ζ value is related to trailing edge and is set to 0.001.

To validate that coefficients found in this step, the Eq. 2 is used, where Z_{par} is the value obtained by the parametrization technique in every coordinate x and is compared to the original position of the airfoil (Fig. 3).

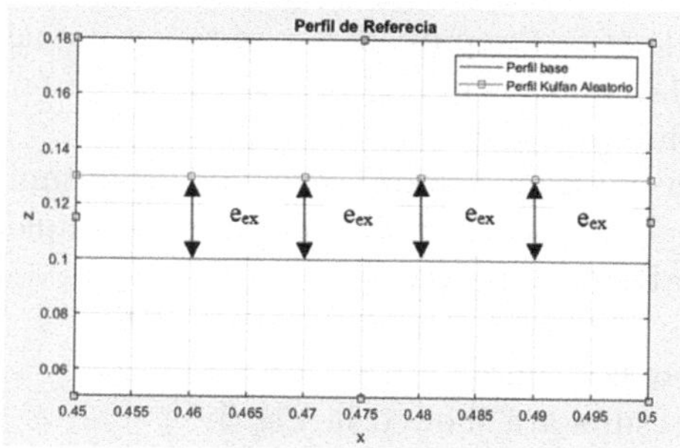

Fig. 3. Graphical representation of function 2 operation

The execution time of search require nearly 1–3 min taking into account the maximum iterations number which enables MatLab, the maximum vector size and the computational resource. The final equation to define the CST parametrization method is as follows:

$$\begin{aligned} S(x) = & \left((x)^{0.5}[1-x]^{1.0}\right)\left(A_1(1-x)^5 + 5A_2(1-x)^4 x + 10A_1(1-x)^3 x^2 \right. \\ & \left. + 10A_4(1-x)^2 x^3 + 5A_5(1-x)x^4 + A_6 x^5\right) + x\zeta \end{aligned} \tag{5}$$

In the other hand, the objective functions are set as follows so that they can satisfy the goals of the above section.

$$f(1) = 1 - C_L \tag{6}$$

$$f(2) = C_D \tag{7}$$

$$f(3) = 1 + C_P \tag{8}$$

The Thoma coefficient (Eq. 9), dimensionless quantity, link local absolute pressure, steam pressure and the kinetic energy by volume. This amount (expressed in negative values) allows to predict the existence of the cavitation phenomena in the fluid.

$$\sigma = \frac{p_0 - p_v}{0.5\rho W^2} = \frac{P_{ATM} + \rho g h - p_v}{0.5\rho W^2} \tag{9}$$

The cavitation effect occurs when the lower negative pressure coefficient (C_P) is greater that the Thoma coefficient, for example

$$-Cp > thoma$$

$$-32 > -20$$

From a strict view point, the C_P amount lower that Thoma coefficient ensures that phenomena cavitation will be present. Therefore, the objective function related to C_P is maximize his rate to avoid cavitation.

The coefficient evaluation was made through XFoil software. This code works mainly with the Reynolds number, this term helps the operations because only is needed such number in each designed station to improve. For this paper is selected 10 stations along to the propeller span to get a quasi-three-dimensional optimization. Nevertheless, the tip and root stations are omitted because those give a negligible propulsion work.

For the above reason, only 8 stations remain to optimize of the which are the conditions of the Table 1. All topics above were implemented in the code NSGA-II through MatLab language.

The results to the parametrization as shown in Fig. 4. The solution way is related with the follow flowchart.

Parametrization results are showed in Fig. 5:

Table 1. Reynolds number table for the different stations chosen

Station	Reynolds number
20%	50620.3900
30%	58952.7626
40%	64505.6269
50%	66193.2478
60%	66301.9375
70%	62530.9932
80%	56496.3187
90%	44574.3631

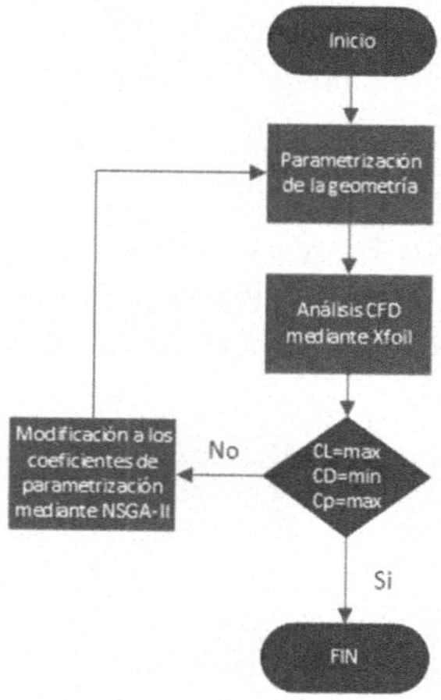

Fig. 4. Implementation flowchart of the used methodology

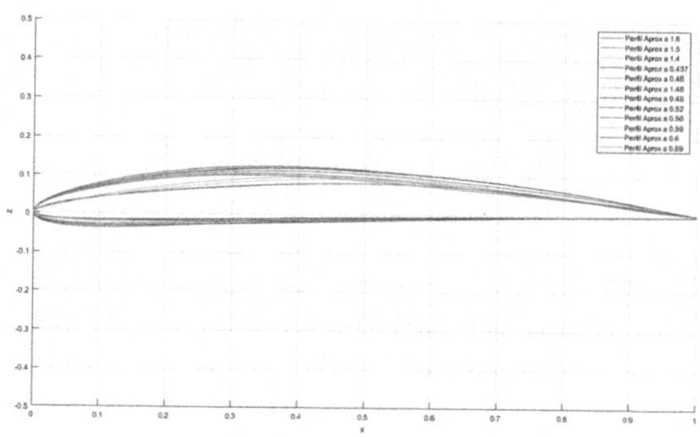

Fig. 5. Generated geometries in the parametrization

For the other hand, analyzing before and after the parametrization all remain stations is presented in the Table 2.

Table 2. Initial C_L values

Station	C_L
20%	1.052
30%	0.962
40%	1.394
50%	1.330
60%	1.351
70%	1.354
80%	1.204
90%	0.824

The Fig. 6, shows a Pareto Front, which is a graph where are compared the solutions in which that the optimal is the solution 1 and cannot exits other solution that improve an objective and worsen other.

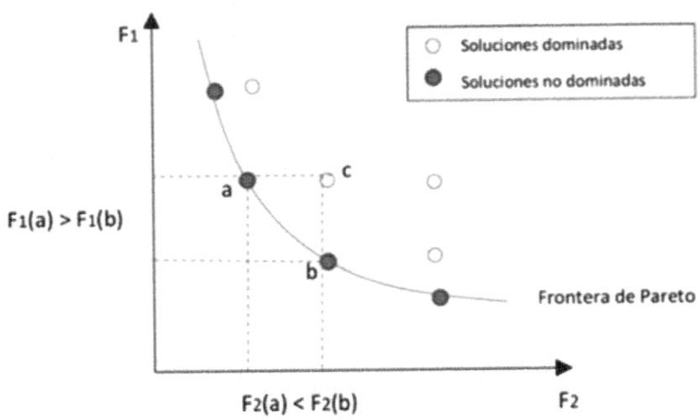

Fig. 6. Pareto's front example

To get a general point of view about the optimization is chosen randomly some individuals from the outer and the middle of the Pareto front curve to define how in the far zones only and objective is improved meanwhile in the intermediate zone is obtained a balance between the objective functions.

The objective functions are next

$$f(1) = 2 - C_L \tag{10}$$

$$f(2) = C_D \tag{11}$$

$$f(3) > -24.72 \tag{12}$$

To help the optimization and analysis through Pareto's front, the problem is sectioned in two parts. The first section is about to maximize the lift to drag ratio; the second part is increasing the C_P value in the selected sections along the propeller span. The sectioned problem helps to analyze the created geometries as well as make easier the graph visualization about the objectives and how was the improvement in coefficients on each environment media work.

The results related to aerodynamical objective, are depicted in the Fig. 7

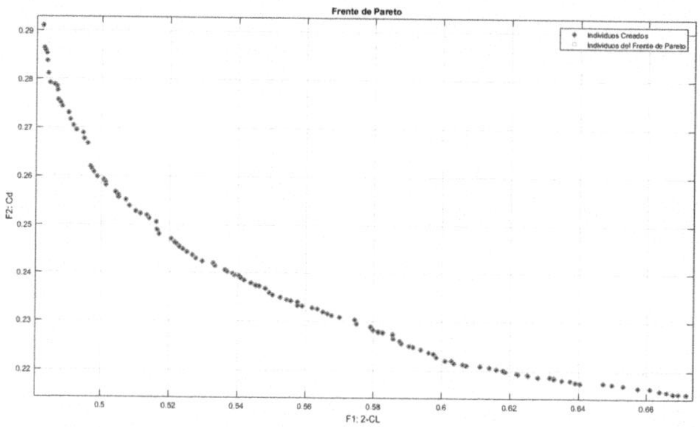

Fig. 7. Pareto's front of station 7.

Some individuals selected from the Pareto's front are given in the Fig. 8

From those airfoils, now is executed the improvement to the pressure coefficient, substituting the Reynolds number associated to the underwater operation. From this new calculation, its obtained the Table 3 which specify the Thoma quantities for each designated station.

Table 3. Table of Thoma coefficients for each station

Station	C_P
20%	−21.0528
30%	−19.1875
40%	−18.3791
50%	−17.3411
60%	−15.4947
70%	−13.6608
80%	−11.8019
90%	−11.3029

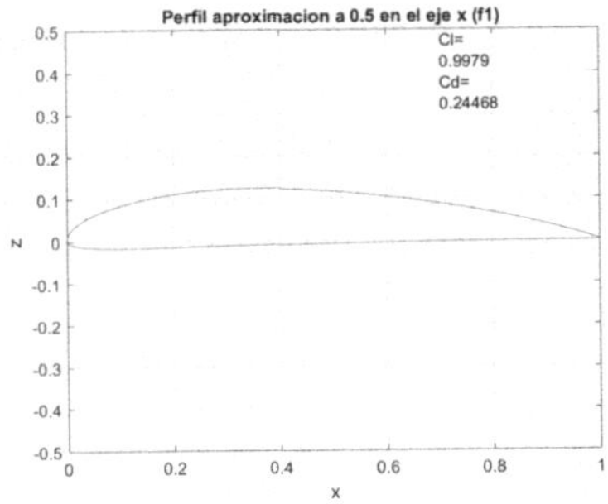

(a) Airfoil in left zone (C_D y C_L menores)

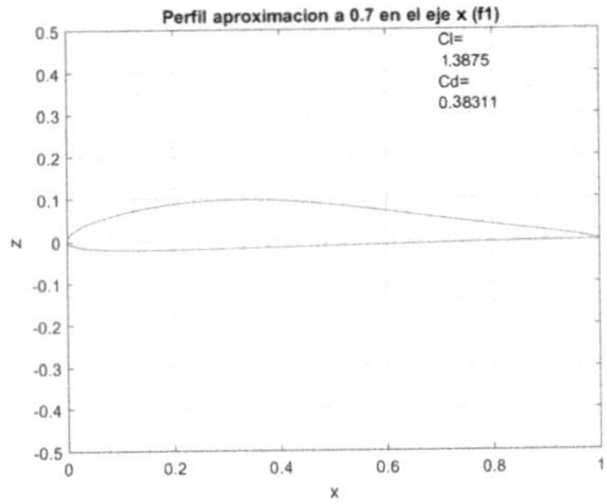

(b) Airfoil in right zone (C_D y C_L mayores

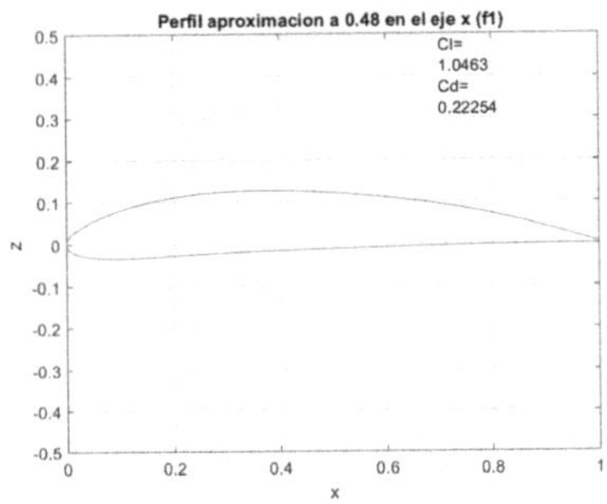

(c) Airfoil in middle zone (Equilibrio entre C_D y C_L)

Fig. 8. Airfoil improved for station 7

Thus, the new airfoils made with this improvement are showed in the Fig. 9 which are the final geometries.

Finally, the comparation between C_P values are showed in the Table 4, before and after geometry optimization.

Table 4. Comparison of pressure coefficient before and after optimization

Station	C_P before	C_P after
20%	−9.392	−6.1
30%	−8.395	−5.35
40%	−7.396	−5.026
50%	−6.401	−4.61
60%	−5.405	−3.87
70%	−4.4	−3.135
80%	−3.411	−2.39
90%	−2.415	−2.19

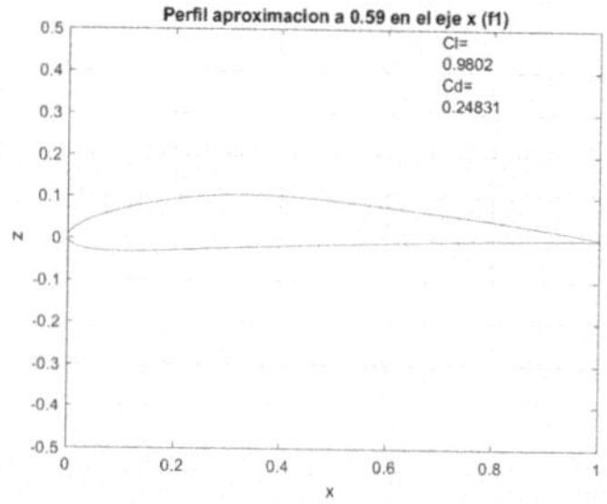

(a) Airfoil in left zone (C_D y C_L menores)

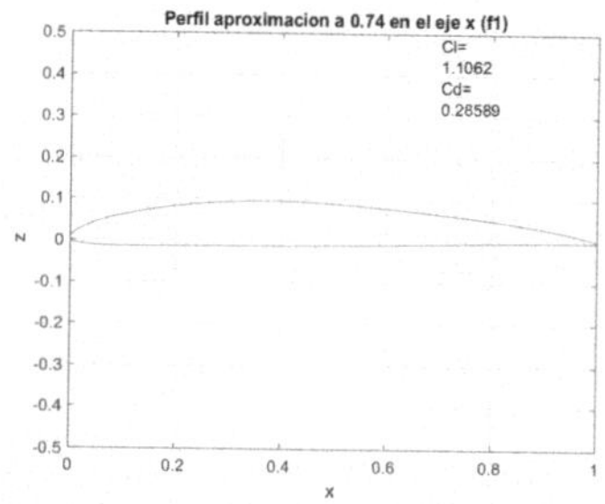

(b) Airfoil in right zone (C_D y C_L mayores

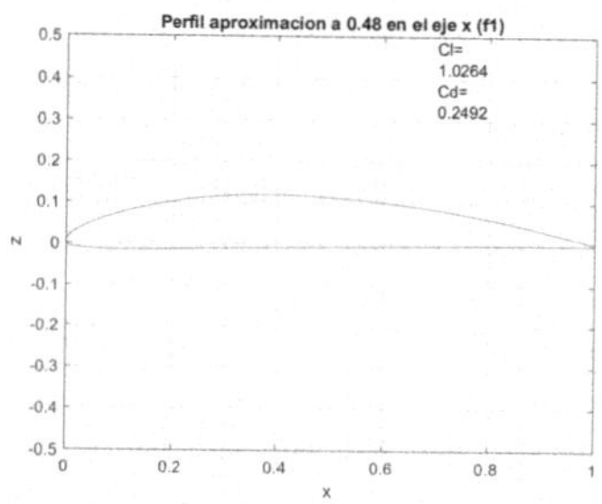

(c) Airfoil in middle zone (Equilibrio entre C_D y C_L)

Fig. 9. Airfoil hydrodynamic improved for station 7

4 Conclusions

When doing this work it was possible to set up a new novel methodology for optimization in biphasic fluid to UAV propellers with the 50-50 ratio work objective. Also, can be proved how, depend which individual for Pareto's front is selected, one of the all objective functions is improved. The best option is located in the middle zone, due to the setting the different objectives. In other way, the objective functions must be set carefully to made a right selection in the Pareto's front. Another important aspect to the optimization is found on the CFD solver. A code faster with the right accuracy to the problem, help to have a biggest variety of geometries, ensuring a bigger search space with a lower calculation time. The Table 4, shows the aerodynamical and hydrodynamical improvement for the airfoils in the different stations. Although a value decrease of aerodynamical improvement, the hydrodynamical operation was benefitedt significantly because the cavitation phenomena was avoided. These results demonstrate the feasibility and effectiveness of the methodology used to improve the amphibious operation in a 50-50 relationship eliminating adverse main effects during both operations.

References

1. Arias-Montano A, Coello CAC, Mezura-Montes E (2011) Evolutionary algorithms applied to multi-objective aerodynamic shape optimization. In: Koziel S, Yang XS (eds) Computational optimization, methods and algorithms. Springer, Heidelberg, pp 211–240
2. Kulfan BM (2008) Universal parametric geometry representation method. J Aircr 45(1):142–158
3. Kulfan BM (2007) Recent extensions and applications of the "CST" universal parametric geometry representation method. In: AIAA paper
4. Kulfan BM, Bussoletti JE (2006) Fundamental parametric geometry representations for aircraft component shapes. In: 11th AIAA/ISSMO multidisciplinary analysis and optimization conference
5. Deb K, Pratap A, Agarwal S, Meyarivan TAMT (2002) A fast and elitist multiobjective genetic algorithm: NSGA-II. IEEE Trans Evol Comput 6:182–197
6. Drela M (1989) XFOIL: an analysis and design system for low Reynolds number airfoils. In: Mueller TJ (ed) Low reynolds number aerodynamic, vol 54. Lecture notes in engineering. Springer, Heidelberg, pp 1–12
7. Hájek J (2009) The generalized PARSEC parameterization for airfoil optimization. In: Aerodynamic Research and Test Institute, Prague, 9 October 2009
8. Ponce LAJ, Montaño AA (2016) Parametrización Geométrica y Optimización Aerodinámica de Alas en Flujo Transónico. Instituto Politécnico Nacional, November 2016
9. Yang X, Wang T, Liang J, Yao G, Liu M (2015) Survey on the novel hybrid aquatic-aerial amphibious aircraft: aquatic unmanned aerial vehicle (AquaUAV). In: ELSEVIER progress in aerospace sciences, pp 131–151

Trajectory Tracking of Multiple Quadrotors While Maintaining Time-Varying Spatial Formations via Synchronization

C.-A. Arteaga-Escamilla(✉), R. Castro-Linares(✉), and J. Álvarez-Gallegos(✉)

Research Center for Advanced Studies (Cinvestav), Mexico City, Mexico
{cruzantonio.arteaga,rcastro,jalvarez}@cinvestav.mx

Abstract. In this paper, the trajectory tracking problem of multiple quadrotors while maintaining time-varying spatial formations is addressed through a synchronization approach. The aim is to control each quadrotor to track its desired trajectory while its motion is synchronized with the others within the formation to keep a desired time-varying formation. First, a way to linearize the translational dynamics of the quadrotor through feedback state is shown, considering the nonlinear dynamic model. An attitude PD controller for each quadrotor is also designed to guarantee the attitude error approaches zero asymptotically. Second, a synchronous controller for each quadrotor translation is developed to guarantee that both position and synchronization errors approach zero asymptotically. Finally, simulations were performed to demonstrate the effectiveness of the proposed synchronization control approach in the formation control tasks.

Keywords: Formation · Multiple quadrotors · Synchronization

1 Introduction

Study on the formation control of multiple quadrotors (Unmanned Aerial Vehicles, UAVs for short) has received increasing attention in the last decade. There are many strategies to solve the multiple UAVs formation control problem, for example, leader-follower approach, virtual structure approach, cluster space strategy, among others. In the leader–follower approach some robots take the role of leaders and aim to track predefined trajectories, while the follower robots follow the leaders according to a relative posture [6–8,13]. A disadvantage is that there is no feedback from the followers to the leaders. In the virtual structure approach, each robot is treated as an element of a rigid virtual structure, each of them has a local control that makes the structure possible to maintain a given arrangement [10,15,16]. This formation scheme is difficult to be implemented in

Supported by Conacyt, under Project CB-254329.

E. E. Hernandez et al. (Eds.): LASIRS 2019, MMS 86, pp. 97–108, 2020.
https://doi.org/10.1007/978-3-030-45402-9_11

some applications. The cluster space strategy conceptualizes the n-robot system as a single entity (cluster), and desired motions are specified as a function of cluster attributes (state variables) such as position, orientation, and geometry. These state variables form the system cluster space. Cluster space state variables may be related to robot-specific state variables through a formal set of kinematic transformations. These transformations allow cluster commands to be converted to robot specific commands to achieve the individual trajectory tracking [1,2,4].

In this paper we propose to use a synchronization control strategy to address the multiple quadrotors formation control problem. The main idea is that a team of quadrotors tracks each individual desired trajectory while synchronizing motions among the others to keep relative kinematics relationship for maintaining the desired time-varying shape. The main contribution of this formation scheme is that the desired shape of the formation can take any spatial shape, unlike [11,12,14,17,18], where the desired shape of the formation must have a planar ellipse form. Besides, this formation scheme is easy to be implemented in applications where the formation is spatial.

2 Dynamic Model of Quadrotor

Consider an inertial coordinate frame $G = \{X_G, Y_G, Z_G\}$ fixed to ground and a body fixed coordinate frame $B = \{X_B, Y_B, Z_B\}$. A picture of a quadrotor with the corresponding frames is shown in Fig. 1. The generalized coordinates are given by $\boldsymbol{\xi} = \begin{bmatrix} x \; y \; z \end{bmatrix}^T$ and $\boldsymbol{\eta} = \begin{bmatrix} \phi \; \theta \; \psi \end{bmatrix}^T$, where $\boldsymbol{\xi}$ denotes the spatial position coordinate of the center of gravity respect to the inertial frame G, and $\boldsymbol{\eta}$ denotes the Euler angles. These angles have the following names and they represent a rotation around their corresponding axes: ϕ is the roll angle around X_B axis, θ is the pitch angle around Y_B axis, and ψ is the yaw angle around Z_B axis. The Newton-Euler approach was used to obtain the dynamical model of the quadrotor, which considers a quadrotor as a rigid body under external forces applied to the center of mass; so the dynamic model of the quadrotor is given by [5,9].

$$m\ddot{\boldsymbol{\xi}} = \boldsymbol{R}\,{}^B\boldsymbol{F} - mg\,{}^G\boldsymbol{D}, \tag{1}$$

$$\boldsymbol{J}\dot{\boldsymbol{\Omega}} = -\boldsymbol{\Omega} \times \boldsymbol{J}\boldsymbol{\Omega} + \boldsymbol{\tau}, \tag{2}$$

where m is the mass of the quadrotor, ${}^G\boldsymbol{D} = \begin{bmatrix} 0 \; 0 \; 1 \end{bmatrix}^T$ is a unit vector in Z_G direction, g is the gravity constant; $\boldsymbol{J} \in \mathbb{R}^{3\times3}$ denotes the inertia matrix respect to B; $\boldsymbol{\Omega} = \begin{bmatrix} \Omega_1 \; \Omega_2 \; \Omega_3 \end{bmatrix}^T$ represents the angular velocity in the body frame B. $\boldsymbol{\tau} = \begin{bmatrix} \tau_\phi \; \tau_\theta \; \tau_\psi \end{bmatrix}^T$ is the control torque defined in the body fixed frame B. ${}^B\boldsymbol{F} = \begin{bmatrix} 0 \; 0 \; u \end{bmatrix}^T$ is the control input vector defined in the body fixed frame B. $u > 0$ is the total thrust force generated by the rotors, this is $u = \sum_{i=1}^4 f_i$, where f_i is the thrust force generated by the ith rotor. $\boldsymbol{R} \in SO(3)$ is the so-called rotation matrix that represents the orientation of the rigid body respect to the inertial frame such that $\boldsymbol{R} : B \rightarrow G$ is orthogonal, which is obtained by making 3 successive rotations using the convention $X_B Y_B Z_B$. Using the compact notation $c\theta = cos\theta$ and $s\theta = sin\theta$, $\boldsymbol{R}$ is given by

$$\boldsymbol{R} = \begin{bmatrix} c\theta c\psi & s\phi s\theta c\psi - c\phi s\psi & c\phi s\theta c\psi + s\phi s\psi \\ c\theta s\psi & s\phi s\theta s\psi + c\phi c\psi & c\phi s\theta s\psi - s\phi c\psi \\ -s\theta & s\phi c\theta & c\phi c\theta \end{bmatrix} \tag{3}$$

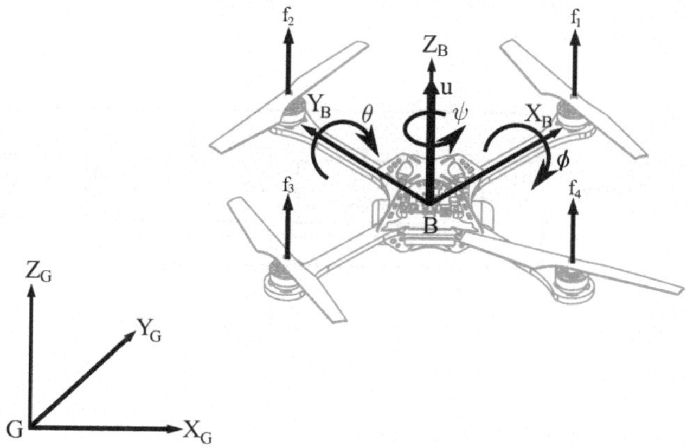

Fig. 1. Coordinate systems and moment/forces acting on the quadrotor.

The kinematic relation of the angular velocity $\boldsymbol{\Omega}$ and the generalized angular velocities $\dot{\boldsymbol{\eta}}$, in the region where the Euler angles are valid, is given by [5,9].

$$\boldsymbol{\Omega} = \boldsymbol{W}_\eta \dot{\boldsymbol{\eta}}, \tag{4}$$

where

$$\boldsymbol{W}_\eta = \begin{bmatrix} 1 & 0 & -sin\theta \\ 0 & cos\phi & sin\phi cos\theta \\ 0 & -sin\phi & cos\phi cos\theta \end{bmatrix} \tag{5}$$

Substituting (4) in (2), the dynamic model of the quadrotor is obtained, expressed in terms of the generalized coordinates, given by

$$m\ddot{\boldsymbol{\xi}} = \boldsymbol{R}^B\boldsymbol{F} - mg^G\boldsymbol{D}, \tag{6}$$

$$\boldsymbol{J}\boldsymbol{W}_\eta\ddot{\boldsymbol{\eta}} = -\boldsymbol{J}\dot{\boldsymbol{W}}_\eta\dot{\boldsymbol{\eta}} - \boldsymbol{W}_\eta\dot{\boldsymbol{\eta}} \times \boldsymbol{J}\boldsymbol{W}_\eta\dot{\boldsymbol{\eta}} + \boldsymbol{\tau}, \tag{7}$$

with the control input vectors given by ${}^B\boldsymbol{F} = \begin{bmatrix} 0 \; 0 \; u \end{bmatrix}^T$ and $\boldsymbol{\tau} = \begin{bmatrix} \tau_\phi \; \tau_\theta \; \tau_\psi \end{bmatrix}^T$.

3 Control Strategy

Assuming that the closed-loop dynamics of rotation is much faster than the translational one, it is possible to separate the model in two systems [3]. The control strategy block diagram for each UAV is shown in Fig. 2, where the superscript denotes the desired value of any variable. The control strategy consists of the designing of a controller via synchronization for the translational dynamics such that it guarantees the trajectory tracking and solves the formation control problem, providing as its output the desired orientation to be fed as input to the attitude stabilization controller. Such synchronous controller is used for all UAVs within the formation.

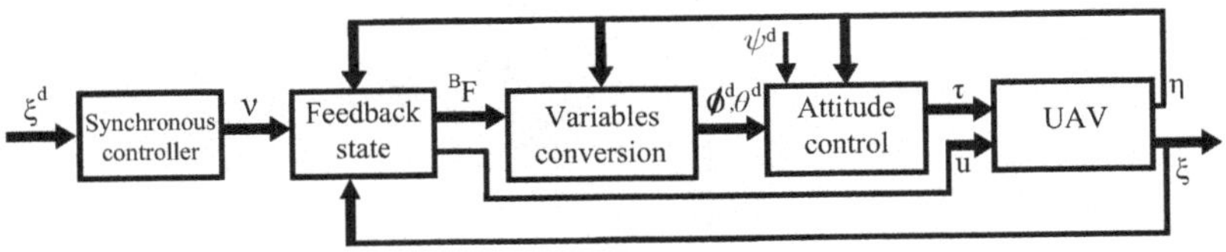

Fig. 2. Control strategy block diagram for each UAV.

3.1 Dynamic Model Linearization

In order to linearize the translational dynamics of the UAV, consider (6), suppose that it is possible to apply forces in all directions, i.e., the translational dynamics of the UAV is given by

$$m\ddot{\boldsymbol{\xi}} = \boldsymbol{R}^B\boldsymbol{T} - mg^G\boldsymbol{D}, \tag{8}$$

where ${}^B\boldsymbol{T}$ is a virtual control input vector. By using a feedback state given by

$${}^B\boldsymbol{T} = m\boldsymbol{R}^{-1}(g^G\boldsymbol{D} + \boldsymbol{\nu}), \tag{9}$$

where $\boldsymbol{\nu} \in \mathrm{I\!R}^3$ is an auxiliary control input vector, which will be defined in Sect. 4; a linear translational dynamics is obtained

$$\ddot{\boldsymbol{\xi}} = \boldsymbol{\nu}.$$

Remark 1: Note that $\boldsymbol{R}$ is singular when $cos\phi cos\theta = 0$.

Obviously, it is not possible to apply forces in all directions (only in Z_B direction). In order to ${}^B\boldsymbol{F}$ force to have the same magnitude, direction and sense of ${}^B\boldsymbol{T}$ force, then the UAV must rotate, for that reason, in (6), $\boldsymbol{R}$ will be replaced by $\boldsymbol{R^d}$, which denotes the desired orientation of the UAV. Therefore, next equations must be satisfied:

$$\boldsymbol{R^d}({}^B\boldsymbol{F}) = \boldsymbol{R}^B\boldsymbol{T}, \tag{10}$$

and

$$\|^{B}\boldsymbol{F}\| = \|^{B}\boldsymbol{T}\| \,. \tag{11}$$

From (11), it is obtained

$$| u | = \|^{B}\boldsymbol{T}\| \,.$$

Since the thrust fcrce generated by the rotors is always positive, then the thrust force that the rotors must generate is given by

$$u = \|^{B}\boldsymbol{T}\| \,. \tag{12}$$

Substituting (9) in (10), it is obtained

$$\boldsymbol{R}^{d}(^{B}\boldsymbol{F}) = m(g^{G}\boldsymbol{D} + \boldsymbol{\nu})$$

$$\begin{bmatrix} s\psi^{d}s\phi^{d} + c\psi^{d}s\theta^{d}c\phi^{d} \\ -c\psi^{d}s\phi^{d} + s\psi^{d}s\theta^{d}c\phi^{d} \\ c\theta^{d}c\phi^{d} \end{bmatrix} u = m(g^{G}\boldsymbol{D} + \boldsymbol{\nu})$$

$$\begin{bmatrix} s\psi^{d}s\phi^{d} + c\psi^{d}s\theta^{d}c\phi^{d} \\ -c\psi^{d}s\phi^{d} + s\psi^{d}s\theta^{d}c\phi^{d} \\ c\theta^{d}c\phi^{d} \end{bmatrix} = \frac{m(g^{G}\boldsymbol{D} + \boldsymbol{\nu})}{u} = \begin{bmatrix} N_x \\ N_y \\ N_z \end{bmatrix} \tag{13}$$

Taking ψ^{d} as a constant value, where

$$\psi^{d} \in [-\pi, \pi] \backslash \{\pm\pi/2\},$$

it is possible to find the angles that the UAV must rotate to satisfy (10), these angles are given by

$$\phi^{d} = arcsin(N_x sin\psi^{d} - N_y cos\psi^{d}), \tag{14}$$

$$\theta^{d} = arcsin\left(\frac{N_x - sin\phi^{d} sin\psi^{d}}{cos\phi^{d} cos\psi^{d}}\right). \tag{15}$$

In order to linearize the rotational dynamics of the UAV consider (7), where $\boldsymbol{\tau}$ is the control input vector. By using a feedback state given by

$$\boldsymbol{\tau} = \boldsymbol{J}\dot{\boldsymbol{W}}_{\eta}\dot{\boldsymbol{\eta}} + \boldsymbol{W}_{\eta}\dot{\boldsymbol{\eta}} \times \boldsymbol{J}\boldsymbol{W}_{\eta}\dot{\boldsymbol{\eta}} + \boldsymbol{J}\boldsymbol{W}_{\eta}\boldsymbol{\Gamma}, \tag{16}$$

where $\boldsymbol{\Gamma} = \left[\Gamma_{\phi}\ \Gamma_{\theta}\ \Gamma_{\psi}\right]^{T}$ is an auxiliary control input vector, a linear rotational dynamics is obtained and given by

$$\ddot{\boldsymbol{\eta}} = \boldsymbol{\Gamma}. \tag{17}$$

Remark 2: The dynamics of translation and rotation of the UAV are not independent; because when applying an input torque given by $\boldsymbol{\tau}$, a rotation occurs in the UAV, and this causes a resultant thrust force that will cause the UAV to accelerate in X_G or Y_G direction.

3.2 Attitude PD Controller

The attitude error of the UAV is defined as

$$\bar{\boldsymbol{\eta}} = \boldsymbol{\eta}^d - \boldsymbol{\eta},$$

where $\boldsymbol{\eta}^d = \begin{bmatrix} \phi^d \ \theta^d \ \psi^d \end{bmatrix}^T$ is the desired attitude and $\dot{\boldsymbol{\eta}}^d = \ddot{\boldsymbol{\eta}}^d = 0$. The proposed attitude control law for the UAV is given by

$$\boldsymbol{\Gamma} = Kd_o \dot{\bar{\boldsymbol{\eta}}} + Kp_o \bar{\boldsymbol{\eta}} \tag{18}$$

where $Kp_o, Kd_o > 0$. Substituting (18) in (17), it is obtained an attitude error dynamics of the UAV given by

$$\ddot{\bar{\boldsymbol{\eta}}} + Kd_o \dot{\bar{\boldsymbol{\eta}}} + Kp_o \bar{\boldsymbol{\eta}} = 0,$$

which is asymptotically exponentially stable.

4 Synchronous Controller Design

Consider a group of n UAVs with their translational dynamics given by

$$\ddot{\boldsymbol{\xi}}_i = \boldsymbol{\nu}_i, \tag{19}$$

where $\boldsymbol{\xi}_i = \begin{bmatrix} x_i \ y_i \ z_i \end{bmatrix}^T$ denotes the spatial position coordinate of the ith UAV and $\boldsymbol{\nu}_i$ is the control input respect to $\boldsymbol{\xi}_i$. The position tracking error of the ith UAV is defined as

$$\boldsymbol{e}_i = \boldsymbol{\xi}_i^d - \boldsymbol{\xi}_i, \tag{20}$$

where $\boldsymbol{\xi}_i^d$ is the desired spatial position coordinate. Similar to [14], the synchronization error of the ith UAV is defined as

$$\boldsymbol{\epsilon}_i = \boldsymbol{e}_i - \boldsymbol{e}_{i+1}. \tag{21}$$

Note that when $i = n$, $i+1$ is denoted as 1. Besides, if the synchronization error $\boldsymbol{\epsilon}_i = 0$ for all $i = 1, 2, ..., n$, then the synchronization goal is achieved.

In a similar manner to [14], in order to link the position and synchronization errors into one, the coupling error of the ith UAV is defined as

$$\boldsymbol{E}_i = \boldsymbol{e}_i + \beta \int_0^t (\boldsymbol{\epsilon}_i - \boldsymbol{\epsilon}_{i-1}) d\zeta, \tag{22}$$

where $\beta > 0$. Note that the coupling error feeds back the information of the two neighboring UAVs $i-1$ and $i+1$. Note that when $i = 1$, $i-1$ is denoted as n. Differentiating $\boldsymbol{E}_i$ with respect to time yields

$$\dot{\boldsymbol{E}}_i = \dot{\boldsymbol{e}}_i + \beta(\boldsymbol{\epsilon}_i - \boldsymbol{\epsilon}_{i-1}), \tag{23}$$

$$\ddot{\boldsymbol{E}}_i = \ddot{\boldsymbol{e}}_i + \beta(\dot{\boldsymbol{\epsilon}}_i - \dot{\boldsymbol{\epsilon}}_{i-1}). \tag{24}$$

The proposed control law for each UAV is given by

$$\boldsymbol{\nu}_i = \ddot{\boldsymbol{\xi}}_i^d + \beta(\dot{\boldsymbol{\epsilon}}_i - \dot{\boldsymbol{\epsilon}}_{i-1}) + \Lambda\dot{\boldsymbol{E}}_i + \Upsilon\boldsymbol{E}_i, \tag{25}$$

where Λ, $\Upsilon > \beta$. Substituting the controller (25) into the UAV translational dynamics in (19) leads to the coupling error dynamics given by

$$\ddot{\boldsymbol{E}}_i + \Lambda\dot{\boldsymbol{E}}_i + \Upsilon\boldsymbol{E}_i = 0, \tag{26}$$

which is asymptotically exponentially stable. Therefore, the proposed synchronous controller (25) leads to asymptotic exponentially convergence of both position tracking and synchronization errors to zero.

5 Simulation

A simulation was performed to verify the effectiveness of the proposed synchronization control approach.

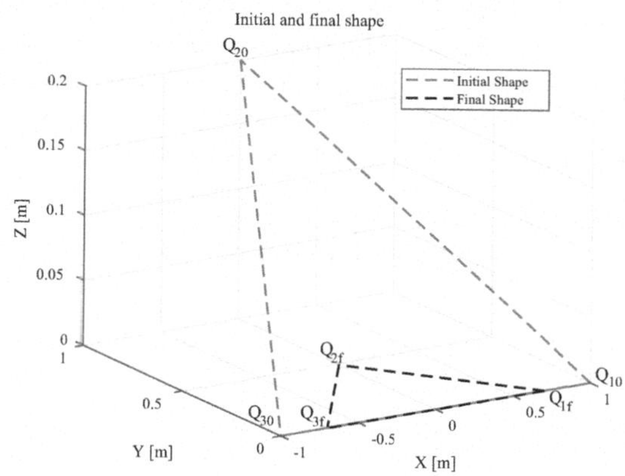

Fig. 3. Initial and final desired shape.

5.1 Build the Formation, Trajectory Tracking and Switch of Shapes

Simulation is divided in three parts, every part takes 10 s. First part consists of each UAV reaches its desired position to build the desired initial shape. The second part consists of trajectory tracking while maintaining the formation shape. Finally, the UAVs have to switch from the initial to the final desired shape. Figure 3 illustrates the initial and final desired shape, where each vertex represents each UAV. The desired shape is defined by specifying distances among

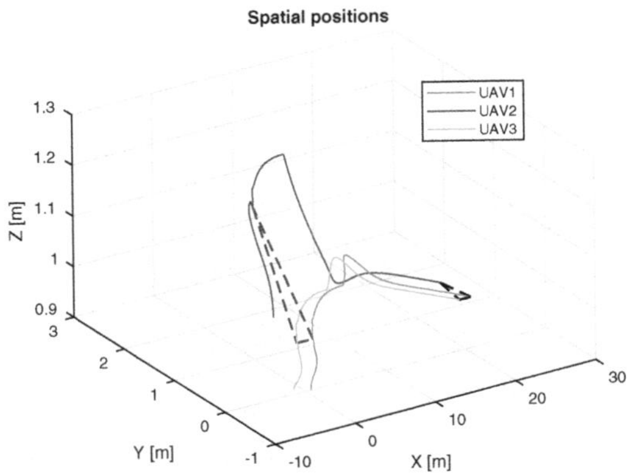

Fig. 4. Spatial positions of UAVs.

UAVs or specifying desired position coordinates respect to a reference point (the origin is the reference point). To build the desired initial shape, the desired coordinates of each UAV are: $Q_{10}(1, 0, 1)$, $Q_{20}(0, 1, 1.2)$ and $Q_{30}(-1, 0, 1)$. For the desired final shape, the desired coordinates of each UAV are: $Q_{1f}(0.7, 0, 1)$, $Q_{2f}(0, 0.5, 1)$ and $Q_{3f}(-0.7, 0, 1)$, as shown in Fig. 3. L_1 denotes the distance between UAV 1 and 2, L_2 denotes the distance between UAV 2 and 3, and L_3 denotes the distance between UAV 3 and 1. The desired trajectory parameters are: $x^d(t) = t$, $y^d(t) = sin(wt)$, $z^d(t) = 1$, with $w = 2\pi/40$. The distance and time are expressed in meters and seconds, respectively.

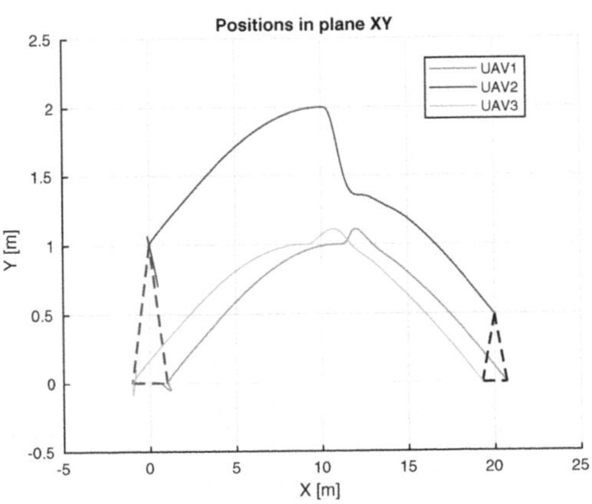

Fig. 5. Positions in plane XY of UAVs.

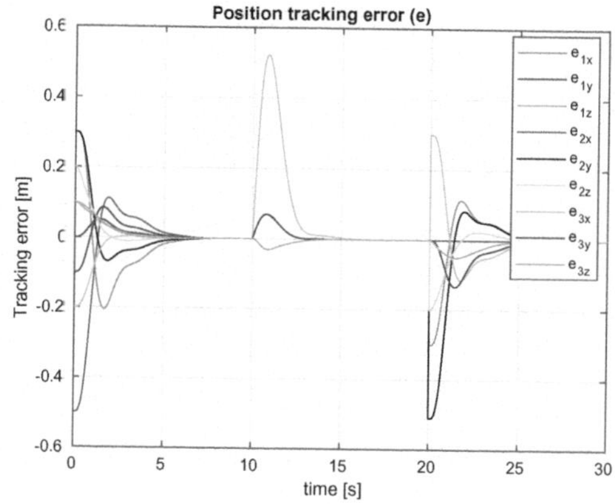

Fig. 6. Position tracking error throughout the time.

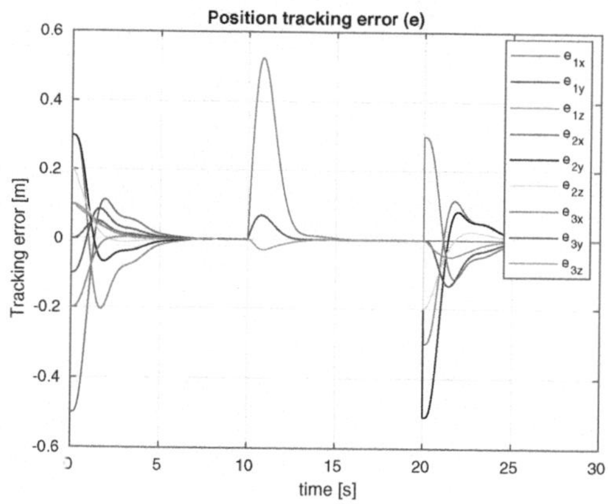

Fig. 7. Coupling error throughout the time.

All UAVs are identical, whose parameters are: mass $m = 100\,\mathrm{g}$, maximum thrust force $u_{max} = 1.05\mathrm{N}$, maximum input torque $\tau_{\phi max} = \tau_{\theta max} = \tau_{\psi max} = 1\times10^{-3}\,\mathrm{Nm}$, inertia matrix $J = diag(2.916\times10^{-4}, 3.075\times10^{-4}, 5.707\times10^{-4})\,\mathrm{kg}*\mathrm{m}^2$, gravity constant $g = 9.81\,\mathrm{m/s}^2$. All UAVs start from hover with the next initial positions and orientations: $\xi_{10} = [0.7\ 0\ 0.9]^T$, $\xi_{20} = [0.5\ 0.7\ 1]^T$, $\xi_{30} = [-0.8\ 0.1\ 0.9]^T$, $\eta_{10} = \eta_{20} = \eta_{30} = [0^o\ 0^o\ 0^o]^T$. The desired yaw angles are: $\psi_1^d = \psi_2^d = \psi_3^d = 0^o$. Control gains are: $kp_o = 2$, $kd_o = 0.5$, $\beta = 0.5$, $\Lambda = 2.2$, $\Upsilon = 1.6$.

Figure 4 illustrates the spatial positions of the UAVs throughout the time. Figure 5 illustrates the positions of the UAVs in plane XY throughout the time. It is seen that there is an impulse when switching between shapes. Figures 6 and 7 depict, respectively, the time evolution of the position tracking and coupling

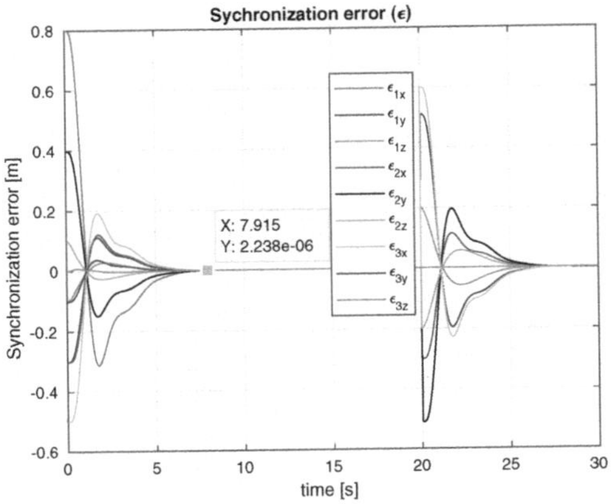

Fig. 8. Synchronization error throughout the time.

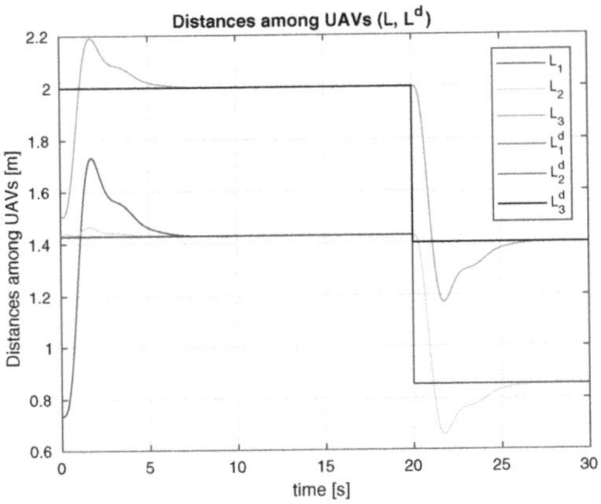

Fig. 9. Real and desired distances among UAVs throughout the time.

error. It is seen that both errors increase their value at the beginning when building the formation, when trajectory tracking (at 10 s), and when switching between shapes (at 20 s), but both errors converge to zero successfully after 6 s approximately. Figure 8 depicts the synchronization error, which increases its value at the beginning and when switching between shapes (at 20 s), but not when trajectory tracking starts (at 10 s), unlike position and coupling errors shown in Figs. 6 and 7, respectively. Figure 9 depicts the time evolution of the real and desired distances among UAVs. It is seen that desired distances are reached after 8 s approximately, and they maintain their value during trajectory tracking from 10 to 20 s. Finally, final desired distances are reached 7 s after switching of shapes.

6 Conclusions and Future Work

In this paper, a synchronous controller is developed for switching between formations and for trajectory tracking of multiple quadrotors while maintaining spatial formations. A synchronization error is defined as differential position error between every pair of two neighboring UAVs. A coupling error is defined to link the position and synchronization errors. It is shown that the developed controller guarantees asymptotic convergence to zero of both position and synchronization errors. Simulation results show the effectiveness of the proposed synchronous controller. It is seen that this controller is not robust, i.e, it is not possible to eliminate effects caused by external disturbances or by parametric uncertainties. Future work will be to add robustness to the proposed controller and its implementation in a real experimental platform.

References

1. Agnew MS (2009) Cluster space control of a 2-robot aerial system. Doctoral dissertation, Santa Clara University, School of Engineering
2. Agnew MS, Dal Canto P, Kitts CA, Li S (2010) Cluster space control of aerial robots. In: 2010 IEEE/ASME international conference on advanced intelligent mechatronics, pp 1305–1310
3. Bertrand S, Guénard N, Hamel T, Piet-Lahanier H, Eck L (2011) A hierarchical controller for miniature VTOL UAVs: design and stability analysis using singular perturbation theory. Control Eng Pract 19(10):1099–1108
4. El Ferik S, Nasir M-T, Baroudi U (2016) A Behavioral Adaptive Fuzzy controller of multi robots in a cluster space. Appl Soft Comput 44:117–127
5. García-Carrillo L-R, Dzul-López A-E, Lozano R, Pégard C (2012) Quad rotorcraft control. Vision-based hovering and navigation. Springer, London
6. Guo C, Qin J, Ge Y, Chen Y (2014) Leader-follower and cascade system based formation control of nonholonomic intelligent vehicles. In: The 26th Chinese control and decision conference (CCDC 2014). IEEE, pp 3931–3935
7. Mercado DA, Castro R, Lozano R (2013) Quadrotors flight formation control using a leader-follower approach. In: 2013 European control conference (ECC). IEEE, pp 3858–3863
8. Peng T, Li S (2014) Formation control of multiple wheeled mobile robots via leader-follower approach. In: The 26th Chinese control and decision conference (CCDC 2014). IEEE, pp 4215–4220
9. Quan Q (2017) Introduction to multicopter design and control. Springer, Singapore
10. Sadowska A, den Broek T-V, Huijberts H, van de Wouw N, Kostić D, Nijmeijer H (2011) A virtual structure approach to formation control of unicycle mobile robots using mutual coupling. Int J Control 84(11):1886–1902
11. Sanhoury I-M, Amin S-H, Husain A-R (2012) Multiple nonholonomic wheeled mobile robots trajectory tracking while maintaining time-varying formation via synchronous controller. Procedia Eng 41:1044–1050
12. Sanhoury IM, Amin SH, Husain, AR (2012) Switching between formations for multiple mobile robots via synchronous controller. In: 2012 IEEE 8th international colloquium on signal processing and its applications, pp 352–357

13. Soorki MN, Talebi HA, Nikravesh SKY (2011) A robust dynamic leader-follower formation control with active obstacle avoidance. In: 2011 IEEE international conference on systems, man, and cybernetics. IEEE, pp 1932–1937
14. Sun D, Wang C, Shang W, Feng G (2009) A synchronization approach to trajectory tracking of multiple mobile robots while maintaining time-varying formations. IEEE Trans Rob 25(5):1074–1086
15. Tan KH, Lewis MA (1996) Virtual structures for high-precision cooperative mobile robotic control. In: Proceedings of IEEE/RSJ international conference on intelligent robots and systems, IROS 1996, vol 1, pp 132–139
16. Van den Broek TH, van de Wouw N, Nijmeijer H (2009) Formation control of unicycle mobile robots: a virtual structure approach. In: Proceedings of the 48th IEEE conference on decision and control (CDC) held jointly with 2009 28th Chinese control conference. IEEE, pp 8328–8333
17. Wang C, Sun D (2008) A synchronous controller for multiple mobile robots in time-varied formations. In: 2008 IEEE/RSJ international conference on intelligent robots and systems. IEEE, pp 2765–2770
18. Zhao D, Zou T (2012) A finite-time approach to formation control of multiple mobile robots with terminal sliding mode. Int J Syst Sci 43(11):1998–2014

Modeling, Simulation and Analysis of the Characteristics of the Non-conventional Vertical Axis Wind Turbine with Variable Pitch Angle that Allows to Establish Different Control Strategies

Jesús García-Barrera[1,2](✉), Jorge Díaz-Salgado[1](✉), Oliver M. Huerta-Chávez[1](✉), and Rafael Sánchez-López[1](✉)

[1] Tecnológico de Estudios Superiores de Ecatepec, 55210 Ecatepec de Morelos, Edo. De México, Mexico
j3susgb90@gmail.com, {jorgesalgado,rsanchez}@tese.edu.mx, lecram_21@hotmail.com

[2] Tecnológico de Estudios Superiores de Tianguistenco, 52650 Tianguistenco, Edo. De México, Mexico

Abstract. This work presents the modeling, simulation and analysis in stationary response of the principal characteristics of a vertical axis wind turbine (VAWT) with variable pitch angle in order to establish different control techniques. The mathematical static model implemented for the study is the so called Double Multiple Stream Tube. For different initial conditions, the model implementation in a specialized software permits to obtain: (i) position vs output torque plots of the wind turbine and of each blade, and (ii) 3D angular position vs average moment vs attack angle plots. The analysis of these plots permits the establishment of an adequate pitch angle at each angular position (from 0 to 2π) for the improvement of the output generated torque of the wind turbine.

Keywords: VAWT · Variable pitch angle · Numerical simulation · Double multiple stream tube model · Momentum models

1 Introduction

Actually, wind energy is exploited by means of wind parks conveniently placed in high wind resource rural areas. Commonly these parks are a set of horizontal wind turbines positioned at high elevation from the ground to reduce the air turbulence. The Horizontal Axis Wind Turbine (HAWT) technology is mature and commercially exploited in this wind parks. This is very far in the case of Vertical Axis Wind Turbine (VAWT) since a lot of technological developments and improvements are necessary. Despite a lower efficiency from respect to the HAWT, VAWT mechanisms offers, by the way, many advantages: low starting torque, omnidirectionality, high turbulence operating capability,

E. E. Hernandez et al. (Eds.): LASIRS 2019, MMS 86, pp. 109–118, 2020.
https://doi.org/10.1007/978-3-030-45402-9_12

and low cost [1, 2]. These features make these turbines to be ideal to be used in urban areas with low wind resource [3].

The reported works of VAWT turbines have been primary concern to: (i) the analysis of different aerodynamical airfoils (ii) the proposition of different mathematical models to predict the behavior of the turbine in steady state, (iii) the calculus of theoretical efficiency based in these proposed models, the number of blades and the type of aerodynamical airfoil and (iv) the effect of blade pitch angle with active and passive control on aerodynamic performance [4–7].

In this work, a static model of a two blade VAWT is proposed and simulated in a specialized numerical software. The static model of a wind turbine is a fundamental tool for the analysis and design of this mechanism, with that in mind here the modeling, simulation and analysis of the behavior of a VAWT is done. The model permits to simulate the effect of a variable pitch angle with a desired trajectory along a rotation cycle.

The methodology followed consists in: (i) the proposition of a Double Multiple Stream Tube mathematical model, (ii) the implementation of this mathematical model in a numeric simulation software, (iii) the realization of various simulations in order to establish a suitable pitch angle to improve the output torque, and (iv) variable pitch simulations to determine the turbine behavior with these proposed trajectory. Section 2 presents the mathematical model and its characteristics, Sect. 3 deal with the implementation of the model. In Sect. 4 the simulated results are presented and discussed. Finally some conclusions are drawn in Sect. 5.

2 Mathematical Model

The theoretical analysis of the VAWT is based in the so called Double Multiple Stream Tube mathematical model. This model has been widely used to predict the total torque and aerodynamic forces of a VAWT with straight or curved blades [8]. The model is based in other two models: (i) two actuator disc model, and (ii) multiple stream tube model.

2.1 Two Actuator Disk Model

This model is based in the partition of the turbine influence zone in three parts: (i) the upwind high velocity U_i zone, (ii) an equilibrium velocity U_e zone and a (iii) downwind low velocity U_i' zone (Fig. 1).

Equations (1)–(4) describe the induced velocities in each zone and the velocity induction factor (2) [9].

$$U_i = U_\infty(1 - a) \tag{1}$$

$$a = \frac{U_\infty - U_e}{U_\infty} \tag{2}$$

$$U_e = U_\infty(1 - 2a) \tag{3}$$

$$U_i' = U_e(1 - a') \tag{4}$$

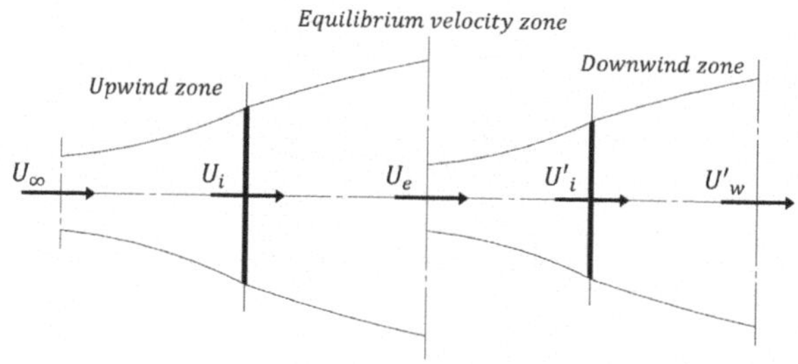

Fig. 1. Two actuator disc model.

2.2 Multiple Stream Tube Model

This model [10] permits to compute the mean turbine torque considering multiple stream tubes along the turbine influence zone parallel to the wind direction. Each considered stream tube has its own interference velocity (Fig. 2). The accuracy of the model depends of the number of streams taken into account.

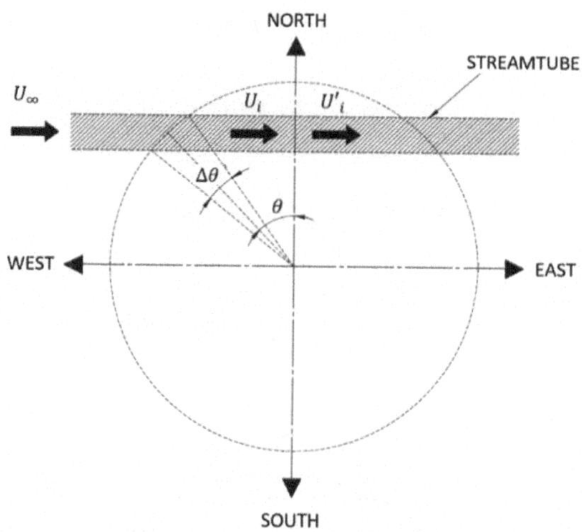

Fig. 2. Multiple stream tube model.

2.3 Double Multiple Stream Tube Model

The Double Multiple Stream Tube Model (Fig. 3) is the combination of the two previous models. Considers a two actuator disc model for the high velocity zone and another for the low velocity one, in this way two actuator discs are located in series. Using Eqs. (5), (6), (7) it is possible to calculate the induced velocities in each streams tube and zones (U_i, U_e, U_i').

$$U_i = U_\infty a \tag{5}$$

$$U_e = U_\infty(2a - 1) \tag{6}$$

$$U_i' = U_e a' \tag{7}$$

$$f_{up} = \frac{N}{8\pi R}\int_{-\frac{\pi}{2}}^{\frac{\pi}{2}} |\sec(\theta)|(C_n\cos(\theta) - C_t sen(\theta))d\theta \tag{8}$$

$$f_{dw} = \frac{N}{8\pi R}\int_{\frac{\pi}{2}}^{\frac{3\pi}{2}} |\sec(\theta)|(C_n\cos(\theta) - C_t sen(\theta))d\theta \tag{9}$$

$$a = \frac{\pi}{f_{up} + \pi} \tag{10}$$

$$a' = \frac{\pi}{f_{dw} + \pi} \tag{11}$$

$$f_{nup(\theta)} = \frac{1}{2}\rho c L W^2 C_n \tag{12}$$

$$f_{tup(\theta)} = \frac{1}{2}\rho c L W^2 C_t \tag{13}$$

$$Q_{up(\theta)} = \frac{1}{2}\rho c R L C_t W^2 \tag{14}$$

$$Q_{AVUP} = \frac{N}{2\pi}\int_{-\frac{\pi}{2}}^{\frac{\pi}{2}} Q_{up(\theta)}d\theta \tag{15}$$

$$Q_{AVDW} = \frac{N}{2\pi}\int_{\frac{\pi}{2}}^{\frac{3\pi}{2}} Q_{dw(\theta)}d\theta \tag{16}$$

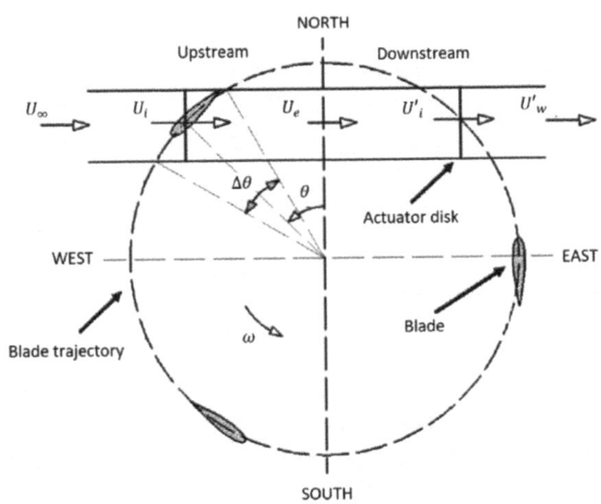

Fig. 3. Double multiple stream tube model.

Equations (10) and (11) permit to calculate low and high velocity induction factors for each stream tube (a, a') by iterating Eqs. (8) and (9). Normal $f_{nup(\theta)}$ and tangential $f_{tup(\theta)}$ forces and output torque $Q_{up(\theta)}$ for each angular position are calculated by Eqs. (12), (13) and (14). Finally, the high Q_{AVUP} and low Q_{AVDW} velocity average torque can be obtained by Eqs. (15) and (16).

3 Model Software Implementation

The proposed model was implemented in a specialized software for numerical simulation. The implementation is based in the reported iterative algorithm presented in [11]. This iterative algorithm computes: (i) the average torque, (ii) tangential forces and (iii) normal forces that the turbine generates for each angular position in steady state. In order to get those results the algorithm must satisfy a convergence criteria. This criteria consist in an error defined as the difference between the actual and previous interference factor. Nevertheless, the reported algorithm lacks an interpolate subroutine to calculate the lift and drag coefficients of the turbine. The problem was surpassed with: (i) the addition of an interpolation method and (ii) the inclusion of experimental data tables of the considered NACA0012 airfoil [12]. Another improvement of the original algorithm is the incorporation of a variable attack-pitch angle.

The modified algorithm permits to: (i) test and analyze the static behavior with variable attack-pitch angle, and (ii) generate curves of the angular position-output torque and angular position-attack angle. These improvements permit to analyze the average output torque with a variable pitch angle with fixed T.S.R.

The proposed algorithm (Fig. 4) needs three input parameters: (i) air velocity V_∞, (ii) angular velocity ω and (iii) VAWT radius R in order to obtain the average output torque.

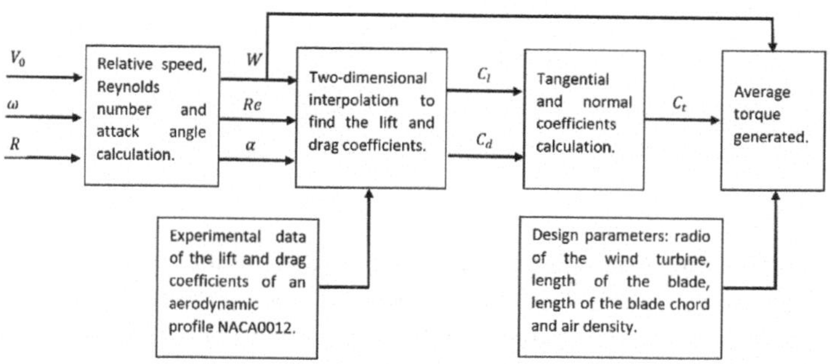

Fig. 4. Implemented static algorithm block diagram.

Employing these input parameters, the proposed algorithm computes the average torque by a bi-dimensional interpolation that relies on: Reynolds number, relative velocity and experimental lift and drag coefficients. For each stream tube an iterative process computes: (i) the interference factors a y a', (ii) Reynolds number, (iii) attack-pitch angle, and (iv) normal and tangential forces.

The drawbacks of the proposed method are: (i) the algorithm requires an iterative process in order to compute the interference factor for each stream tube and (ii) the computed average torque is the steady state value response.

4 Simulation Results

Four different speed velocities from 5 to 20 m/s are simulated with the parameters depicted in Table 1 and conditions of Table 2 with a constant T.S.R. of 4.2.

Table 1. Physical parameters.

Parameter	Value	Units
Blade number	2	Unity
Blade chord	0.3	m
Blade length	2.0	m
Turbine radius	2.0	m

Table 2. Simulation conditions.

Simulation conditions	Test 1	Test 2	Test 3	Test 4
Attack angle	0° [rad]	0° [rad]	0° [rad]	0° [rad]
Number of stream-tubes	20	20	20	20
Wind velocity [m/s]	5	10	15	20
Angular velocity [rad/s]	10.5	21	31.5	42
T.S.R. [/]	4.2	4.2	4.2	4.2
Turbine solidity	0.15	0.15	0.15	0.15

The angular position θ *vs* output torque (Fig. 5), and angular position *vs* attack angle *vs* output torque of each blade (Fig. 6) is showed for each air velocity.

These results show that the tendency of the curves is similar for different air velocities, according with the theory, a quadratic relationship between output torque and air velocity is evident (Figs. 5 and 6). As expected, the maximum torque appears in the high velocity zone. In particular this occurs between 0° and 20° that corresponds to an attack angle between 9° and 10° (Fig. 6). Meanwhile, the minimum torque arises in an angular position around 270° that corresponds to an attack angle between −2° and 0°.

The steady state simulated results permit to propose that a constant attack angle equal to 10° will upgrade the output torque. Test 1 is repeated with this constant attack angle (Fig. 7).

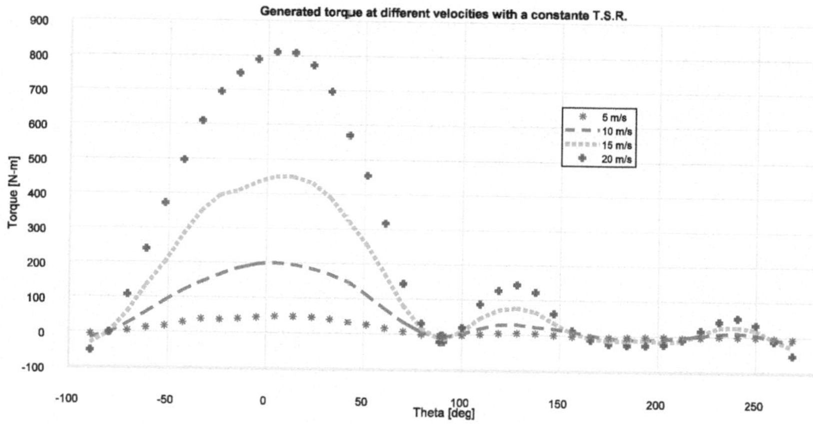

Fig. 5. Angular position vs output torque.

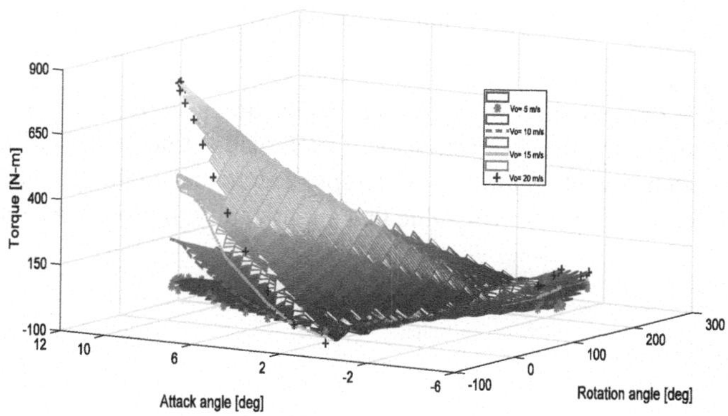

Fig. 6. Rotation angle vs attack angle vs output torque 3D plot.

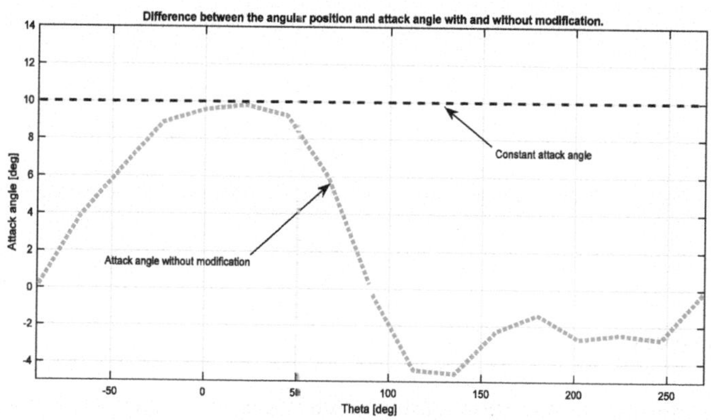

Fig. 7. Difference between the angular position and free and constant attack angle for Test 1.

The output torque with and without constant attack angle shows a significant output torque improvement of about four times (≈3.75) the original output (Fig. 8).

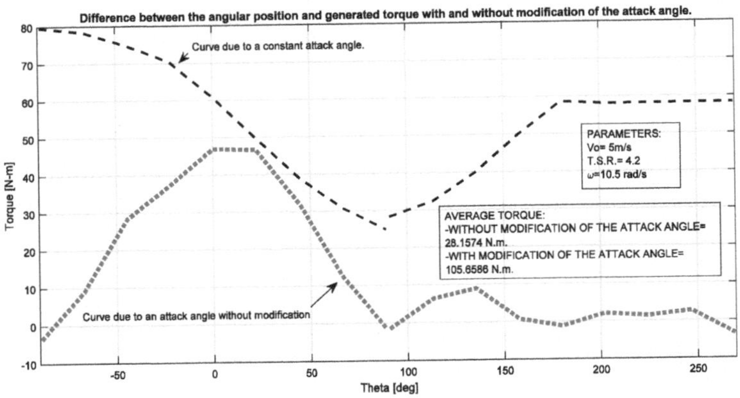

Fig. 8. Difference between the output torque generated with free and constant attack angle.

The respective desired pitch angle (for a constant attack angle of 10°) is presented in Fig. 9, and establishes the desired trajectory of each blade along a revolution cycle. The tracking of this pitch angle trajectory would imply a technical challenge, and establishes a first comparison with a theoretical value that significantly improves the output torque.

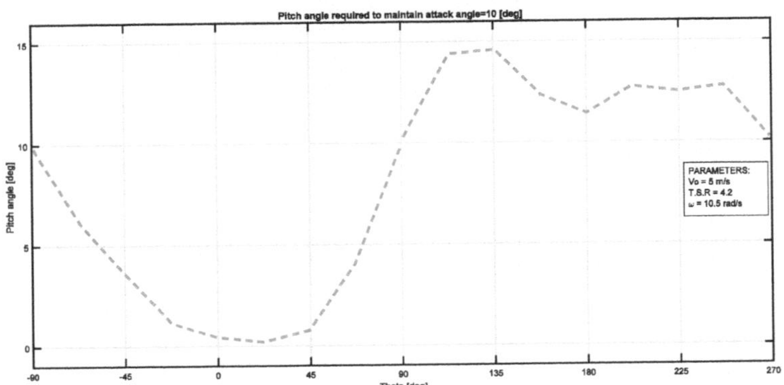

Fig. 9. Pitch angle required to maintain a constant attack angle.

5 Conclusions

The Double Multiple Stream Tube mathematical model was successfully implemented in a numerical simulation software to shape the attack angle and output torque of a two blade vertical axis wind turbine (VAWT) with variable pitch angle.

The implemented model permits to obtain: (i) angular position *vs* output torque plots of the wind turbine and of each blade, and (ii) 3D angular position *vs* average moment *vs* attack angle plots. The steady state simulated results permit to conclude that the output torque can be upgraded with an appropriated attack angle. Moreover, the analysis of these plots permits to establish a desired pitch angle trajectory at each angular position (from 0 to 2π) for the improvement of the average generated torque of the wind turbine.

The proposed pitch angle trajectory was obtained considering a constant attack angle specified near the maximum output torque. This yields a theoretical significant output torque improvement of about four times the original output. Indeed, the desired pitch angle trajectory would imply a great technical challenge, and its implementation would be a difficult task, it establishes a first comparison with a theoretical value that significantly improves the output torque.

There is a lot of additional work to be done: (i) first the experimental implementation of the desired pitch angle trajectory, (ii) the consideration of different T.S.R.'s and blade solidity, (iii) greater air velocities with turbulent flow, (iv) a greater number of blades, and (v) the consideration of a symmetrical central blade, normally used in this types of turbines to automatic start up. Finally, the numerical model must include a greater number of table entries for the Reynolds number computing in order to outperform the numerical precision.

Meanwhile with the desired pitch angle also many different trajectories and control strategies can be proposed and tested as future work.

References

1. Musgrove PJ (1979) Vertical axis wind turbine. 0–5. https://doi.org/10.1016/b978-1-4832-8407-1.50354-8
2. Aslam Bhutta MM, Hayat N, Farooq AU, Ali Z, Jamil SR, Hussain Z (2012) Vertical axis wind turbine - a review of various configurations and design techniques. Renew Sustain. Energy Rev 16:1926–1939. https://doi.org/10.1016/j.rser.2011.12.004
3. Chaichana T, Chaitep S (2010) Wind power potential and characteristic analysis of Chiang Mai, Thailand. J Mech Sci Technol 24:1475–1479. https://doi.org/10.1007/s12206-010-0415-3
4. Zhang LX, Liang YB, Liu XH, Guo J (2014) Effect of blade pitch angle on aerodynamic performance of straight-bladed vertical axis wind turbine. J Cent South Univ 21:1417–1427. https://doi.org/10.1007/s11771-014-2080-7
5. Zhang L, Pei Y, Liang Y, Zhang F, Wang Y, Meng J, Wang H (2015) Design and implementation of straight-bladed vertical axis wind turbine with collective pitch control. In: 2015 IEEE international conference on mechatronics and automation (ICMA), pp 1258–1263. https://doi.org/10.1109/ICMA.2015.7237666
6. Zhang L, Liang Y, Li E, Zhang S, Guo J (2012) Vertical axis wind turbine with individual active blade pitch control. In: Asia-Pacific power and energy engineering conference, APPEEC. https://doi.org/10.1109/APPEEC.2012.6307108
7. Pawsey NCK (2002) Development and evaluation of passive variable-pitch vertical axis wind turbines
8. Paraschivoiu I (2002) Wind Turbine Design With Emphasis on Darrieus Concept. École Polytechnique de Montréal

9. Mohammed AA, Ouakad HM, Sahin AZ, Bahaidarah HMS (2019) Vertical axis wind turbine aerodynamics: summary and review of momentum models. J Energy Resour Technol Trans ASME 141:1–10. https://doi.org/10.1115/1.4042643
10. Paraschivoiu I (1988) Double-multiple stream tube model for studying vertical-axis wind turbines. J Propuls Power 4:370–377
11. Castillo J (2011) Small-Vertical Axis Wind Turbine Design. 69
12. Sheldahl RE, Klimas PC (1981) Aerodynamic characteristics of seven symmetrical airfoil sections through 180-degree angle of attack for use in aerodynamic analysis of vertical axis wind turbines. Chem SAND 80-211, 118. https://doi.org/10.2172/6548367

Numerical Simulation of the Formation, Evolution and Control of a Methanol Spray Generated by a Pressure Swirl Atomizer During the Combustion Process

G. Guevara-Morales[1,2], O. M. Huerta-Chavez[1,2](✉), A. Arias-Montaño[1,2], M. Zirion Flores[1,2], and I. Castorena Solache[2,3]

[1] SEPI-ESIME Ticomán-IPN, Gustavo A. Madero, CDMX, Mexico
Lecram_21@hotmail.com
[2] Catedras Conacyt TESE Posgrado de Mecatronica, Santiago de Querétaro, Mexico
[3] ISAE ENSMA, Chasseneuil du Poitou, France

Abstract. In this study the injection of liquid fuel by a pressure swirl atomizer during the combustion process, is reproduced and analyzed using numerical simulation (Computational Fluid Dynamic). A 2D model along with an Eulerian-Lagrangian method has been use to simulate the presence of the spray and its interaction with the continuous phase. The injection and atomization of the fuel are reproduce using the LISA ("Linearized Instability Sheet Atomization") model together with a Rosin-Rammler distribution, while the dispersion (evolution) of the droplets is simulated through the governing equations of the discrete phase. On the other hand, the continuous phase is simulated using a model based on the Reynolds-averaged Navier-Stokes equations (RANS). The combustion model use is the steady-flamelets. The results of the dispersion (number density), size (Sauter mean diameter) and velocity of the droplets are compare with the experiments, showing good agreement. Finally, in order to study the control of the injection process, the operating parameters of the atomizer have been modified, analyzing the response in the characteristics of the spray.

Keywords: Simulation · RANS · Injection and control

1 Introduction

For research and development, the combustion process, numerical tools such as CFD are implemented. CFD tools can emulate the behavioral details of the flow inside the turbulent combustion chamber without the need of sensors, therefore not invading and not altering the behavior of the flow. It is then very useful to choose the experiments that will be carried out based on the results that the model yields. This is possible thanks to the development of models ever more precise and ever more complex use for the study of the aforementioned phenomena, besides this can be done at a fraction of the cost of the experimental setup.

E. E. Hernandez et al. (Eds.): LASIRS 2019, MMS 86, pp. 119–137, 2020.
https://doi.org/10.1007/978-3-030-45402-9_13

Regarding the combustion modelling, the non-linear nature of the chemical reaction rate (ω), expressed by an Arrhenius law, the mean ω- cannot be described as function of the means of the variables that define it (mass concentration, density and temperature). Such being the case, the transport equations for different chemical species cannot be expressed in the same way as the ones that deal with turbulence. To deal, different models have been developed, based on relatively easy to model variables. Veynante and Vervisch [3] feature an exhaustive revision of combustion models, these can be separated into two categories depending on how they treat chemical rate: either an infinitely fast chemical rate or a finite one. The main way to determine which type of model suits the needs of one specific problem is to compare the chemical rate characteristic time (τc) and the characteristic time of turbulence of the flow (τt). Damköhler's number is meant to characterize the relation between these times, as $Da = \tau t/\tau c$. If $Da \gg 1$ then one can consider an infinitely fast reaction and the turbulence will characterize the behavior of the flow [3].

Models that consider infinitely fast chemical reactions assume that reagents are burned as soon as the right proportions of mixture are reached. The chemical process is fast compared to the transport process (diffusion and convection). As a result, the knowledge of the mixing profile of the chemical species provides us with knowledge on their oxidation and therefore combustion. Among these chemical equilibrium-driven models in diffusion flames we find models that are based on determining mixture fraction (f) inside the combustion chamber. (f) values ranges between 1 and 0 where 1 represents a flow region in which only fuel is present and 0 represents a region in which only oxygen is present. The rest of the field unknowns are determined by exploiting linear relations between them and mixture fraction (f). The model used in this paper (flamelets) is based on the mixture fraction (f) [3–5], it assumes that the turbulent flame is a superposition of laminar flames that vary in position through time. It can be observed that the position of the flame and reaction of the species is directly affected by the relative velocity and distance between the oxidizer and fuel flows and we can clearly identify the regions where the flame is present and its extinction once the mixing proportions do not allow for a flame to form anymore. These flow characteristics are described in a variable called scalar dissipation (χ). Because f varies along the distance between fuel and oxidizer flows, χ is defined based on f's gradient, therefore representing the shape of the flame. The flamelets model takes this narrowing down of the flame into account thanks to the aforementioned variables unlike a simple chemical equilibrium model. To couple the combustion model with the turbulence model we have an Eddy-Breakup model that considers that the reactions take place in the smaller scales of turbulence [3], in which the molecular level mixing occurs. Given that we consider infinitely fast reactions, the flow of reagents and the turbulent mixing process are the limiting process inside the combustion chamber and determines the actual evolution of the flame.

The present paper a 2D model along with a Eulerian-Lagrangian method had been used to simulate the presence of the spray and its interaction with the continuous phase. The injection and atomization of the fuel are reproduced using the LISA ("Linearized Instability Sheet Atomization") model together with a Rosin-Rammler distribution, while the dispersion (evolution) of the droplets is simulated through the governing equations of the discrete phase. On the other hand, the continuous phase is simulated using a model based on the Reynolds-averaged Navier-Stokes equations (RANS). The combustion model used is the steady-flamelets. The results of the dispersion (number density), size (Sauter mean diameter) and velocity of the droplets are compared with the experiments, showing good agreement. Finally, in order to study the control of the injection process, the operating parameters of the atomizer have been modified, analyzing the response in the characteristics of the spray. The model will be evaluated when compared to the experimental setup of combustion chamber carried out by the NIST (National Institute of Standards and Technology) [1]. The experimental parameters and results have been used to validate the data yielded by the coupled models.

2 Methodology

2.1 Experimental Setup

The NIST experimental along with the database by Widmann and Presser [1] were selected as the reference values for the present article. The NIST model, as illustrated in Fig. 1, consists of a cylindrical reservoir made of steel and kept at ambient pressure. Inside the cylindrical container a diffusion flame is generated via the injection of liquid methanol (CH3OH) as a spray and air with a swirling motion. The spray is created by using a pressure injector which forms a hollow cone of fuel. Around this system there's a vortex generator (swirler) with 12 adjustable blades.

This experimental setup was used to create a database that could validate a combustion model. The spray measurements were obtained in a non-intrusive fashion using Phase Doppler Interferometry (PDI). The burnt gases temperature at the combustion chamber's outlet as well as the temperature of the chamber walls were measured by thermocouple probes. To characterize the turbulent flow Particle Image Velocimetry (PIV) was employed. The results reported in [1], measured during the stable status of the process are listed below:

- **Spray**: Sauter's diameter, droplet density, volumetric flow, and, axial and radial mean velocities of drops. This set of data was measured multiple times at different altitudes of the combustion chamber counting form the injection point with their respective error margins. The altitudes were: (5, 15, 25, 35, 45, 55 and 65 mm from the injection point.
- **Flow**: Axial, radial and tangent velocities as well as their fluctuations, measured in function of their radial position and at three axial planes from the injection point (1.4, 9.5 and 17.6 mm) the data is reported when the spray is present as well as when it isn't. It also includes measurement error margin.
- **Temperature**: Mean temperature of the container's Wall measured at 10 different altitudes from its base (99, 165, 231, 297, 429, 561, 693, 826, 958 and 1090 mm). Mean temperature of outlet gases measured at 13 different points distributed along the outlet.

The NIST model was chosen as a starting point for the present paper mainly because the experimental values were extensively reported and readily available alongside the operating parameters for the experiment. This allows for easier definition of boundary conditions in the simulations and for meaningful comparisons to validate the model. Besides, this models can be simplified to focus computer resources on the analysis of the fundamental phenomena intervening in turbulent combustion. The preceding results and considerations of the RANS simulations based on this same NIST experimental setup (Zhu et al. [10], Crocker et al. [11] and Collazo et al. [12]) form the angular stone in which this simulation is built on.

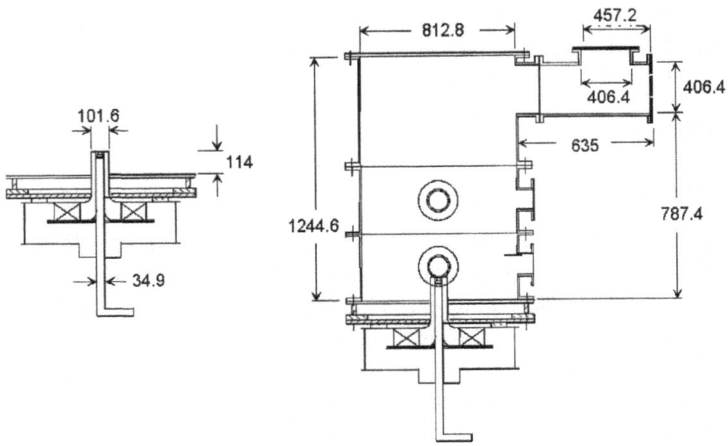

Fig. 1. Experimental Model from NIST, units: mm [1].

2.2 Computer Simulation RANS 2D

As shown "Experimental Model (Fig. 1)", the burnt gas outlet is placed at the side of the combustion chamber, this makes the setup asymmetrical. However, according to Zhu et al. [10] and Crocker et al. [11] the effect of having non-centered outlet is irrelevant to the flow in the vicinity of the injector, zone that will be studied in this paper. This means that the combustion chamber's geometry can be simplified by considering the upper face of the reservoir as the outlet for the gases, giving the combustion chamber axial symmetry and making a 2D simplification of the chamber and flow possible. As a result, only half of the mean plane of the combustion chamber is modelled. This has great effects in the overall simulation time. Even though there are no tangential components in the gradients of an axisymmetric flow, we must still consider the radial fluctuation in velocity of the flow due to turbulence. In a spray's particular case, keeping the fluctuation of velocity in mind implies that injected particles and their properties are also axisymmetric. Although the spray's 3D simulation is able to emulate these fluctuations and the 2D is not, the 2D simulation is capable of rendering the general characteristics of the spray [10]. In "Computational Model for RANS Simulation (Fig. 2), a 2D model for the combustion chamber is shown, the inlet and outlet for the gases are outlined. The outlet's area is the same as in the experimental setup, with a 203.2 mm radius.

Taking into account the simple geometry of the combustion chamber a structured mesh of 338 058 square elements was used, the mesh was refined around the air and fuel inlets. The mesh's parameters were determined thanks to a mesh independence study. In the aforementioned study the mean temperature in relation to the mass contained in each of the mesh's elements was compared between the elements. In contrast with a classical transient simulation in which the solution converges to a steady, invariant state, the coupling of the spray's transient model modifies the results. In this case, the injected particles' properties into the chamber such as size, velocity, trajectory, position and collision modes vary randomly during each time step. As a result, the liberated heat coming from the chemical reaction varies and consequently the global temperature of the combustion chamber varies too. As this is the case the simulation's result will converge to a statistic stability state instead of a steady value. In this state, the solution will be in the vicinity of a predicted value (statistical mean) and will have a given variance. Therefore, it is the statistical mean of the temperature obtained by the convergence of the simulation that will be used to compare the different mesh elements in the mesh independence study.

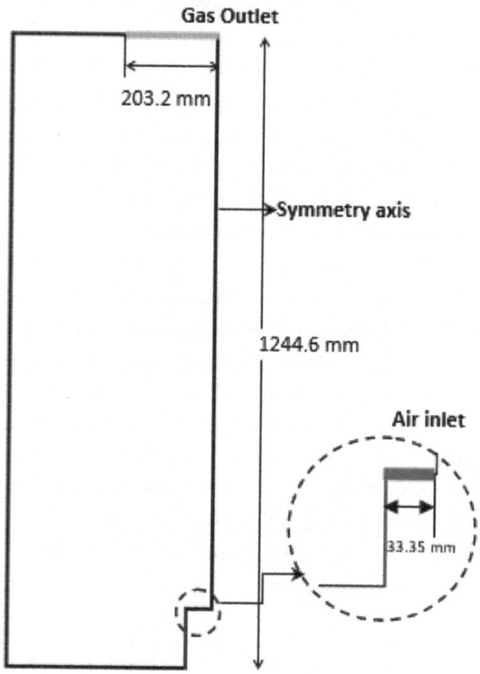

Fig. 2. Computational Model for RANS Simulation.

2.3 Turbulence Model

In a RANS model, the flow calculation is obtained thanks to the mean transport equations of mass Eq. (1) and momentum Eq. (2). If the density of the flow is constant the Reynolds mean (noted with the bar "‾‾") is used in these equations. However, if the density is not constant, (which is the case in reactive flows with great temperature gradients) the Favre mean (noted with the tilde "∼") is adapted to resolve the equations. The equations obtained with this mean are analog to the equations obtained with the Reynolds mean and the mean density of the products is taken into account. In Eqs. 1 and 2, variables m_S and F_S represent source terms for mass and momentum respectively, this terms translate the vapor and momentum transferred from the fuel drops to the gaseous mix.

$$\frac{\partial \bar{\rho}}{\partial t} + \frac{\partial \bar{\rho} \tilde{u}_j}{\partial x_j} = m_S \quad (1)$$

$$\frac{\partial}{\partial t}(\bar{\rho}\widetilde{u}_t) + \frac{\partial}{\partial x_j}(\bar{\rho}\widetilde{u}_t\widetilde{u}_j) = -\frac{\partial \bar{p}}{\partial x_t} + \frac{\partial \tau_{tj}}{\partial x_j} - \frac{\partial}{\partial x_j}(\bar{\rho}\widetilde{u'_t u'_J}) + F_S \quad (2)$$

The turbulent flow was modelled used a standard Low Reynolds κ-ε model. This model was chosen because of its two-equation approach and versatility in this type of problems and the low computational cost that it represents. Even though, in general, the Reynolds Stress Model is more common in problems where recirculation is present, the feeble swirl number inside the flow justifies the hypotheses for a standard κ-ε model.

The κ-ε model is based on Boussinesq's hypothesis, which replaces the unknown residual stress tensor $(\widetilde{u'_t u'_j})$ that results from applying the Reynolds mean to the Navier-Stokes Eq. (2) equations with a "turbulent viscosity" (μ_t) term that models the effects of the residual tensor. The turbulent viscosity term is obtained via the turbulent kinetic energy (κ) and the energy dissipation (ε) with the following expression, in which is a model's constant. The model resolves two different transport equations, one for the turbulence's kinetic energy Eq. (3), and one for energy dissipation, both equations present the classical form of a transport equation, which is: **Rate of change + Transport by convection = Transport by diffusion + Production – Destruction**. The production and destruction rates, as well as the diffusive transport general coefficients are modelled by empirical observation of various turbulent flows over the years and adjusted with the different constants that make up the standard model, as shown in Eqs. 3 and 4.

In this study, the gaseous phase is modelled as mixture of ideal gases which density only varies in function of temperature thanks to the fact that we consider the combustion to happen at subsonic speeds and having therefore an isobaric flame. By using the ideal gas law, the software is capable thereby capable of relating the density of the mix and the temperature yielded by the simulation.

$$\frac{\partial \bar{\rho}\kappa}{\partial t} + \frac{\partial \bar{\rho}\widetilde{u}_t\kappa}{x_i} = \frac{\partial}{\partial x_j}\left[\left(\frac{\mu_t}{\sigma_k}\right)\nabla\kappa\right] + \bar{\rho}P_k + \bar{\rho}\varepsilon \quad (3)$$

$$\frac{\partial}{\partial t}(\bar{\rho}\varepsilon) + \frac{\partial}{\partial xj}(\bar{\rho}\widetilde{u}i\varepsilon) = \frac{\partial}{\partial xj}\left[\left(\frac{\mu t}{\sigma\varepsilon}\right)\nabla\varepsilon\right] + C1\varepsilon\frac{\bar{\rho}Pk\varepsilon}{k} - C2\varepsilon\frac{\bar{\rho}\varepsilon^2}{k} \quad (4)$$

2.4 Spray Model

The injection and atomization of fuel are not directly simulated. The sub-model of a pressure atomizer available in Fluent that was used on this paper is based on the LISA atomizing model (Linearized Instability Sheet Atomization) developed by Schmidt et al. [13]. The LISA model determines the statistical mean of the spray's drops using the atomizer's known parameters. The model is directly based on the classical atomization of a fuel layer illustrated on "Atomization phenomena modelled by LISA (Fig. 3).

The model calculates, first hand, the dimension of the fuel layer (t_s) as a function of mass, pressure, injection diameter, and spray angle. The parameters used for this particular injector were a hollow cone type injector with a nominal angle of 60°. However, when comparing the experimental results to the simulation this paper found that the drop dispersion was better approximated by an angle of 58°. According to the NIST database [14] the atomizer used on the experiment was a "Delavan WDA-1.0" with an outlet diameter of 0.28 mm. The pressure and mass flow were reported by Widmann and Presser [1] and are condensed, along the other injector operation parameters on Table 1.

Table 1. xx

Parameter	Value
Air Temperature	300 K
Fuel Temperature	300 K
Fuel Flow	3 kg/hr
Injection Pressure	690 kPa
Injection Angle	60°
Operation Pressure	1 atm

After considering the aforementioned parameters, the model applies a perturbations specter to the fuel layer with a depth of t_s moving at the velocity U_s through a static, oxidant atmosphere, the most unstable perturbation, and consequently the one which will break up the fuel film into drops, will be the one that reaches the highest growth rate (Ω). Following this reasoning, the model will calculate Ω by studying the upper limits of the growth rate equation. The equation translates the fact that growth rate is a function of the wavelength "k", therefore Ω will be yielded by the maximum wavelength "k_s". When the fuel layer perturbation with k_s wavelength and growth rate "Ω", reaches a critical wave amplitude (η_b), the fuel layer will break up into ligaments of "d_L" diameter. This diameter is expressed in Eq. 5 which presents two possibilities: a) in the first case, the wavelength is negligible compared with t_s and therefore the diameter d_L will depend only of k_s and the ligament constant C_L or b) In this case, the wavelength is larger than t_s thus the ligament diameter d_L will depend on k_s and h_f, which represents the width of the fuel layer were the atomization occur. The distance (L_b) between this point and the injection point is calculated thanks to the fuel layer's velocity and the time elapsed before atomization (τ_s), i.e. $L_b = U_s \tau_s$. Time (τ_s) is given by the critical wave amplitude equation Eq. (6). Considering the former, L_b is expressed in Eq. 7 as a function of U_s, Ω and the term $ln(\eta_b/\eta_0)$, also known as layer constant. The software affects a value of 12 to this constant thanks to the fuel's feeble viscosity (when compared to water).

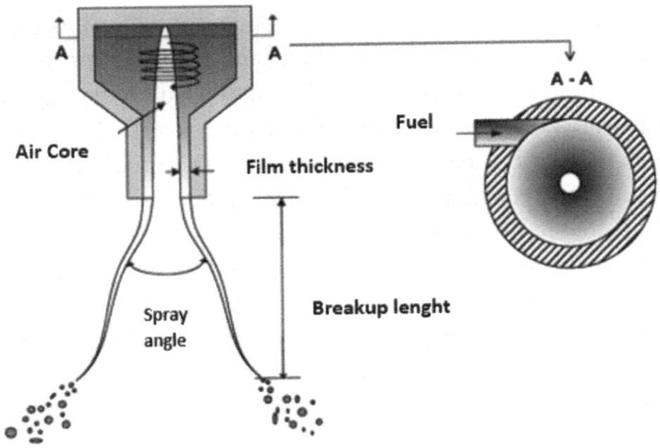

Fig. 3. Atomization phenomena modelled by LISA [15].

The value of 12 was corroborated by experimental results with different fuels which Weber's numbers varied between 2 and 200 [16]. The width (h_f) is computed thanks to L_b and the initial fuel layer width. The initial drop diameter (d_0) of drops resulting from the atomization of the ligaments with a d_L diameter is determined thanks to Eq. 8. Said equation is based on the wavelength of the wave caused by the breakup of the ligament (λ_{opt}), equating the volume of the ligament and the resulting spherical drop. The value of d_0, yielded by the LISA model is equated to the mean diameter ($\bar{d}$) in a Rosin-Rammler drop distribution. In this function the mass fraction of drops that have a diameter superior to a certain (d) diameter, is given by Eq. 8. Said equation yields Y_d, the mass fraction and n, a parameter that translates the dispersion in drop size. For most practical applications, n varies between 1.8 and 3 [17].

$$d_L = \left\{ \sqrt{\frac{2\pi C_L}{k_s}}, \sqrt{\frac{8h_f}{k_s}} \right\} \tag{5}$$

$$\eta_b = \eta_0 e^{\Omega \tau_s} \tag{6}$$

$$L_b = \frac{U_s}{\Omega} ln\left(\frac{\eta_b}{\eta_0}\right) \tag{7}$$

$$d_0 = 1.88 d_L (1 + 3Oh)^{1/6} \tag{8}$$

Given the value of $\bar{d}$ and n, the injected drops diameter "d" is calculated by randomly studying the mass fraction Y_d in Eq. 9. For this paper's purposes the value of n is not calculated as it is a statistical value. The actual value of the parameter was studied in a number of cases to determine the best match for the experimental database result, after this empirical approach it was found that the best match for spray drop dispersion is $n = 4.5$. Furthermore, the trajectory of injected drops is also random. However, considering the spray angle of 58° and the dispersion angle presented by Zhu et al. [10] it was concluded that the dispersion angle should be around 10°. After a certain number of iterations, it was concluded that an angle of 15° has a greater degree of fidelity towards the atomization distances presented in the experimental results. The number of particles injected into the domain was determined to be 8000 per time step, this enabled the mass flow to be realistic and therefore went well to the rest of the spray mean parameters.

Once the drops are injected into the domain, thanks to the LISA sub-model, the second atomization mode is simulated. The sub-model TAB ("Taylor Analogy Breakup) that is used to simulate this second mode is based on Taylor's analogy, which compares a system composed of a spring and a mass and a drop's deformation and oscillation. It is given then that the drops will breakup into smaller drops as soon as their deformation rate reaches a critical value. To calculate the drops' trajectory the simulation takes into account the mean values of flow velocity ($\tilde{u}$), we need however to consider the whole velocity when determining trajectories, which creates a need to consider the random fluctuations of the mean velocity value throughout time, in order to do this the simulation used the "random walk" probabilistic model which introduces an eddy whose proper velocity is noted u$'$, the eddies are introduced throughout the particle´s trajectory. By using this model, the velocity considered before integrating the position equation of the drop is: $\tilde{u} + u'(t)$ The value for $u'(t)$ is calculated using a random probability variable. Furthermore, collisions between particles are also taken into account in the spray model and are accounted for with O'Rourke's probabilistic algorithm. The algorithm considers that two drops may only collide if both are inside of the same mesh element. To determine the probability of a collision happening a collision volume is computed. The probability of collision is then computed by calculating the ratio between the collision volume and the mesh element volume. Once the algorithm has determined a collision, the results of this collision were reduced to only two outcomes: either coalescence or an elastic collision. The outcome of collision is defined by the relative position of both drops during the collision and a random variable. The drop's properties after collision are calculated thanks to conservation laws.

2.5 Combustion Model

The turbulent diffusion flame is modelled through a steady non-adiabatic flame model also known as flamelets, this is available in the non-premixed combustion models in Fluent. This model was designed to simulate turbulent non-premixed flames. Using the fundamental properties of a non-premixed flame whose reaction zone can be assimilated to a surface, it is inferred that the flame divides the domain into two zones, a zone with oxidant and combustion products outside the flame and a zone with fuel and combustion products, inside the flame. Therefore, we can add the transport equations for mass fraction and enthalpy. Besides, the hypothesis of a Lewis number equal to 1 throughout the flow lets us assume that the diffusion coefficients for chemical species and for heat are equal.

To avoid having to calculate the chemical reaction rate and thanks to the previously made hypothesis we use the Shvab-Zeldovitch formalism to write the transport equations in terms of Zeldovtich's variables. We norm these variables by using the mixture fraction f defined using Zeldovitch's variables. The software then solves for a transport equation of f which is only a lineal combination of a Zeldovitch variables and therefore satisfies the same transport equations. Applying Favre's mean to the transport equation yields Eq. 9. It is important to note that the production term for mixture fraction comes solely from the evaporation rate of the drops within the domain, same as mass and momentum in previous equations. Furthermore, the diffusion coefficient for mixture fraction is considered to be defined by turbulence which has a greater effect on chemical species diffusion than molecular diffusion by itself. Therefore, we consider the molecular diffusion coefficient to be negligible in favor of a turbulent diffusion coefficient defined by the "turbulent viscosity" that models Reynolds Stress Tensor. Besides a transport equation for the variance of fraction mixture ($\widetilde{f'^2}$), given in Eq. 10. Since the simulation takes consideration of the steady flow the variation of mixture fraction and its variance (first term for Eqs. 9 and 10) is zero. We define scalar dissipation (χ) which characterizes the narrowing of the flame from the mixture fraction gradient, i.e. $\chi = 2D|\nabla f|^2$. To represent flames with different narrowing degree we can express scalar dissipation in the stoichiometric zone of the mixture (χ_{st}). The Favre's mean $\widetilde{\chi_{st}}$ is then assimilated through a linear regression of the transport equation of χ [3], yielding Eq. 11.

$$\frac{\partial}{\partial t}(\bar{\rho}\tilde{f}) + \frac{\partial}{\partial xj}(\bar{\rho}\tilde{u}i\,\tilde{f}) = \frac{\partial}{\partial xj}\left[\left(\frac{\mu t}{\sigma f}\right)\nabla\tilde{f}\right] + fs \tag{9}$$

$$\frac{\partial}{\partial t}(\bar{\rho}\,\widetilde{f'^2}) + \frac{\partial}{\partial xj}(\bar{\rho}\tilde{u}i\,\widetilde{f'^2}) = \frac{\partial}{\partial xj}\left[\left(\frac{\mu t}{\sigma f}\right)\nabla\,\widetilde{f'^2}\right] + Cg\mu t\left(\nabla\tilde{f}\right)^2 - \bar{\rho}Cd\frac{\varepsilon}{k}\widetilde{f'^2} \tag{10}$$

$$\widetilde{\chi_{st}} = C_\chi\frac{\varepsilon}{\kappa}\widetilde{f'^2} \tag{11}$$

Before the simulation, the model generates a number of tables (one for each flame with a certain narrowing degree) that resolve the steady transport equations (a flamelets model considers the turbulent flame to be a superposition of laminar, steady flames that vary in position and form over a defined step of time) and yield results for temperature and mass fraction of each of the chemical species involved in the combustion problem, these results are given in function of the mean values of f, χ_{st} and enthalpy (H), even though the latter does not affect the chemical reaction directly. The relation between temperature T and mass fraction of the chemical species $n(Y_n)$ in function of the variables f, χ_{st} y H is yielded by Eqs. 12 and 13. This equations are obtained via the transport equations of chemical species and enthalpy for a laminar flame, in which the coordinate system can be assimilated to a 1D problem in which f varies (Shvab-Zeldovitch formalism), therefore the convection transport is negligible in favor of the other terms.

$$\rho\frac{\partial Y_n}{\partial t} = \frac{1}{2}\rho\chi\frac{\partial^2 Y_n}{\partial f^2} + \omega_n \tag{12}$$

$$\rho\frac{\partial T}{\partial t} = \frac{1}{2}\rho X\frac{\partial^2 T}{\partial f^2} - \frac{1}{Cp}\sum_n Hn\omega n + \frac{1}{2Cp}\rho X\left[\frac{\partial Cp}{\partial f} + \sum_n Cp,n\frac{\partial Yn}{\partial f}\right]\frac{\partial T}{\partial f} \tag{13}$$

The thermochemical information needed to solve for these equations is obtained by the reaction mechanism of the fuel, in this case, methanol. The reaction mechanism used in the simulation includes 167 elementary chemical reactions between 32 chemical species. This kinetic mechanism has been developed and approved by Lindstedt et al. [18] and is consultable in [19]. Using Eqs. 12 and 13 the software will calculate the weighted average between the combustion variables and the probability density function of f.

The function of the probability $p(f)$ is calculated using a beta PDF, which means it'll be calculated using the values of $\tilde{f}$ and $\widetilde{f'^2}$. In the case of $p(\chi_{st})$ a delta function is considered, meaning that the variance won´t be necessary for χ_{st}. This way a relation exists between combustion variables and $\tilde{f}$, $\widetilde{f'^2}$ $\widetilde{\chi_{st}}$ and $\tilde{H}$ (The latter are the variables yielded by the simulation).

In the same way, as previous simulations, flame-emitted radiation will be taken into account by the Discrete Ordinates model ("DO"). This models discretizes the whole range of solid angles that are necessary to calculate the radiation in the heat transfer term of energy transport. In contrast with other models such as Rosseland's, DO deals with the whole optical depth specter, this makes it adequate to simulate the radiation coming from combustion.

3 Presentation and Discussion of Results

3.1 Comparison with Experimental Results

First, the flow velocity profiles are compared with the experimental results (reported at 9.5 and 17.6 mm in height with respect to the injection point). Figures 4, 5 and 6 show the axial, radial and tangential velocity profiles respectively. The discrepancy between the velocity values of the three components and the experimental results in the range of approximately 0 to 15 mm of the radius is notorious. A similar difference is reported by Zhu et al. [17] and Collazo et al. [2]. This discrepancy is due to the increase in speed produced by the spray on the gaseous medium. The software determines this increase by calculating the sum of momentum lost by the drops that cross an element of the mesh, later this moment is introduced as a source term to the equations of the flow of the corresponding element. This method takes the point effect of the particles and distributes it "on the mesh element, overestimating the amount of fluid that is affected by the drops. On the contrary, in the PIV measurement the measured velocity corresponds to the fluid found between the drops of fuel, and that does not reach to be affected by its displacement, so that in the experiment the punctual effect of the drops is not relevant, while in the simulation it increases. As can be seen in "Axial velocity at 9.5 and 17.6 mm of the injector (Fig. 4)" the difference in the area near the axis of the injection is greater in the axial velocity profile, a component that is mostly affected by the speed of the spray, while the tangential velocity profile (Fig. 6) has the smallest difference in this area.

After this discrepancy interval, the results obtained from the simulation consistently reproduce the evolution of the flow. The increase in radial velocity in the axial direction (Fig. 5), due to the swirling expansion in the chamber, is adequately reproduced by simulation. On the other hand, the increase in the maximum axial speed in the simulated profiles (Fig. 4) when increasing the axial distance with the injector is due to the increase in temperature. Considering the uncertainty of the experimental data, including the uncertainty of the 1.4 mm velocity profile of the injector based on which the input profile was defined, the velocity results given by the simulation are acceptable.

The results of the number of drops density (number of drops per cubic centimeter) along the radial axis and in different axial planes are compared in "Number of drops density (drop/cm3) at 5, 25, 45 and 65 mm (Fig. 7)" and "Sauter diameter at 5, 25, 45 and 65 mm (Fig. 8)". As noted, the dispersion of the drops increases along the axial axis, as the spray expands. Although the increase in the dispersion of the drops and their displacement along the radius is adequately captured by the simulation, the discrepancy in the peaks of maximum concentration is notorious. This suggests, in addition to the lack of calculated dispersion of the drops, that the software considers more drops, which means that a smaller average diameter is calculated. Compared with the results of the number of drops density reported by Zhu et al. [2], the results of the present work show greater similarity with the experiment.

The average diameter of Sauter is an important characteristic of the spray when the interface between the drops and the medium is relevant, as in the case of combustion. Reproducing this characteristic properly is necessary to make the spray behavior similar to the experiment. In Fig. 9 and 10 it is observed that in general the total surface of the drops is overestimated in the simulation, which suggests that the diameter of the drops calculated in general is smaller than the experimental one. This would imply that the number of drops is greater, since the same volume of fuel must be distributed in smaller drops, which is consistent with the increase in the number of drops density. However, the diameter trend is adequately reproduced by simulation. The slope observed in the distribution of the Sauter diameter can be explained considering that the diameter of the drops near the flame is smaller (due to the increase in evaporation) which translates into a smaller volume/area ratio. On the other hand, the increase of the diameter with the axial distance is possibly due to the fact that the drops of smaller diameter and greater surface evaporate with greater speed, so that the drops of greater diameter remain at a greater distance from the injector.

The "Temperature contours (K) (Fig. 11)" shows the temperature profile inside the combustion chamber, where the candle-shaped flame is observed, which is congruent with the conical shape of the spray and similar to the results of the previous simulations [2–4]. The maximum temperature reached, according to the simulation, is approximately 1790 K. No experimental data on the temperature inside the chamber is reported, however

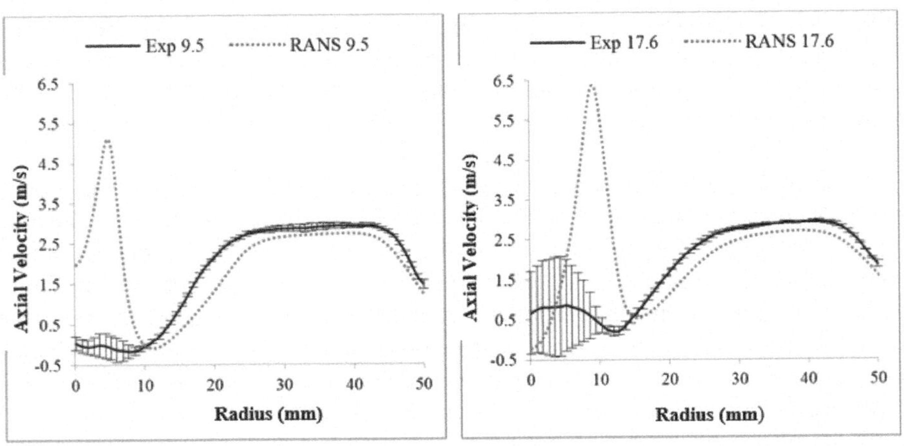

Fig. 4. Axial velocity at 9.5 and 17.6 mm of the injector

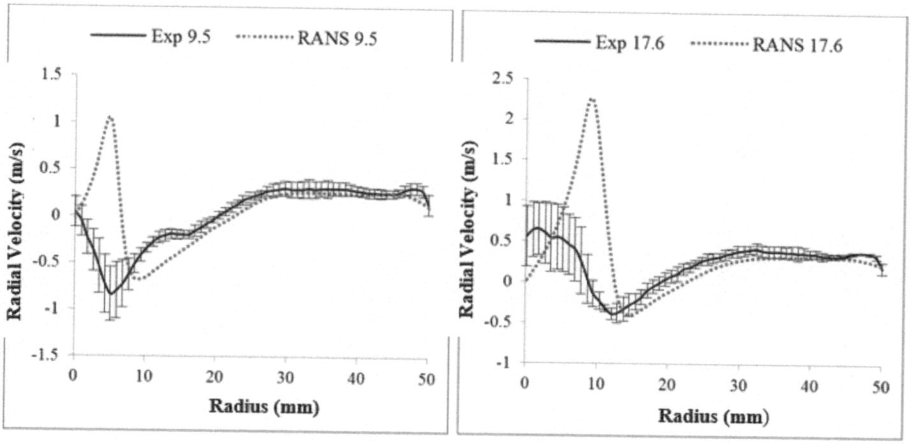

Fig. 5. Radial velocity at 9.5 and 17.6 mm of the injector

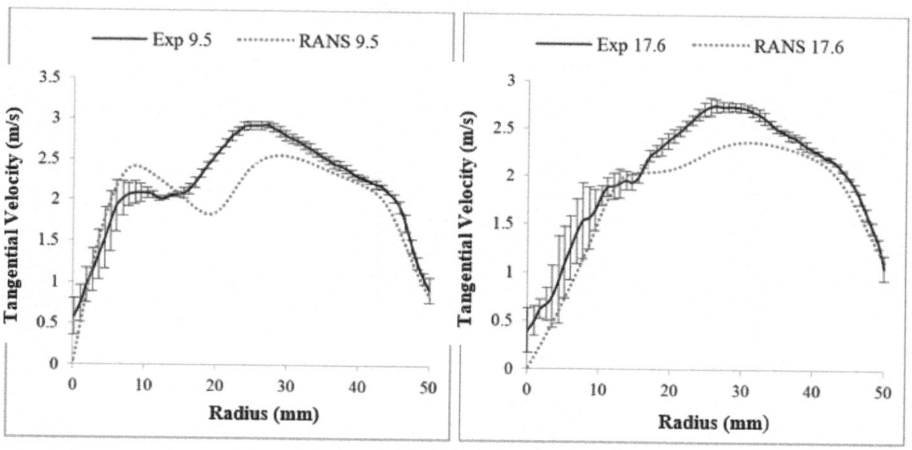

Fig. 6. Tangential velocity at 9.5 and 17.6 mm of the injector

the result is similar to the previous simulations, where Zhu et al. [2] report a maximum temperature of 1850 K, Collazo et al. [3] of 2100 K and Crocker et al. [4] of 1900 K.

The temperature at the exit of the chamber is approximately 750 K. Since the output of the chamber was modified it is not possible to directly compare this result with the experiment. However, and as indicated by Zhu et al. [2], considering that the average output temperature is 550 K according to the experiment, the output being located at a point farther than in the simulation, the temperature of 750 K given by the simulation is congruent.

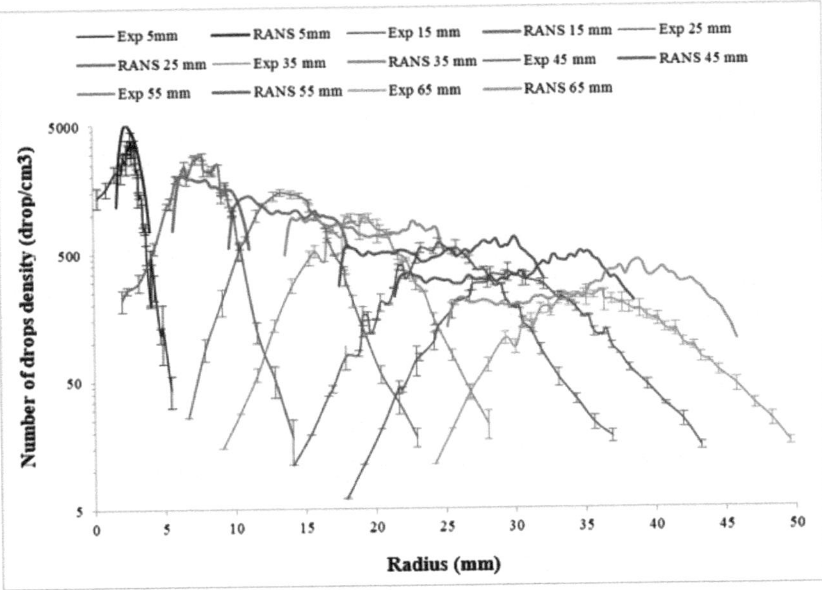

Fig. 7. Number of drops density (drop/cm^3) at 5, 25, 45 and 65 mm

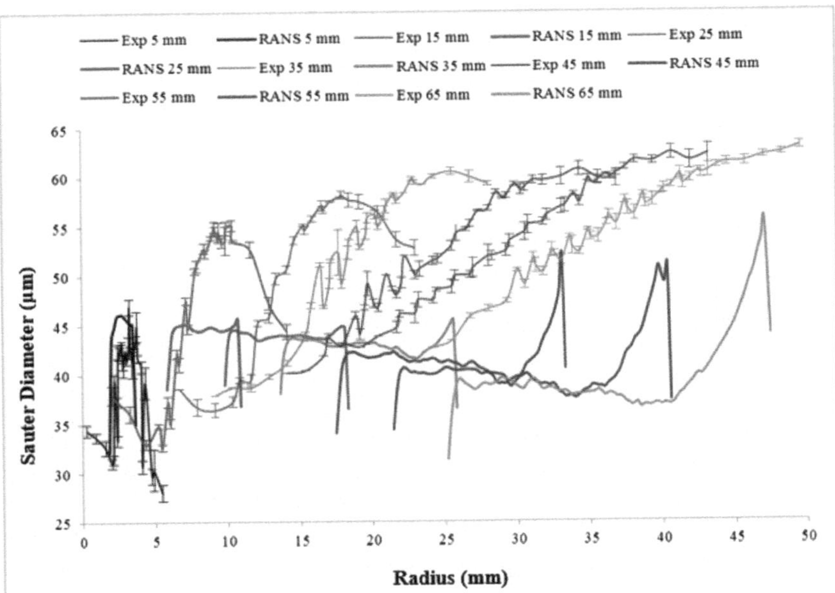

Fig. 8. Sauter diameter at 5, 25, 45 and 65 mm

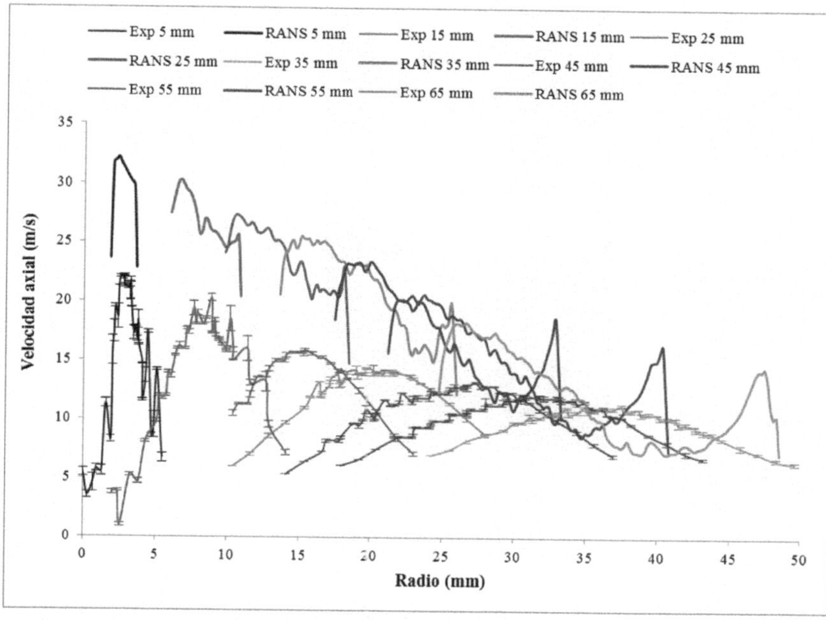

Fig. 9. Axial Velocity at 5, 25, 45 y 65 mm of the injector.

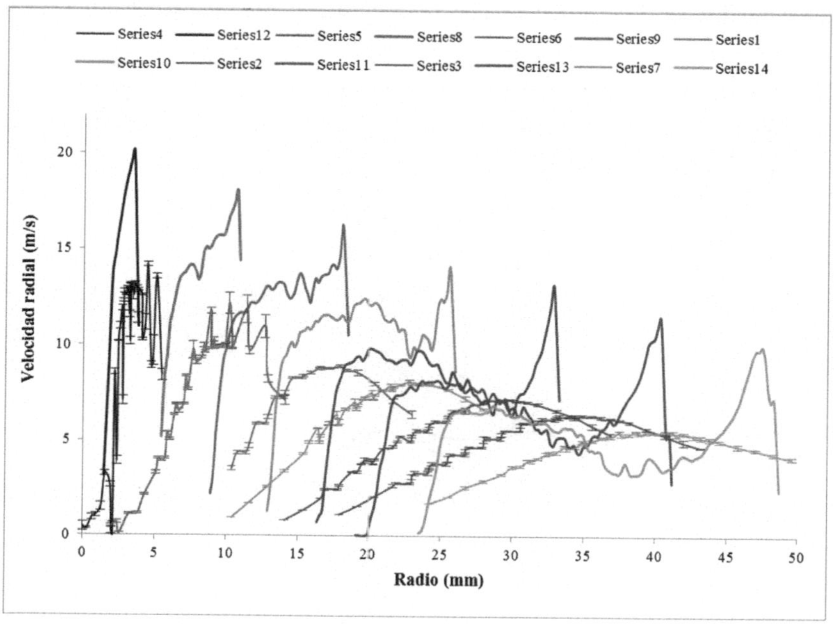

Fig. 10. Radial Velocity (15, 35 y 55 mm).

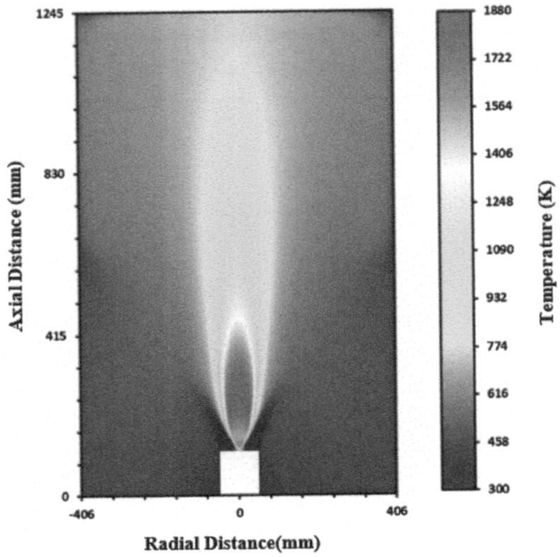

Fig. 11. Temperature contours (K).

The velocity vectors are plotted in "Flow velocity vectors (m/s) (Fig. 12)", where three zones with low intensity recirculation are observed. This recirculation is due to the pressure gradient between the inlet air flow and the fuel flow. The fuel flow, which is faster than the incoming air flow, produces a decrease in static pressure, which attracts part of

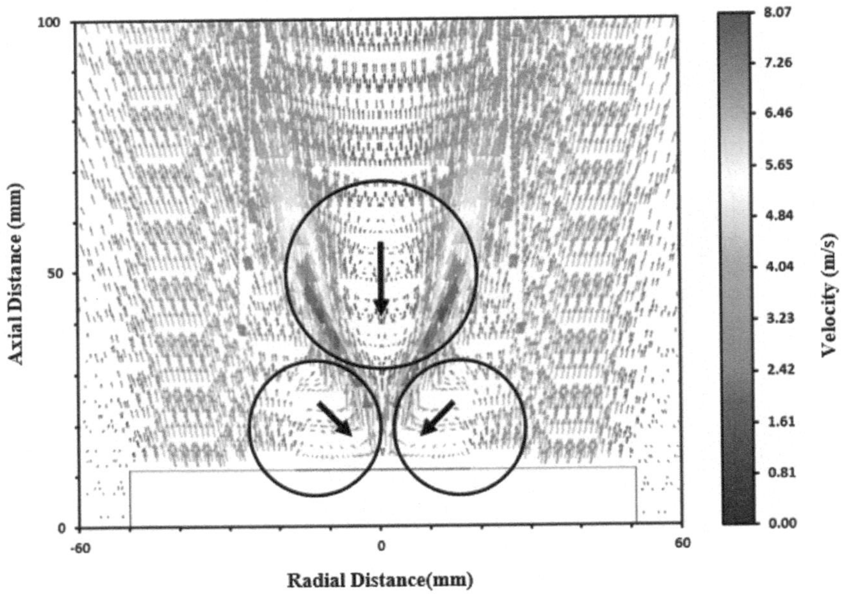

Fig. 12. Flow velocity vectors (m/s).

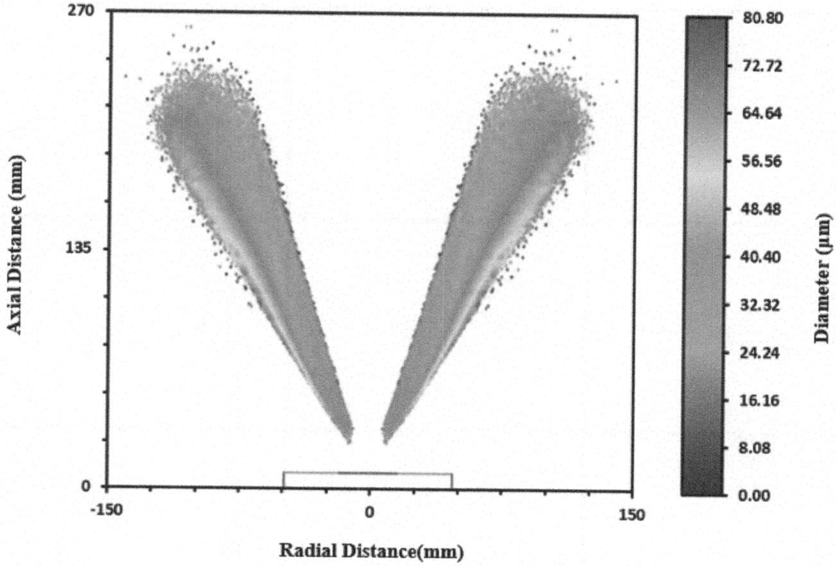

Fig. 13. Spray droplet diameter (μm).

the inlet air flow. The small lateral recirculation zones are congruent with the axial velocity profile reported experimentally. On the other hand, the expected vortex rupture does not occur, since an abrupt loss of the eddy structure or a larger recirculation zone is not observed.

The "Spray droplet diameter (μm) (Fig. 13)" shows the distribution and properties of the fuel particles (droplet packages) that are in the computational domain during the last iteration of the simulation (the spray is transitory). This figure shows the distribution of the size of drops, where it can be seen that inside the cone the drops are smaller (due to the proximity with the flame). This is consistent with the Sauter diameter distribution discussed above. All particles evaporate completely and incomplete trajectories are not reported (particles that escape the domain or collide with its walls). Comparing the distribution of the drops with the temperature profile, it is observed that some drops cross the high temperature zone, which is congruent with the percentage of unburned methanol experimentally reported (although in the simulation it is assumed that this vapor also reacts). The maximum distance to the injection point reached by the spray is approximately 297 mm.

Once the numerical technique to reproduce turbulent spray combustion phenomena was valid, the injection process control was studied, setting the atomizer operating parameters and varying the Reynolds number, the graphs of temperature contours of Fig. 14, show us the different events if the injection velocity increases, so that the dispersion of the drops are affected during the combustion process in control of the size of the drops decreases, the recirculation of the flow intensifies, the temperature decreases until the flame goes out.

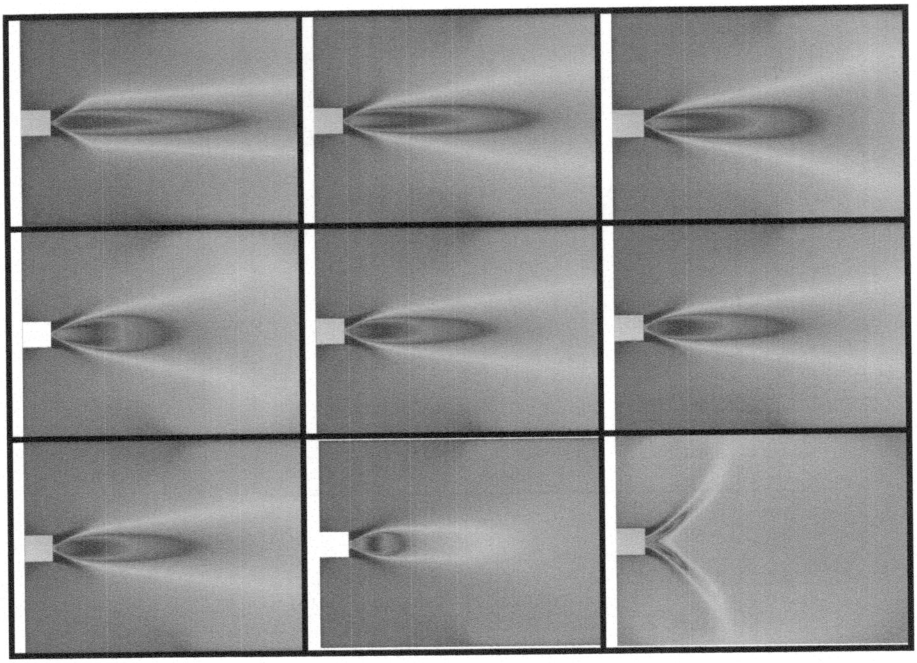

Fig. 14. Temperature Contours at different Reynolds

4 Conclusions

In the present work, the turbulent spray combustion process was simulated in the experimental combustion chamber model developed by NIST. Experimental data have been reported in other works, as well as the boundary conditions used in this simulation. The combustion model used is the flamelet model, which demands few computational resources considering that the reaction is rapid compared to the turbulent processes of the flow. To simulate the spray, the LISA model was used, in addition collisions between drops and their turbulent dispersion are considered. The data suggests that the droplet size determined by the LISA model is smaller than the actual spray droplet size. The results of the RANS simulation were used to estimate the turbulence scales present in the flow.

The control of the injection of drops, the atomized or pulverized both are of the utmost importance during combustion, since they depend on the quality of the burn, the maximum temperature in the chamber, the number of oxidants and the amount of contaminant products that are discarded to the atmosphere, so specialized studies in the control of injectors are able to deliver adequate information to keep the flame on.

References

1. Widmann JF, Presser C (2002) A benchmark experimental database for multiphase combustion model input and validation. Combust Flame 129:47–86

2. Pope SB (2000) Turbulent flows. Cambridge University Press, Cambridge
3. Veynante D, Vervisch L (2002) Turbulent combustion modeling. Prog. Energy Combust. Sci. 28:193–266
4. Peters N (1986) Laminar flamelet concepts in turbulet combustion. In: Twenty-First Symposium (International) on Combustion, pp. 1231–1250
5. Warnatz J, Maas U, Dibble RW (2006) Combustion, physical and chemical fundaments, modeling and simulation, experiments, pollutant formation, 4 ed
6. Magnuseen BF (1989) Modeling of NOx and soot formation by the Eddy Dissipation Concept. International Flame Research Foundation
7. Magnussen BF (1981) On the structure of turbulence and a generalized eddy dissipation concept for chemical reaction in turbulent flow. In: 19th American Institute of Aeronautics and Astronautics Aerospace Science Meeting
8. Zhou LX et al (2016) A review on studies of a SOM combustion model for single-and-two-phase combustion. Int. J. Heat Mass Transf. 96:154–163
9. Zhou L, Li K, Wang F (2012) Advances in large-eddy simulation of two-phase combustion (I) LES of spray combustion. Chin. J. Chem. Eng. 20(2):205–211
10. Zhu S, Roekaerts D, Meer Tvd (2011) Numerical study of a methanol spray flame

Control of a Stewart-Gough Platform for Earthquake Ground Motion Simulation

Erick Alvarado Requena(✉), Antonio Estrada, Gengis Toledo Ramírez, Noe Reyes Elias, Jorge Uribe, and Berenice Rodríguez

Centro de Ingeniería y Desarrollo Industrial, 76125 Querétaro, QRO, Mexico
e.alvarado@posgrado.cidesi.edu.mx

Abstract. The Stewart-Gough platform is a six DOF parallel robot manipulator with higher precision, mechanical stiffness and load capacity compared with a conventional serial link arm. In this work, a CAD model of CIDESI Stewart Platform HxCf1 is imported to Simscape as a dynamic model and an application of an inverse dynamics control scheme is proposed; full kinematic equations of the robot are presented. A ground motion movement "profile" consisting on position (x, y, z) and orientation (α, β, γ) based on real earthquake records is loaded and simulated; simulation results illustrate the performance of the HxCf1 robot using a standard PID controller.

Keywords: Parallel manipulator · Robotics · Control

1 Introduction

Parallel manipulators offer some advantages such as precision, mechanical stiffness and higher load capacity, due to its parallel structure if compared with serial manipulators; on the other side, a disadvantage of parallel manipulators is the workspace limitations due to the link lengths and the mechanical limits of the passive joints [1]. Generally, because of the non-linearity and the complexity of the equations, forward kinematics of parallel manipulators is very complicated and difficult to solve; this is also in contrast with serial manipulators. Dynamical analysis of parallel robots is very important to develop a model based controller but also complicated due to the existence of multiple closed-loop chains. Dynamic equations of a Stewart-Gough platform can be derived based on Lagrange's formulation [2]; in this paper, application of an inverse dynamics control scheme is proposed for trajectory tracking on a Stewart-Gough platform.

The CIDESI Hexapod robotic system is a project that aims to develop an industrial grade robotic system that provides functionalities similar to those usually found in an industrial robot; the CIDESI Hexapod has been designed to be operated by a human operator (User), or by another system or device (Application) e.g., a vision system.

Generally, a Hexapod robot system consists of a Robot (mechanical structure), controller, Teach Pendant, computer, user documentation and accessories.

E. E. Hernandez et al. (Eds.): LASIRS 2019, MMS 86, pp. 138–146, 2020.
https://doi.org/10.1007/978-3-030-45402-9_14

These are the basic components that are generally included in any serious commercial robotic system. The "Hexapod CIDESI project" is divided at least in three phases, with the development of one prototype with increased capabilities in each phase; the current prototype corresponds to the phase I, which is called "Hexapod CIDESI phase 1", better known by its nickname: "HxCf1" [3,4]. In the Phase I, the robot is mainly intended as a low cost industrial-like experimental prototype [5].

2 Stewart-Gough Platform Model

The Stewart - Gough platform is a robot designed in a closed chain, in which a fixed platform in space, considered as the base of the robot, and a mobile platform are connected; both joined by six linear extensible legs. The points of union with the platforms are mechanical joints with more than two degrees of movement; this manipulator is actuated by motors located on the prismatic joints [6]. The design provides high performance in precision and strength compared to serial manipulators.

In Fig. 1, the HxCf1 [7] prototype is shown. For modeling purposes two coordinate frames are defined, a coordinate frame, B(x, y, z), attached to the fixed base containing the universal joints A_i and a frame T(x, y, z) attached to the mobile platform containing the universal joints GT_i being $i = 1$ to 6. Extension of each leg is denoted by l_i [8].

(a) (b)

Fig. 1. CIDESI Hexapod HxCf1.

Hexapod positioners are based on 6 DOF motions; these are the three linear movements x, y, z (lateral, longitudinal and vertical), and the three rotations roll (x-axis), pitch (y-axis) and yaw (z-axis) of frame B. The following position vectors of the connection points on its corresponding coordinate system are derived from Fig. 1.

$$GT_i = \begin{bmatrix} r_p cos(\lambda_i) \\ r_p sin(\lambda_i) \\ 0 \end{bmatrix} ; B_i = \begin{bmatrix} r_b cos(\upsilon_i) \\ r_b sin(\upsilon_i) \\ 0 \end{bmatrix} \qquad (1)$$

The coordinate values for each point calculated using Eq. 1 are shown in Table 1.

Table 1. Point coordinates for B_i and GT_i.

Point	Coordinates	Point	Coordinates
B1	$(160.1226, 189.3407, 0)$	GT1	$(189.2780, 36.9977, 0)$
B2	$(83.9125, 189.3407, 0)$	GT2	$(-62.5981, 182.4184, 0)$
B3	$(-244.0351, 43.9999, 0)$	GT3	$(-126.6799, 145.4207, 0)$
B4	$(-244.0351, -43.9999, 0)$	GT4	$(-126.6799, -145.4207, 0)$
B5	$(83.9125, -233.3407, 0)$	GT5	$(-62.5981, -182.4184, 0)$
B6	$(160.1226, -189.3407, 0)$	GT6	$(189.2780, -36.9977, 0)$

We refer to the origin of the moving platform as G. The position vector of G w.r.t. the B_{XYZ} coordinate system is denoted as $P = [P_x P_y P_z]^T$. The rotation matrix of the moving platform relative to the fixed base is given by:

$$^{B}R_T = R_Z(\gamma)R_Y(\beta)R_X(\alpha) =$$

Thus, the vector L_i corresponding to the link i is obtained as follows (in B_{XYZ} coordinates).

$$L_i = R_{XYZ}GT_i + P - B_i, i = 1, ..., 6 \tag{2}$$

and their lengths, l_i, correspond to the actuator positions

$$l_i = \|L_i\|$$

After obtaining inverse kinematics, a dynamic model of the Stewart-Gough platform may be derived using the Lagrange formulation [2]. Kinetic and potential energies of the moving platform and the legs, each one consisting of a piston and a cylinder, are obtained in compact expressions. Lagrangian formulations are taken in compact symbolic forms in order to make the expression understandable. The matrix equation is showed below:

$$M'(X_p)\ddot{X}_p + V'_m(X_p, \dot{X}_p)\dot{X}_p + G'(X_p) = J^T(X_{p)}\tau, \tag{3}$$

Where $M'(.)$ is the mass matrix, $V'_m(.)$ is the velocity product term, $G'(.)$ is the gravitational term, J(.) is the Jacobian matrix and $\tau \in R^6$ is the generalized actuator force at each actuator all with respect to the based frame.

Dynamic formulation of the hexapod in the joint space coordinates can be described by transforming the dynamics written in the task-space coordinates starting with the Jacobian Matrix $J(.)$:

$$\dot{x} = J^{-1}\dot{q}, \tag{4}$$

where $\dot{q} \in R^6$ is the velocity vector for the six legs. A dynamic matrix equation in the joint space coordinates will be obtained as follows:

$$M(q)\ddot{q} + V_m(q, \dot{q})\dot{q} + G(q) = \tau, \tag{5}$$

where

$$\begin{aligned} M(q) &= J^{-T}M'(X_p)J^{-1}, \\ V_m(q, \dot{q}) &= J^{-T}M'(X_p)J^{-1}\dot{J}J^{-1} + J^{-T}V_m'(X_p, \dot{X}_p)J^{-1}, \\ G(q) &= J^{-T}G'(X_p), \end{aligned} \tag{6}$$

Where $q \in R^6$ represents the six link displacement vector.

3 Control Scheme Proposal

In order to linearize and decouple the robot manipulator dynamics, inverse dynamic control is used; possible non-linearities such as Coriolis and centrifugal terms as well as gravity terms can be compensated by adding these terms to the control input. However, this type of feedback linearizing controller is much more complicated and it's conception is not straightforward for the robot manipulator dynamics [9].

In order to cancel non-linear terms and decouple the dynamics of each link, it is used an inverse dynamics control of the form:

$$u = M(q)u_0 + C(q, \dot{q})\dot{q} + G(q), \tag{7}$$

applied to (3). The mass matrix, $M(q)$, is a symmetric positive definite matrix which verifies:

$$\lambda_m I \leq M(q) \leq \lambda_M I, \tag{8}$$

where $\lambda_m(\lambda_M < \infty)$ denotes the strictly positive minimum (maximum) eigenvalue of $M(q)$ for all configuration q. These properties show the regularity of the inertia matrix. The controller of (7) and the regularity property of $M(q)$, yield a set of n decoupled linear systems.

$$\ddot{q} = u_0 \tag{9}$$

where u_0 is an auxiliary control input to be designed. A typical choice for u_0 is:

$$u_0 = \ddot{q}_d + K_D(\dot{q}_d - \dot{q}) + K_P(q_d - q), \tag{10}$$

which leads to the error equation

$$\ddot{\tilde{q}} + K_D\dot{\tilde{q}} + K_P\tilde{q} = 0, \tag{11}$$

where $\tilde{q} = q_d - q = -e$.

The error equation, (11), is exponentially stable by a suitable choice of the matrices K_D and K_P; Eq. 11 can be rewrite in its state space form, $\dot{\xi} = A\xi$, where the matrix A is Hurwitz provided K_D and K_P were suitably chosen.

4 Dynamic Model

The dynamic model of the CIDESI HxCf1 Hexapod was made taking as base the CAD model in Simscape software. Simscape enables to create models of physical systems within the Simulink environment; it is possible to build physical component models based on physical connections that directly integrate with block diagrams and other modeling paradigms.

Simscape helps to develop control systems and test system-level performance. The block diagram modeling environment for the engineering design and simulation of rigid body machine and their motions, allows the user to model and simulate mechanical systems with a suite of tools to specify bodies and their mass properties, their possible motions, kinematics constraints and coordinate systems, and to initiate and measure body motions. The full model with its components is shown in Fig. 2.

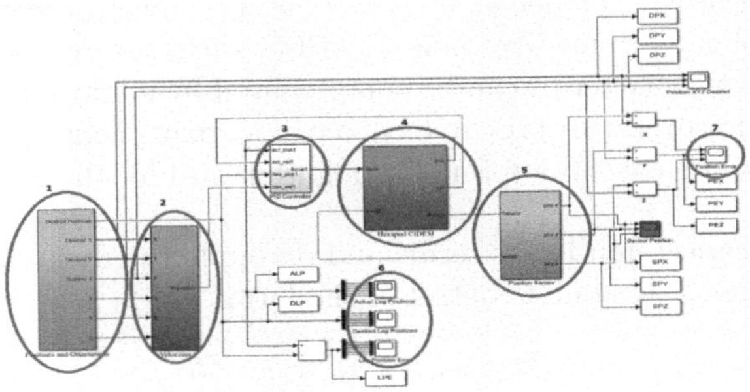

Fig. 2. CIDESI HxCf1 Hexapod Simscape model. 1. Kinematic blocks, 2. Jacobian blocks, 3. Configurable PID controller, 4. Assembly exported from SolidWorks, 5. Position sensor of the mobile platform, 6. Leg position and error, 7. Mobile platform position error.

The exported CAD model from SolidWorks to Simscape using SimMechanics add-on generate one XML file and a set of geometry files that were converted in order to create the platform part by part by linked blocks. After the conversion was complete, the geometry blocks of the platform looks as depicted in Fig. 3 and the visual output of the platform is shown in Fig. 4.

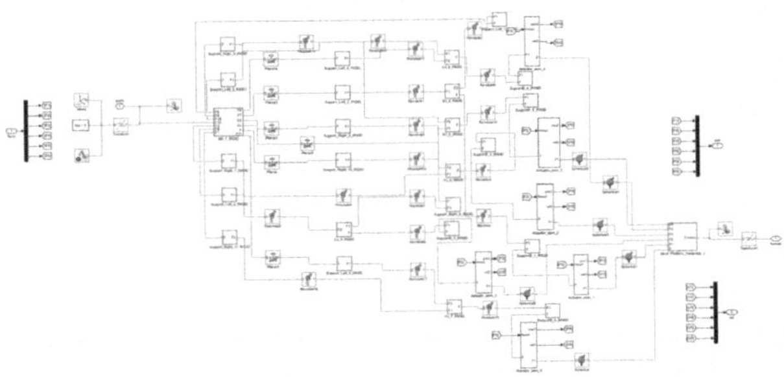

Fig. 3. CIDESI HxCf1 Hexapod exported geometry.

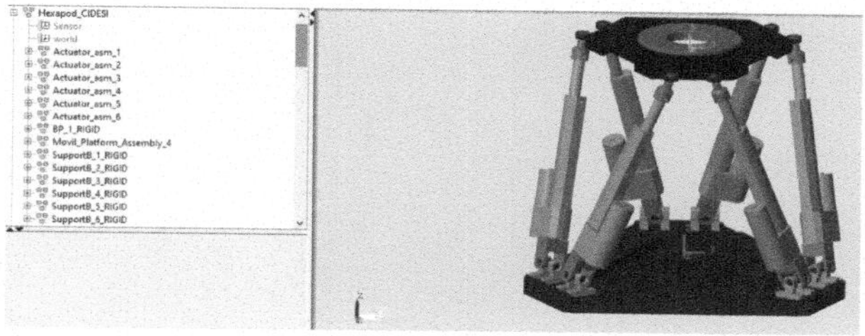

Fig. 4. CIDESI HxCf1 Hexapod visual output.

5 Simulation

A simulation study will be performed to illustrate the defined Simscape model of the HxCf1 Hexapod, using an earthquake ground motion "profile", see Fig. 2, consisting on 3 positions (x, y, z) and 3 orientations (α, β, γ) based on real ground motions that was obtained from PEER database [10] and loaded as input. In the simulation, the reference positions are updated at a 50 ms rate and a classic PID controller is used, setting the gains as follows: $Kp = 100$, $Ki = 20$ and $Kd = 2$.

In order to perform a simulation, an input file containing the required positions and orientations for the Stewart Platform is used; for this simulation a real earthquake ground motion obtained from PEER database [10] was used as input. In the simulation, the reference positions are updated at a 50 ms rate and a classic PID is used; after running the simulation, the values given by the position sensor in the mobile platform are compared with the desired values and the length errors of the actuators 1, 3 and 5. The results are shown in Fig. 5. For the mobile platform, the position sensor data was plotted against the desired position data in the x, y, z axis; plots are shown in Fig. 6 having the error plots in Fig. 7.

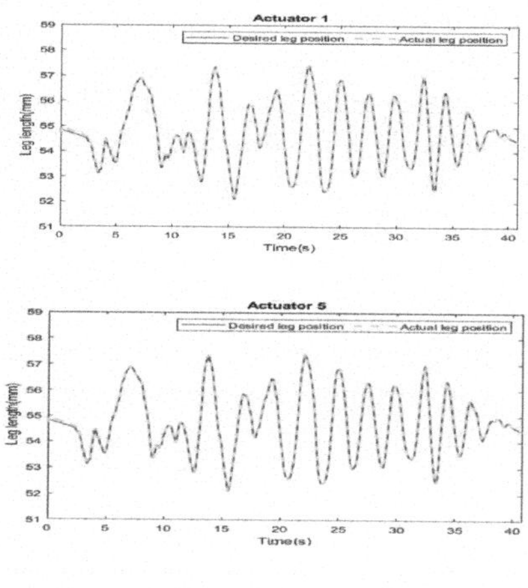

Fig. 5. Actuator length errors

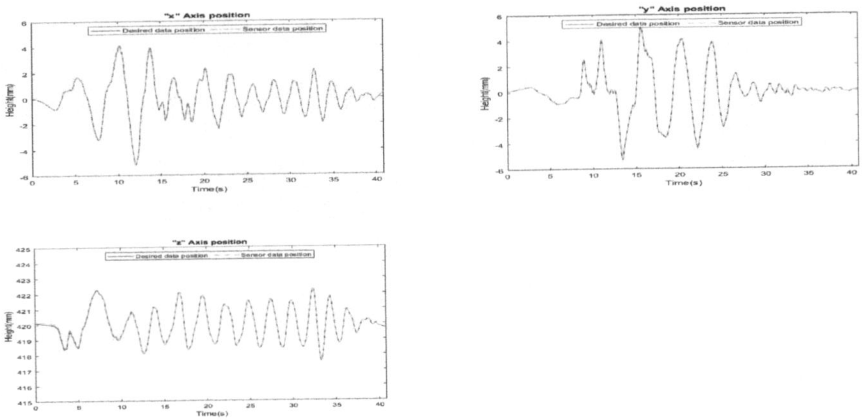

Fig. 6. Mobile platform axis position.

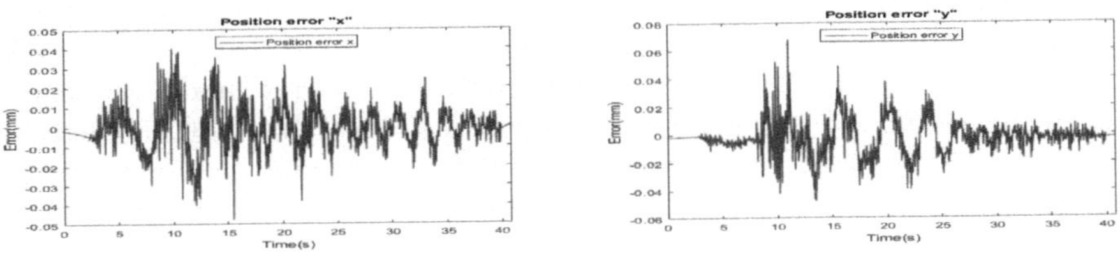

Fig. 7. Mobile platform axis error.

6 Experimental Results

In order to perform experimental tests and be able to compare the results with the Simulink dynamic model, an Arduino program using FreeRTOS library was written; the program allow to read the measurement of the position of the actuators, using their own potentiometer, at a specified time.

With the measurements done, the real position of the actuators can be compared with the desired position that was used as input to the Stewart Platform at different points of the routine. The plots shown in Fig. 8 display the performance of the experimental readings against the desired positions in the x, y, z axis for the actuators numbered 1, 3 and 5; the plots shown from Fig. 9 show the error in the same referred axis.

The measured response of the actuator potentiometers is very noisy for each measurement. It was necessary to use a filter in order to obtain a more realistic result. After passing the data trough a first order filter with a time constant of 1 s, the corresponding plots are shown in Fig. 10.

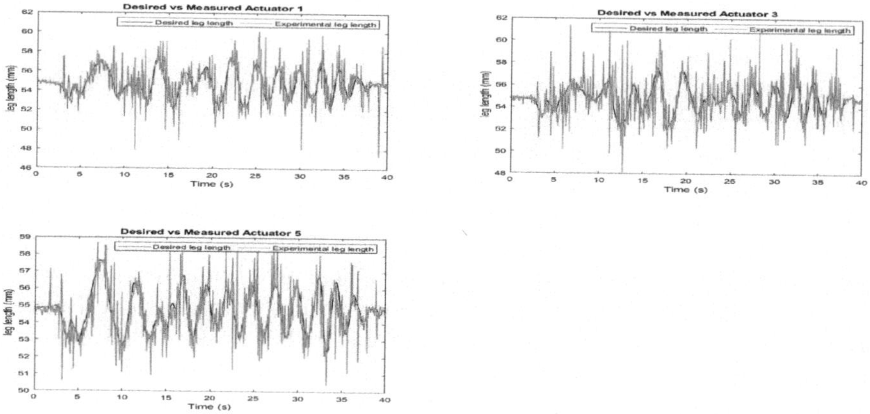

Fig. 8. Measured positions against desired positions.

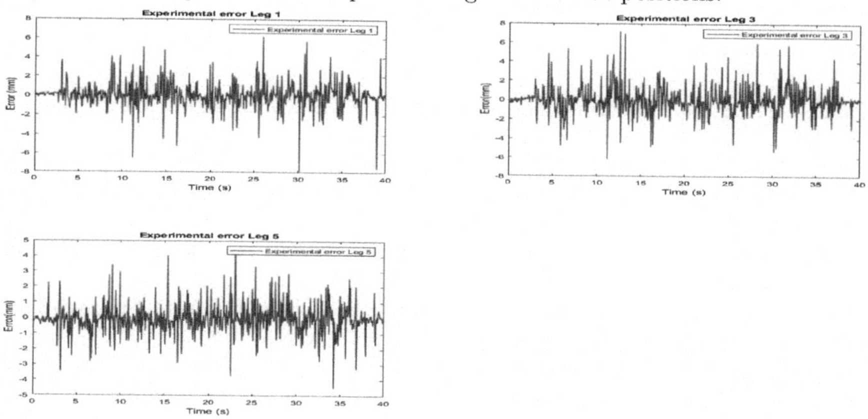

Fig. 9. Mobile platform axis position.

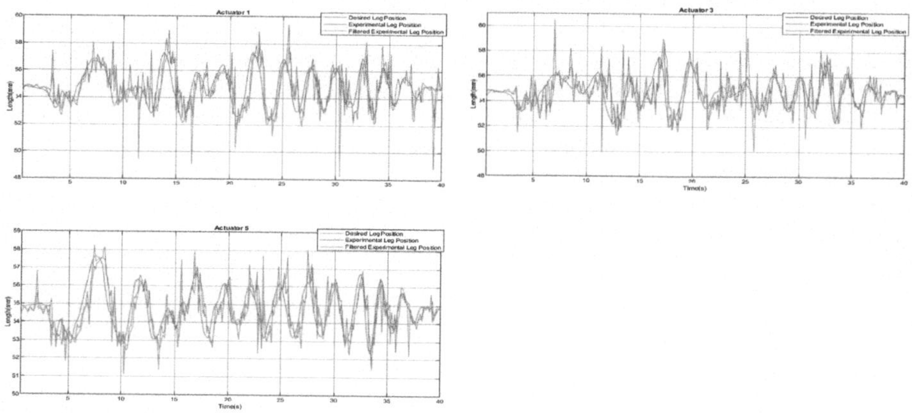

Fig. 10. Filtered experimental error.

7 Conclusions

The simulation of the earthquake profile was very fast and the resulting error is very low in every axis. For the experimental tests, it is evident that the measured position values at each point are following the reference values. Nevertheless, the error is quite larger than the results in simulation. Additionally, there is the problem of the noise in measurements of positions, this could be due to parasitic current that are returning to the voltage source when the routine is being performed; the filter that was applied help to obtain results closer to reality since the spikes on measured positions do not represent a real feasible motion of the platform. It is important that for further work to eliminate most of the noise generated in the measurements in order to be able to experiment with more earthquake "profiles" with different frequencies and have a better evaluation of the HxCf1 performance.

References

1. Merlet J-R (1995) Determination of the orientation workspace of parallel manipulators. J Intell Rob Syst 13:143–160
2. Ghobakhloo A, Eghtesad M (2006) Closed form solution to inverse dynamics problem of a parallel robot (Stewart platform) by Lagrangian approach. In: Tehran international congress on manufacturing engineering (TICME 2005)
3. Toledo-Ramírez GK (2018) Hexápodo-CIDESI, fase 1. conceptos de operación y requerimientos. Technical report, CIDESI-CONACYT, August 2018
4. López GAD (2018) Desarrollo de prototipo alfa de plataforma Stewart-Gough "Hexápodo CIDESI". Technical report, CIDESI-CONACYT
5. Germán Félix JL (2018) Robot controller for a parallel robot, from conception to implementation. Master's thesis, CIDESI-CONACYT & FH Aachen
6. Stewart D (1965) A platform with six degrees of freedom. In: Proceedings of the institution of mechanical engineers
7. Barasja Fletes JR, Toledo Ramírez GK (2018) Diseño de detalle de prototipo alfa Hexápodo CIDESI - Fase I. Parte Mecánica. Technical report, CIDESI-CONACYT
8. Arciniega E (2018) Cinematic and dynamic simulations of CIDESI Hexapod HxCf1. Technical report, CIDESI
9. Ghobakhloo A, Eghtesad M (2006) Position control of a Stewart-Gough platform using inverse dynamics method with full dynamics. In: Advanced motion control
10. Users manual for the PEER ground motion database web application. https://ngawest2.berkeley.edu/peer_ground_motion_database_users_manual.pdf. Accessed 19 Apr 2019

Flow Control on a Wing Section Through the Implementation of the SDBD Plasma Actuator

R. A. Bernal-Orozco(✉), A. Arias-Montano, and O. Huerta-Chávez

IPN-ESIME Unidad Ticoman, SEPI. Av. Ticoman 600, San José Ticoman, C.P. 07340 Gustavo A. Madero, Ciudad de México, Mexico
rulozco@hotmail.com, aarias@ipn.mx, lecram_21@hotmail.com

Abstract. This work presents a Computational Fluid Dynamics (CFD) wind flow analysis technique applied to a Wing Section using a two-dimensional Reynolds-Averaged Navier-Stokes Method. The behavior of the flow has been studied from different angles of attack of the aerodynamic section until the angle of stall is found. An active flow control by means of the SDBD plasma actuator is implemented numerically, through body force source terms in the Navier-Stokes equations to reattach the boundary layer of the flow. Then, altering the position and frequency of the SDBD plasma actuator results in the delay of the stall beyond 16° and an increase of lift with reduction of the drag for new critical angles of stall.

Keywords: Flow-control · SDBD · Dielectric barrier discharge · Airfoil with plasma actuator · RANS

1 Introduction

The lift of an airfoil depends greatly on the attachment of the flow. One way to keep the flow attached is through the use of flow control methods such as Single Dielectric Barrier Discharge (SDBD) actuators. A flow control method as defined by Flatt [4] is "any mechanism or process, which alters the boundary layer so it behaves different from it will naturally do". An SDBD actuator is a device that through an electrical discharge ionizes the air and by the acceleration of the ions by the electromagnetic forces, produces a transfer of momentum to the flow, thus diminishing its separation, Fig. 1 shows a schematic diagram of a plasma actuator.

The use of SDBD actuators in aerodynamic applications such as reattachment of flow in laminar separation bubbles was proved by Corke [2] who successfully reattached the flow in a NACA 66_3-018 airfoil up to 8° beyond the stall. Corke also tested them in blades from a low pressure turbine of a gas

E. E. Hernandez et al. (Eds.): LASIRS 2019, MMS 86, pp. 147–158, 2020.
https://doi.org/10.1007/978-3-030-45402-9_15

turbine engine. Greenblatt [6] proposed its use in wind turbines, with the use of an actuator he obtained a 7% increase in the net power delivered by the wind turbine.

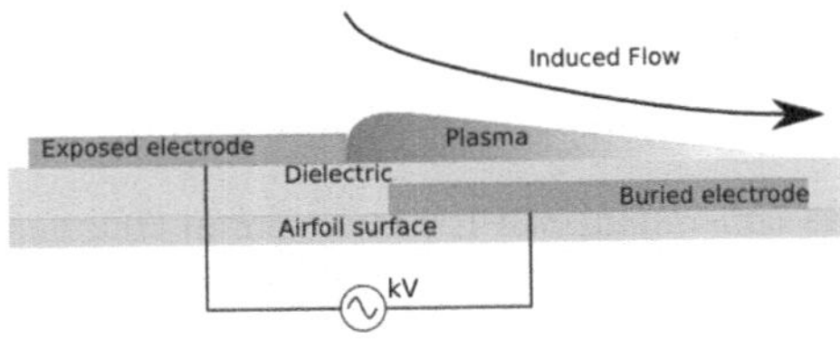

Fig. 1. SDBD scheme.

In the present study, two-dimensional Reynolds-averaged Navier–Stokes equations (RANS) simulations of an airfoil with an SDBD at a Reynolds number (Re) of 2.0×10^5 are conducted, to analyze the effect of changing the position and frequency of the actuator. While most previous studies used commercial software, in this work the simulations are performed with the open source software OpenFOAM. The actuator was implemented as a source term body force so it can be directly introduced in the flow equations, as explained in Sect. 3. The position of the actuator is analyzed by placing the actuator at 3% and then at 10% of the airfoil chord for several angles of attack, as shown in Sect. 6.1. For the frequency analysis the base frequency remain fixed and the burst frequency was changed in a range from 3 Hz to 45 Hz, this is presented in Sect. 6.2. In Sect. 7 the methodology used in this study was validated through comparison with experimental data.

2 Flow Equations

To simulate the flow around the airfoil. The RANS equations are used. Taking the flow as incompressible and by the fact that most of the electric energy consumed by the actuator is transformed into kinetic energy, the change in temperature can be neglected, so it is only required the continuity equation, Eq. (1), and the momentum equation Eq. (2).

$$\frac{\partial \bar{u}_j}{\partial x_j} = 0 \tag{1}$$

$$\frac{\partial \bar{u}_i}{\partial t} + \bar{u}_j \frac{\partial \bar{u}_i}{\partial x_j} = -\frac{1}{\rho}\frac{\partial \bar{p}}{\partial x_j} + \frac{\partial}{\partial x_j}\left[(\nu + \nu_t)\frac{\partial \bar{u}_i}{\partial x_j}\right] + \frac{f_b}{\rho} \tag{2}$$

where p is the pressure, u the speed, ρ the density, ν the kinematic viscosity, ν_t the eddy viscosity, and f_b the body force of the actuator.

From the turbulence models available in OpenFOAM the $k - \omega$SST was selected. It is a lineal model, in which the Reynolds stress are modeled by the

Boussinesq hypothesis which is a linear relationship: $-\rho\langle u_i u_j\rangle = 2\mu_t S_{i,j} - \frac{2}{3}\rho k\delta_{ij}$, where μ_t is the eddy viscosity, k the turbulence kinetic energy and $S_{i,j}$ the strain rate tensor. Then the transport of energy from the turbulence fluctuations is modeled through the eddy viscosity.

The $k-\omega$SST model is characterized by being a two equations model, one for turbulent kinetic energy Eq. (3), and another for specific dissipation Eq. (4). This model uses a $k-\omega$ formulation in the inner part of the boundary layer from the wall up to the viscous sub-layer. This formulation changes the behavior to $k-\epsilon$ model in the free stream, thus avoiding the problem of over sensitivity of the $k-\epsilon$ model to the turbulence intensity in the free stream. It is a model capable of capturing separation.

$$\frac{\partial k}{\partial t} + U_j \frac{\partial k}{\partial x_j} = \tau_{ij}\frac{\partial u_i}{\partial x_j} - \beta^* \omega k + \frac{\partial}{\partial x_j}\left[(\nu + \sigma_{k1}\nu_t)\frac{\partial k}{\partial x_j}\right] \tag{3}$$

$$\frac{\partial \omega}{\partial t} + U_j \frac{\partial \omega}{\partial x_j} = \frac{\gamma_1}{\nu_t}\tau_{ij}\frac{\partial u_i}{\partial x_j} - \beta^* \omega^2 + \frac{\partial}{\partial x_j}\left[(\nu + \sigma_\omega \nu_t)\frac{\partial k}{\partial x_j}\right] + 2\left(1 - F_1\right)\sigma_{\omega 2}\frac{1}{\omega}\frac{\partial k}{\partial t x_i}\frac{\partial \omega}{\partial x_i} \tag{4}$$

The turbulent viscosity ν_t in the model $k-\omega$SST is defined by the Eq. (5).

$$\nu_t = \frac{a_1 k}{\max\left(a_1\omega, b_1 F_{23} S\right)} \tag{5}$$

3 Modeling of the Actuator

To simulate an SDBD plasma actuator its computational model is required. Various models have been developed, those seek to accurately reproduce the properties of plasma, such as ion formation rate, collisions. But them entails an increase in the computational cost making them impractical for engineering applications. Alternatively engineering models such as Suzen model [11] or Shyy model [10] have been developed, these models are based on the idea that the body force is directly proportional to the product of the charge density by the electric field. However, these models have some drawbacks such as that the electrodes should be strictly modeled in the domain since the thickness and position are required by the equations that determine the electric field. Besides that in the solver the electrical equations must be included.

That is why in works such as those carried out by Dörr and Kloker [3] the body forces are modeled as source terms. Kloker states that a SDBD can be modeled as a body force parallel to the wall. In the work performed by Sato [9] he used the Suzen model to calculate the magnitude of the force that is applied as a source term directly in the momentum equation.

In this work an actuator with a length of 0.03x/c was modeled, by solving the Eq. (6) from the Kloker model, the extension of the body was obtained, the body force extends from the wall up to a height of 0.003x/c, in the Fig. 2 the extension of body force is shown.

$$f(x,y) = c_x \left[\left(a_0 a_1 x + a_0^2 a_2 x^2\right) e^{-a_0 x}\right] \cdot \left[\left(b_1 y + b_2 y^2\right) e^{-b_0 y^{2/5}}\right] \tag{6}$$

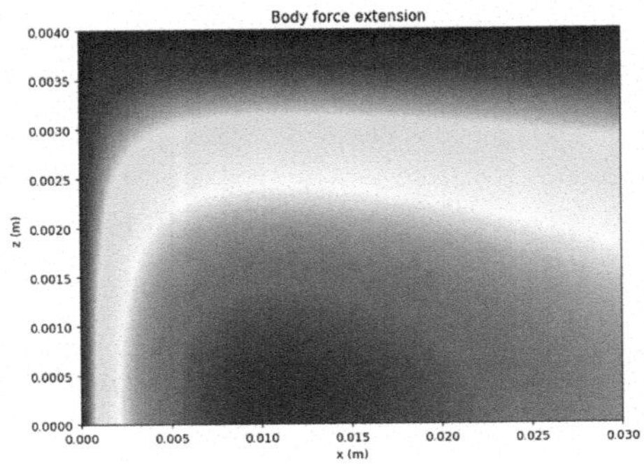

Fig. 2. Extension of the body force for an actuator 0.03x/c length.

The magnitude of the body force was selected from the results obtained by Hofkens [7] in which he determined the strength of a plasma actuator from velocity field data. The results obtained by Hoefkens are shown in Table 1.

Table 1. Body force at different voltages for $f_{base} = 2$ kHz.

Voltage kV	Force N/m^3
8	416
10	2038.8
12	2308.6
14	3258.8
16	3302.6

Plasma actuators can operate both in a continuous mode or through pulses as in the burst mode. It has been proven that an actuator operating in burst mode is more energy efficient [1,5,9,12].

The burst mode is generated multiplying a low frequency square signal with a high frequency sinusoidal base signal. In Fig. 3 is shown a scheme of the burst

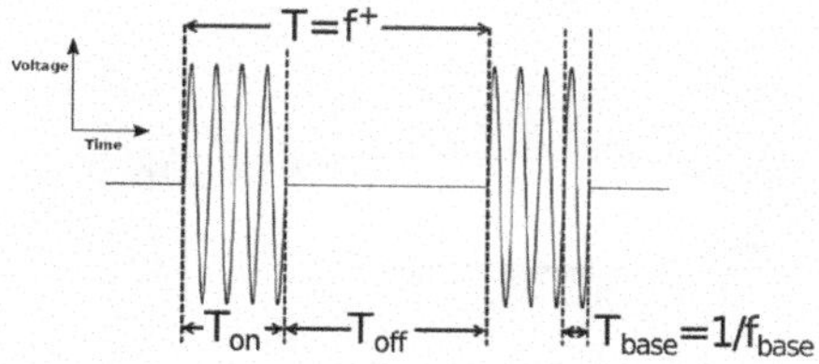

Fig. 3. Signal scheme for burst mode.

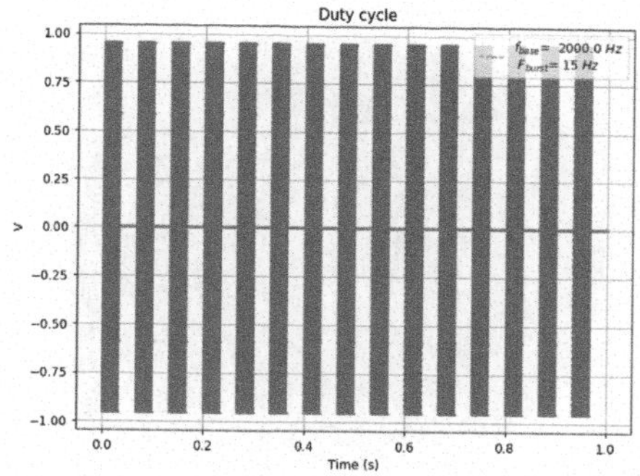

Fig. 4. Duty cycle for F_{burst} = 15 Hz and f_{base} = 2 kHz.

mode, and in Fig. 4 the graph generated by a burst frequency of 15 Hz and a base frequency of 2 kHz is shown.

The burst mode depends on the burst ratio Eq. (7), the burst frequency Eq. (8) and the base frequency Eq. (9).

$$BR = \frac{T_{on}}{T} \tag{7}$$

$$F^{+} = F_{burst} L/U_{\infty} \tag{8}$$

$$f^{+} = f_{base} L/U_{\infty} \tag{9}$$

where L is the length reference (airfoil chord) and U_{∞} is the free stream velocity.

A body force of 416 N/m^3 was selected, which corresponds to a voltage of 8 kV and a base frequency of 2 kHz, the burst frequency will have values of 3 Hz, 15 Hz and 45 Hz.

4 Geometry and Meshing

A structured C-type mesh was generated with `blockMesh`, Fig. 5 and Fig. 6 show the mesh. The airfoil has a unit chord of c = 1.0. Three meshes were used to perform a mesh independence analysis, the properties of the meshes are shown in the Table 2.

For the mesh study the values of the lift coefficient (C_l) and drag coefficient (C_d) were compared for a range of angles of attack from 0° to 12° for a Re number of 2.0×10^5. Table 3 and Table 4 show the results of the mesh independence study. Mesh 3 was selected to perform the simulations for the study of the plasma actuator, the y^+ was kept under a value of 5, and the mesh has a length of 25 × c and a height of 20 × c.

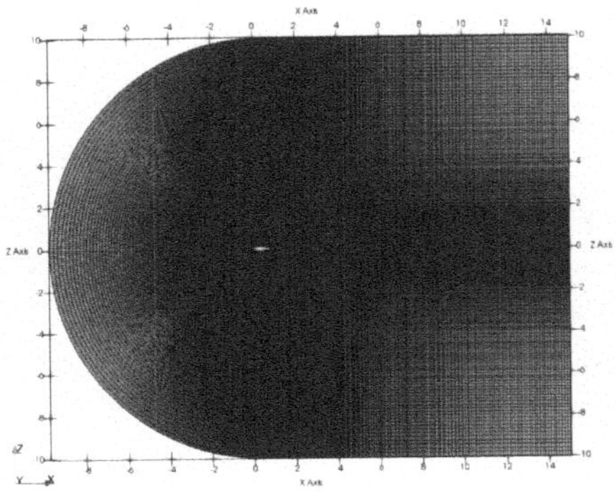

Fig. 5. General aspect of the mesh.

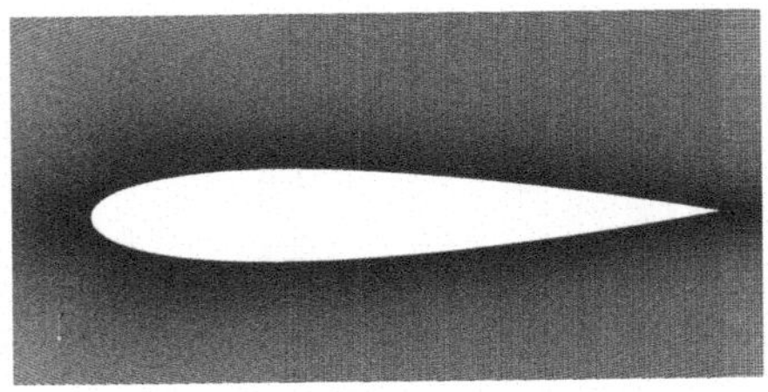

Fig. 6. Mesh detail.

Table 2. Mesh properties.

Parameter	Mesh 1	Mesh 2	Mesh 3
Nodes	629,700	775,500	104,7340
Cells	313,650	386,400	522,000
Max aspect ratio	306.773	309.543	277.867
Max non-orthogonality	29.171	29.5681	29.6446
Max skewness	0.45114	0.45267	0.45037

Table 3. C_l values for the mesh independence study.

α	C_l Mesh 1	C_l Mesh 2	C_l Mesh 3
0°	1.29E–07	–3.81E-08	–4.30E–07
4°	0.3717	0.3739	0.3739
8°	0.7189	0.7223	0.7273
12°	0.9751	0.9793	0.9859

Table 4. C_d values for the mesh independence study.

α	C_d Mesh 1	C_d Mesh 2	C_d Mesh 3
0°	0.0156	0.0154	0.0145
4°	0.0088	0.0093	0.0096
8°	0.0787	0.0795	0.0804
12°	0.1711	0.1724	0.1742

5 OpenFoam Setting up

The actuator was implemented in OpenFOAM as a source term using the utility `CodedSource` in which through `C++` language several types of source terms can be programmed. Then the source term is introduced into a set of specific cells (cell zone). The selected cell zone has a length of $0.03 \times c$ and height of $0.003 \times c$, which corresponds to the height established with the Eq. (6).

For the boundary conditions, a condition of type `fixedValue` is applied to the inlet to set the free stream velocity, a condition `inletOutlet` which provides a generic outlet when applying zero gradient for positive flows (outside the domain). A boundary condition of type `noSlip` sets the velocity with a value of zero at the wall. Also over the airfoil sruface wall functions are applied for turbulent kinetic energy `kLowReWallFunction`, turbulent viscosity `nutkWallFunction` and for the specific turbulent dissipation `omegaWallFunction`. To ensure a two-dimensional flow, a boundary condition of type `empty` is applied to the boundaries perpendicular to the flow.

The solver `pimpleFoam` was used and second order numerical schemes were established, for the time a backward scheme was applied, for the gradient, divergence and laplacians Gauss linear schemes with limiters, the interpolation were lineal. The time step was modified in each iteration automatically based on a maximum Courant number.

6 Results

The simulations were performed at $Re = 2.0 \times 10^5$ with the following initial conditions $U_\infty = 3.0 \quad \text{m/s}$, $\rho_\infty = 1.0 \quad \text{kg/m}^3$ and $\nu = 1.5 \times -5 \quad \text{m}^2/\text{s}$. The performance of the actuator position and frequency were studied through its effect over the lift and drag.

6.1 Actuator Position Analysis

Simulations of a NACA 0015 profile were performed at a Re of 2.0×10^5 for a range of angles of attack from 0° to 16° to study the effect of the actuator along of the chord, the actuator was placed at 3% (x/c = 0.03) and 10% (x/c = 0.1) of the chord. In this case the actuator operated in its continuous mode.

Figure 7 shows the velocity field without actuator for an angle of attack $\alpha = 16°$, at this angle the airfoil is stalled. When the actuator is active the stall is avoided and delays the detachment up to x/c = 0.5, this can be seen observed in Fig. 8.

Fig. 7. Velocity field for $\alpha = 16°$ without actuator.

Fig. 8. Velocity field for $\alpha = 16°$ whit actuator at x/c = 0.03.

The actuator reduces the detachment and prevent the stall, because the electric discharge ionizes the air and due to potential difference between the electrodes, an electromagnetic force moves those ions and that motion transfers momentum within the boundary layer. The increase of momentum in the boundary layer manifests itself as a jet within the boundary layer, see Fig. 9, that jet modifies the velocity profile of the boundary layer to a positive velocity gradient.

The plasma actuator also increases the vorticity in the nearby region where it is placed, this is evident in the graph of the Fig. 10 that shows two vorticity peaks in the place where the actuator is located in each case x/c = 0.03 and x/c = 0.1. The vorticity promotes the transition of the laminar boundary layer to a turbulent boundary layer, in the turbulent boundary layer there is a greater interaction between the flow outside it and that which is inside, so there is an addition of momentum from the external flow to the boundary layer which helps the flow to remain attached for longer.

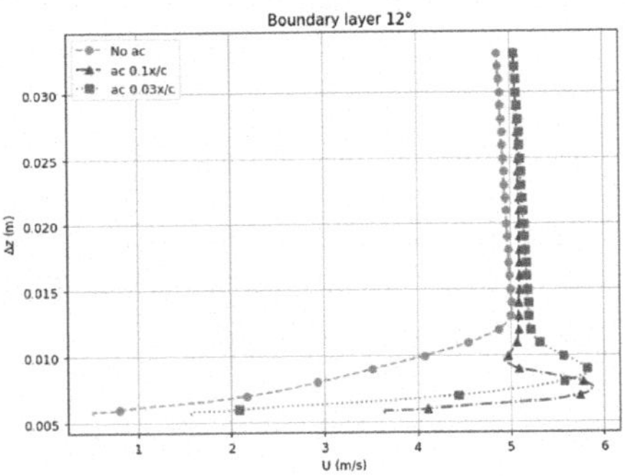

Fig. 9. Boundary layer at x/c = 0.1 for $\alpha = 12°$.

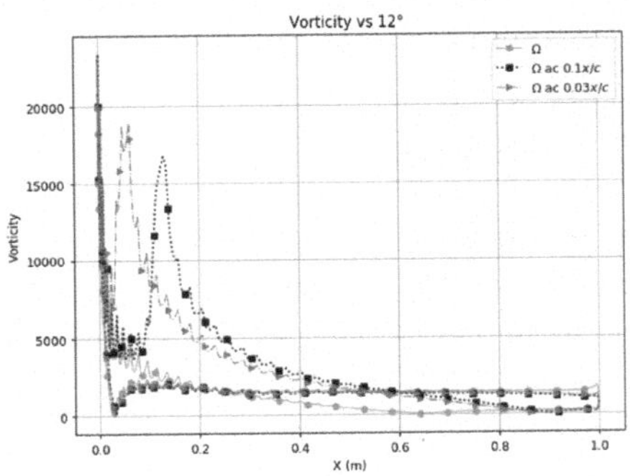

Fig. 10. Vorticity (Ω) distribution for $\alpha = 12°$.

The pressure coefficient (C_p) distribution is also affected, when there is a low pressure peak near the leading edge (x/c = 0.0) as seen in Fig. 11 and Fig. 12, this is more noticeable for the case at $\alpha = 16°$. As consequence of the low pressure increment the C_l increases.

The plot in Fig. 13 corresponds to the lift airfoil curve, in this graph is evident that the actuator produce an increase of the C_l. Taking into account the effect of the position of the actuator at different angles of attack, it is observed that the actuator placed at x/c = 0.1 achieves the greater C_l for angles previous α_{maxCl}, and from $\alpha = 13°$ to $\alpha = 16°$ the actuator at x/c = 0.03 is more effective.

From the plot in Fig. 14 it is observed that there is an increase of C_d in angles of attack smaller than $\alpha = 8°$, within the range of $\alpha = 10°$ to $\alpha = 15°$ the actuator is able to reduce the drag by achieving a notable difference with respect to the C_d that is present without an actuator, finally at the angle of stall $\alpha = 16°$ due to the actuator effect the C_d decreases significantly because the flow reattachment. It can be mentioned that there is no significant difference in the

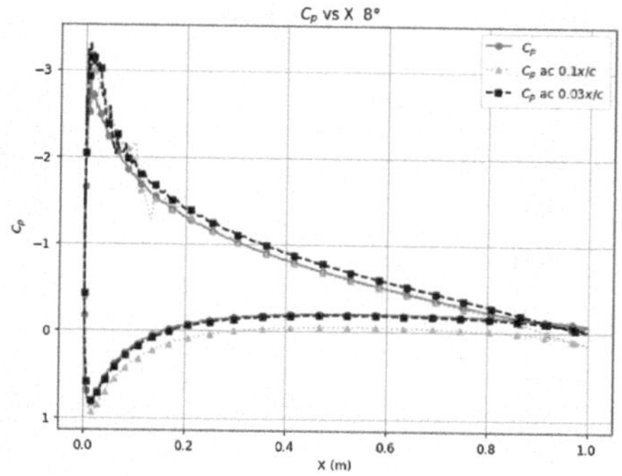

Fig. 11. C_p distribution for $\alpha = 8°$.

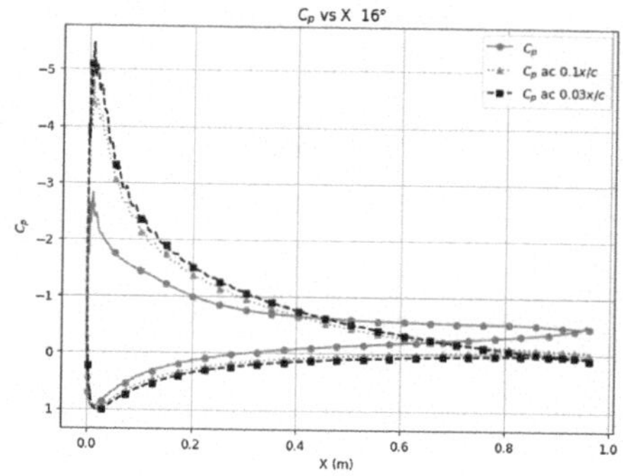

Fig. 12. C_p distribution for $\alpha = 16°$.

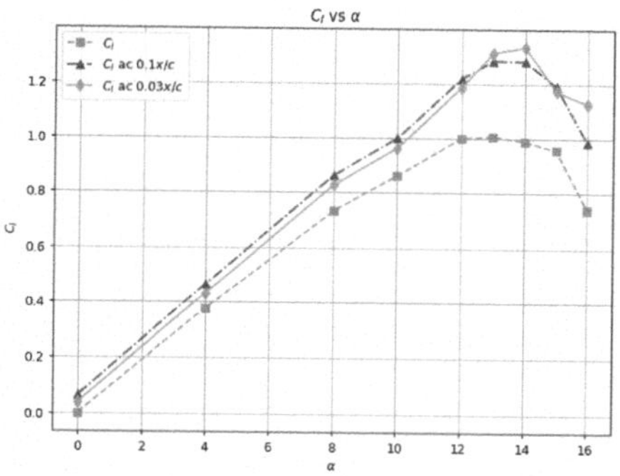

Fig. 13. C_l polar for Re $= 2.0 \times 10^5$.

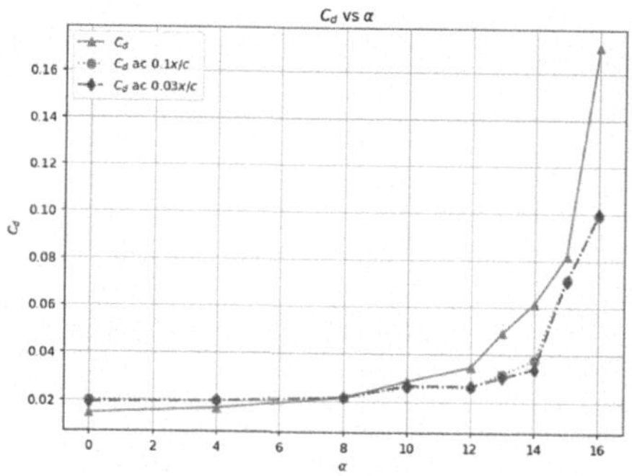

Fig. 14. C_d polar for Re $= 2.0 \times 10^5$.

effect of the actuator position over the C_d, having the same behavior regardless of the actuator is at x/c = 0.03 or x/c = 0.1.

6.2 Actuator Frequency Analysis

For the analysis of the actuator frequency, 4 cases were studied for an angle of attack of 16°, this angle of attack is of great interest because is when the airfoil stalls, the cases are listed in the Table 5.

The plots in Fig. 15 compare the effect of the different burst frequencies over the C_l for an actuation period of 1.5 s. It is observed that all the cases behave like a step input, and with a burst frequency of 3 Hz, the highest C_l is achieved additionally the increase is continuous and without any drops or over shoots.

The net effect of the actuator at different burst frequencies can be seen through the lift and drag coefficients, in Fig. 16 the coefficients with and without actuator are compared. In all cases with actuator there is an increase of the C_l of more than 50% in comparison to the case without actuator, the maximum C_l was achieved in continuous mode and with a burst frequency of 3 Hz. with respect to C_d this remains virtually unchanged.

Table 5. Cases for the study of the frequency of the actuator.

α○	F^+	F_{burst} Hz	f^+	f_{base} kHz
16	1	3	660	2
16	5	15	660	2
16	15	45	660	2
16	Continuous	–	660	2

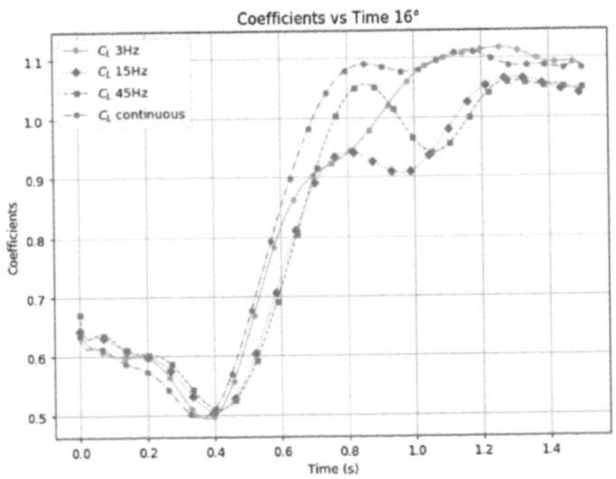

Fig. 15. Lift coefficient versus actuation time, for $\alpha = 16°$.

In Fig. 17 the aerodynamic efficiency (C_l/C_d) is compared, the highest aerodynamic efficiency was achieved with the actuator operating at a burst frequency of 45 Hz, although it is evident that the variation in the efficiency of one case or another is small.

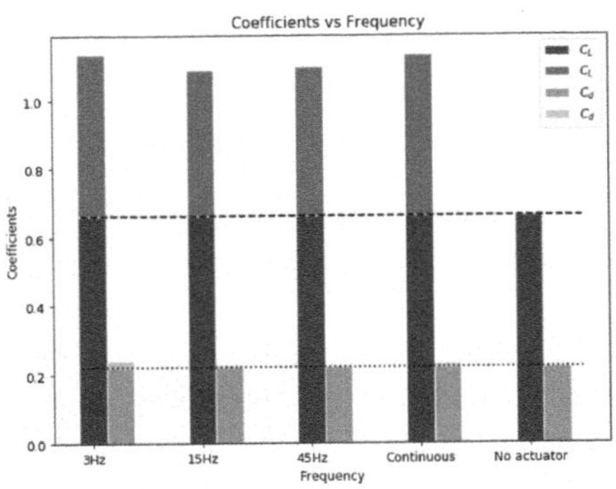

Fig. 16. Aerodynamic coefficients vs f_{burst}, for $\alpha = 16°$.

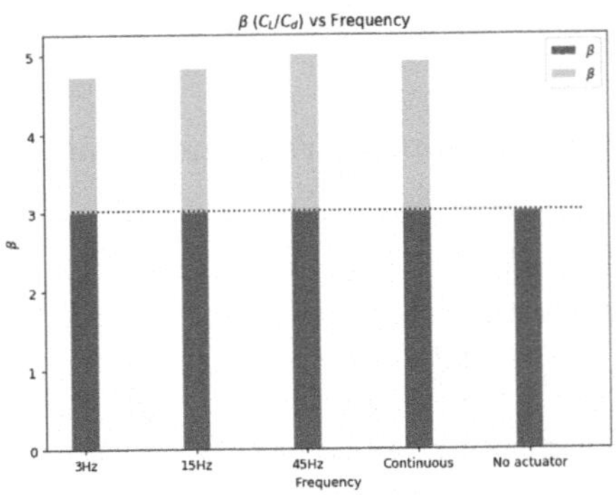

Fig. 17. Aerodynamic efficiency vs f_{burst}, for $\alpha = 16°$.

7 Validation

To validate the methodology implemented in this work, simulations with and without actuator were performed at a Re of 1.58×10^5. The Re was used as similarity parameter to compare the C_l from the simulations with the C_l reported by Post [8], in which he used an SDBD in a NACA 0015 to attach the flow and prevent the stall.

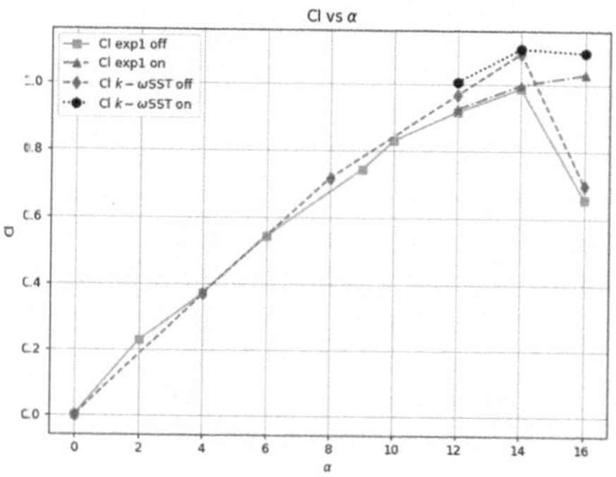

Fig. 18. Simulation and experimental C_l validation plot.

In Fig. 18 the C_l polar shows the results from Post and from the simulations with the $k - \omega$SST turbulence model. It is observed that the C_l behavior from the simulations is similar to the one showed by the experimental data, so there is a good correlation between the experimental data and the one from the simulations.

8 Conclusions

The position of the actuator study showed that the actuator is more effective when it is closer to the detachment point, when the separation occurs closer to the leading edge the actuator placed at x/c = 0.03 performs better, if the separation is closer to the trailing edge or x/c = 0.5 the actuator at x/c = 0.1 is better. If the detachment occurs before of the actuator it wont be able to reattach the detached flow before it.

The graph in Fig. 13 shows that for angles of attack $\alpha \geq 10°$ the C_l increase is higher than for $\alpha \leq 10°$. For angles of $\alpha < 8°$ there is an increase of the C_d, at higher angles with greater separation regions the drag is reduced, as seen in Fig. 14.

From the frequency study it was identified that the 3 Hz burst frequency performs better as it achieves the highest C_l and the behavior of C_l during the period of actuation shows a smooth and continuous increase. It was identified that frequency variations do not have a noticeable impact on C_d. As in Fig. 16

and Fig. 17 there is no significant magnitude variations over the C_l, C_d and β regardless the frequency we suggest that a more detailed and sensitive frequency study should be performed.

For future works it is intended to analyze the three-dimensional effects of the actuator through Large Eddy Simulation (LES), and the implementation of arrays with actuators placed at different positions of the airfoil in a way that they can be controlled to be activated depending of the angle of attack in a way that the actuator acts closer to the separation point.

References

1. Aono H, Sekimoto S, Sato M, Yakeno A, Nonomura T, Fujii K (2015) Computational and experimental analysis of flow structures induced by a plasma actuator with burst modulations in quiescent air. Mech Eng J 2(4):15–00233
2. Corke T, Post M (2005): Overview of plasma flow control: concepts, optimization, and applications. In: 43rd AIAA aerospace sciences meeting and exhibit, p 563
3. Dörr P, Kloker M (2015) Numerical investigation of plasma-actuator force-term estimations from flow experiments. J Phys D Appl Phys 48(39):395203
4. Flatt J (1961) The history of boundary layer control research in the United States of America. Bound Layer Flow Control 1:122–143
5. Fujii K (2014) High-performance computing-based exploration of flow control with micro devices. Phil Trans R Soc A 372(2022):20130326
6. Greenblatt D, Müller-Vahl H, Lautman R, Ben-Harav A, Eshel B (2015) Dielectric barrier discharge plasma flow control on a vertical axis wind turbine. In: Active flow and combustion control 2014, pp 71–86. Springer, Cham
7. Hofkens A (2016) Determination of the body force generated by a plasma actuator through numerical optimization
8. Post ML, Corke TC (2004) Separation control on high angle of attack airfoil using plasma actuators. AIAA J 42(11):2117–2184
9. Sato M, Nonomura T, Okada K, Asada K, Aono H, Yakeno A, Abe Y, Fujii K (2015) Mechanisms for laminar separated-flow control using dielectric-barrier-discharge plasma actuator at low reynolds number. Phys Fluids 27(11):117101
10. Shyy W, Jayaraman B, Andersson A (2002) Modeling of glow discharge-induced fluid dynamics. J Appl Phys 92(11):6434–6443
11. Suzen Y, Huang P, Jacob J, Ashpis D (2005) Numerical simulations of plasma based flow control applications. In: AIAA paper, vol 4633, p 2005
12. West T, Hosder S (2012) Numerical investigation of plasma actuator configurations for flow separation control at multiple angles of attack. In: 6th AIAA flow control conference, p 3053

Experimental Platform for Modeling and Control of Articulated Bodies

Ana Hernandez-Maldonado[1(✉)], Juan G. Diaz-Martinez[1], Hirohisa Kojima[2], Irandi Gutierrez-Carmona[1,2], and Sajjad Keshtkar[1,2]

[1] School of Engineering and Sciences, Tecnológico de Monterrey, 89600 Altamira, Tamaulipas, Mexico
A01410096@itesm.mx, juan.diaz@tec.mx

[2] Department of Aerospace Engineering, Tokyo Metropolitan University, Hino, Tokyo 191–0065, Japan
hkojima@tmu.ac.jp

Abstract. Articulated bodies, like robotic manipulators and tethered satellites, have gained great attention in recent decades due to their application in several space solutions, as active debris removal, momentum transfer systems, electrodynamic reboots of space stations, among others. The dynamics of these mechanisms are usually considered as rigid pendulums or dumbbells. Several studies have been carried out to validate the mathematical modeling and control either of tether satellites with a massless tether and thus, no inertia; or rigid pendulums under earth conditions. The aim of this paper is to combine the prior knowledge of both proposals and present a mathematical model of the articulated space system in orbital conditions. Numerical and experimental results, based on the ground-based experimental setup, are considered to validate this model. This report also includes technical information about the hardware and software components used in the communication and connections of an experimental platform designed to emulate different orbital trajectories.

Keywords: Kane formulation · Articulated space structures · Tethered satellites · Orbital ground based experiment · Space robotics · Space pendulum

1 Introduction

Articulated satellite systems, consisting of two or more artificial satellites connected to each other in space varies from traditional spacecraft with some aspects. One of the main characteristics of these systems is a large length providing a stable orientation along the local vertical and the creation of a small artificial gravity. Another important feature is changeable configuration flexibility: the ability to change the length and orientation of the connecting body allows the realization of different operations. This phenomenon allows adjusting the relative position and orientation of spacecraft, attaching and detaching objects from tethers, moving payloads along the connecting rod, and so on [1, 2].

One of the possible areas of application of these architectures is space debris removal. Nowadays, pollution isn't an exclusive problem of the earth, this phenomenon has

E. E. Hernandez et al. (Eds.): LASIRS 2019, MMS 86, pp. 159–169, 2020.
https://doi.org/10.1007/978-3-030-45402-9_16

reached dimensions from outer space, it's estimated that there are more than 17,000 objects in LEO, including rocket bodies, spacecraft, and fragmentation debris, however because of the high complexity of track small debris is impossible to know the real amount for sure [3]. Many solutions have been proposed to this problem: Harpoons, nets, tether satellites, and articulated bodies, such as robotic manipulators.

An articulated satellite system is a complicated physical system, therefore, most research on dynamics and control of these systems use simplified models to make predictions about its behavior. These simplifications, however, may introduce a lack of accuracy since important phenomena, as inertia, may not be considered. In many recent studies, the TSS structure has been considered as a dumbbell model (Fig. 1). The proposed model assumes that two satellites are connected by a rigid tether and can be simplified by considering the satellites as point masses and the tether is a massless rod whose bending, and stretching are negligible [4, 5]. In some other researches, the connecting rod is considered as a massive rigid body that provides more accurate dynamic analysis [6, 7]. Although in the last years many interesting and theoretically reliable results have been obtained, the lack of experimental validation is an important issue that caught the eyes of the authors.

The main concern in this paper is to realize a detailed analysis of the behavior of the articulated systems under space conditions. The obtained results can be served to predict the behavior of the articulated satellite systems in different orbital conditions, develop precise control schemes and supporting subsystems. Following these criteria, it is possible to prevent failures or damages caused by oscillations/vibration during the satellite operation. Although there exist some experimental banks which are used for studying the tethered satellites, flexible space systems and space climber [8, 9] but until now, they haven't been any bench or results related to the experiment the articulated bodies or dumbbell models. After the detailed mathematical modeling of the space articulated systems, this work presents the necessary conditions and methods to carry out the experimentation, validation, and control of articulated space systems in the ground conditions.

2 Mathematical Model

2.1 Simplified Double Pendulum Model in Space Satellite

Consider the space structures as a pair of rigid pendulums with concentrated masses, hence, a moment of inertia. This mechanism can be presented as a three-mass system composed of the mother ship, which rotates around the Earth in an elliptical orbit, and a pair of links (pendulums) connected to the main satellite by rotational joints as can be seen in Fig. 1.

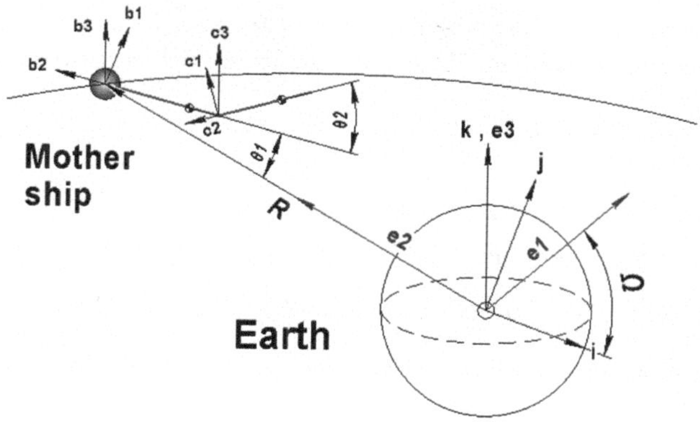

Fig. 1. Orbital model

Hereafter, the following terms and variables will be referred to as follows:

J_m: Joint with the mother ship.
J_p: Joint between both pendulums,
$\theta_1(t)$ and $\theta_2(t)$: Libration angles of the center of mass of each body from the local vertical,
L_1: Distance from j_m to the center of mass of the first pendulum,
L_2: Distance from j_p to the center of mass of the second pendulum,
L_1': Distance between both joints,
L_2': Full length of the second pendulum

The lengths L_1, L_2 and L_1' are constant in this study. The orbital radius of the system $\boldsymbol{R} = R(t)\boldsymbol{e}_2$ is proposed as a variable for the cases of elliptical orders and related to the orbital rate $\boldsymbol{\Omega} = \Omega(t)\boldsymbol{e}_3$ by using Kepler's time equations. The orbital rotation reference frame $\{\boldsymbol{e}_1, \boldsymbol{e}_2, \boldsymbol{e}_3\}$ has its origin in the center of the Earth. The resemblance of this model with the experimental setup will be explained later.

2.2 Equations of Motion

The equations of motion presented in this study are obtained using Kane's method [10]. The position of the center of mass of the first and second pendulum, are given by the vectors $\boldsymbol{R}_1$ and $\boldsymbol{R}_2$, as follows:

$$\boldsymbol{R}_1 = (L_1 \sin\theta_1)\boldsymbol{e}_1 + (R - L_1\cos\theta_1)\boldsymbol{e}_2 \tag{1}$$

$$\boldsymbol{R}_2 = \left(L_1' \sin\theta_1 + L_2 \sin\theta_2\right)\boldsymbol{e}_1 + \left(R - L_1'\cos\theta_1 - L_2\cos\theta_2\right)\boldsymbol{e}_2 \tag{2}$$

The generalized velocities are selected as

$$u_1 = \dot{\theta}_1 \tag{3}$$

$$u_2 = \dot{\theta}_2 \tag{4}$$

Considering the effect of the orbital rate vector $\boldsymbol{\Omega}$, the velocity of the center of mass of the first and second pendulum, $\boldsymbol{V}_1$ and $\boldsymbol{V}_2$, are expressed respectively, as follows:

$$\boldsymbol{V}_1 = \left[(-R + L_1 cos\,\theta_1)\dot{\Omega} + L_1\dot{\theta}_1 cos\,\theta_1\right]\boldsymbol{e}_1 + \left[\dot{R} + L_1 sin\,\theta\left(\dot{\Omega} + \dot{\theta}_1\right)\right]\boldsymbol{e}_2 \quad (5)$$

$$\begin{aligned}\boldsymbol{V}_2 =& [\left(-R + L_1' cos\,\theta_1 + L_2 cos\,(\theta_1 + \theta_2)\right)\dot{\Omega} + \left(L_1' cos\,\theta_1 + L_2\dot{\theta}_1 cos\,(\theta_1 + \theta_2)\right) \\ &+ L_2\dot{\theta}_2 cos\,(\theta_1 + \theta_2)]\boldsymbol{e}_1 + [\dot{R} + \dot{\Omega}\left(L_1' sin\,\theta_1 + L_2 sin\,(\theta_1 + \theta_2)\right) \\ &+ L_1'\dot{\theta}_1 sin\,\theta_1 + L_2\dot{\theta}_1 sin\,(\theta_1 + \theta_2) + L_2 sin\,(\theta_1 + \theta_2)\dot{\theta}_2]\boldsymbol{e}_2\end{aligned} \quad (6)$$

The acceleration vectors of both the center of masses, $\boldsymbol{A}_1$ and $\boldsymbol{A}_2$, are therefore

$$\begin{aligned}\boldsymbol{A}_1 =& [-2\dot{R}\dot{\Omega} + (R + L_1 cos\,\theta_1)\ddot{\Omega} - L_1(\dot{\Omega}^2 sin\,\theta_1 + 2\dot{\Omega}\dot{\theta}_1 sin\,\theta_1 + sin\,\theta_1\dot{\theta}_1^2 - \ddot{\theta}_1 cos\,\theta_1)]\boldsymbol{e}_1 \\ &+ [\ddot{R} + (-R + L_1 cos\,\theta_1)\dot{\Omega}^2 + 2L_1\dot{\Omega}\dot{\theta}_1 cos\,\theta_1 + L_1(\ddot{\Omega} sin\,\theta_1 + \dot{\theta}_1^2 cos\,\theta_1 + \ddot{\theta}_1 sin\,\theta_1)]\boldsymbol{e}_2\end{aligned} \quad (7)$$

$$\begin{aligned}\boldsymbol{A}_2 =& [-2\dot{R}\dot{\Omega} + \left(-R + L_1' cos\,\theta_1 + L_2 cos\,(\theta_1 + \theta_2)\right)\ddot{\Omega} - L_1'\dot{\Omega}^2 sin\,\theta_1 - L_2\dot{\Omega}^2 sin\,(\theta_1 + \theta_2) \\ &- 2L_1'\dot{\Omega}\dot{\theta}_1 sin\,\theta_1 - 2L_2\dot{\Omega}\dot{\theta}_1 sin\,(\theta_1 + \theta_2) - L_1'\dot{\theta}_1^2 sin\,\theta_1 - L_2\dot{\theta}_1^2 sin\,(\theta_1 + \theta_1) \\ &- 2L_2\dot{\Omega}\dot{\theta}_2 sin\,(\theta_1 + \theta_2) - 2L_2\dot{\theta}_1\dot{\theta}_2 sin\,(\theta_1 + \theta_2) - L_2\dot{\theta}_1^2 sin\,(\theta_1 + \theta_2) + L_1'\ddot{\theta}_1 cos\,\theta_1 \\ &+ L_2\ddot{\theta}_1 cos\,(\theta_1 + \theta_2) + L_2\ddot{\theta}_2 cos\,(\theta_1 + \theta_2)]\boldsymbol{e}_1 + [\ddot{R} + (-R + L_1' cos\,\theta_1 + L_2 cos\,(\theta_1 + \theta_2))\dot{\Omega}^2 \\ &+ L_1'\ddot{\Omega} sin\theta_1 + L_2\ddot{\Omega} sin(\theta_1 + \theta_2) + L_1'\dot{\theta}_1^2\, cos\,\theta_1 + L_2\dot{\theta}_1^2 cos\,(\theta_1 + \theta_2) \\ &+ 2L_2\dot{\theta}_1\dot{\theta}_2 cos\,(\theta_1 + \theta_2) + L_2\dot{\theta}_2^2 cos\,(\theta_1 + \theta_2) + 2\dot{\Omega}\left((L_1' cos\,\theta_1 + L_2 cos\,(\theta_1 + \theta_2))\dot{\theta}_1\right. \\ &\left.+ L_2\dot{\theta}_2 cos\,(\theta_1 + \theta_2)\right) + L_1'\ddot{\theta}_1 sin\,\theta_1 + L_2\ddot{\theta}_1 sin\,(\theta_1 + \theta_2) + L_2\ddot{\theta}_2 sin\,(\theta_1 + \theta_2)]\boldsymbol{e}_2\end{aligned} \quad (8)$$

The partial velocities of each body with respect to the generalized velocities are

$$\frac{\partial V_1}{\partial u_1} = L_1 cos\,\theta_1\,\boldsymbol{e}_1 - L_1 cos\,\theta_1\,\boldsymbol{e}_2 \quad (9)$$

$$\frac{\partial V_1}{\partial u_2} = 0 \quad (10)$$

$$\frac{\partial V_2}{\partial u_1} = [L_1' cos\,\theta_1 + L_2 cos\,\,(\theta_1 + \theta_2)]\boldsymbol{e}_1 + \left[L_1' sin\,\,\theta_1 + L_2 sin\,(\theta_1 + \theta_2)\right]\boldsymbol{e}_2 \quad (11)$$

$$\frac{\partial V_2}{\partial u_2} = [L_2 cos\,(\theta_1 + \theta_2)]\boldsymbol{e}_1 + [L_2 sin\,(\theta_1 + \theta_2)]\boldsymbol{e}_2 \quad (12)$$

The gravitational forces affecting both pendulums are given by

$$\boldsymbol{F}_{g1} = -\mu m_1 \frac{\boldsymbol{R}_1}{|R_1|^3} \quad (13)$$

$$\boldsymbol{F}_{g2} = -\mu m_2 \frac{\boldsymbol{R}_2}{|R_2|^3} \quad (14)$$

Employing Kane's equation,

$$K_i + K_i^* = 0, i = \{1, 2\} \quad (15)$$

The generalized active forces are considered as

$$K_i = \boldsymbol{F}_{g1} \cdot \frac{\partial \boldsymbol{V}_1}{\partial u_i} + \boldsymbol{F}_{g2} \cdot \frac{\partial \boldsymbol{V}_2}{\partial u_1} \tag{16}$$

And the generalized inertia forces, are expressed as

$$K_i^* = \left[\frac{\partial \boldsymbol{V}_1}{\partial u_i} \cdot (m_1 \boldsymbol{A}_1) + \frac{\partial \boldsymbol{V}_2}{\partial u_i} \cdot (m_2 \boldsymbol{A}_2)\right] + \left[\frac{\partial \omega_1}{\partial u_i} \boldsymbol{T}_1^* + \frac{\partial \omega_2}{\partial u_i} \boldsymbol{T}_2^*\right] \tag{17}$$

Where ω_1, ω_2 are the angular velocities of the first and second pendulum respectively, while T_1^*, T_2^* are the inertial torque of each body defined as:

$$\boldsymbol{T}_i^* = \boldsymbol{\alpha}_i \cdot \boldsymbol{I}_i - \boldsymbol{\omega}_i \times \boldsymbol{I}_i \cdot \boldsymbol{\omega}_i \tag{18}$$

where $\boldsymbol{\alpha}_i$, $\boldsymbol{I}_i$ are the angular acceleration and the central inertia of the analyzed body.

Finally, the equations of motion can be obtained as

$$\boldsymbol{M}\ddot{\boldsymbol{q}} + \boldsymbol{f} = [0, 0]^T. \tag{19}$$

where $\boldsymbol{M}$ is the mass matrix, $\boldsymbol{f}$ is the nonlinear term vector combining the Coriolis and centrifugal forces and $\boldsymbol{q} = [\theta_1, \theta_2]$ is the state vector. The elements of M and f, are expressed as follows:

$$M_{11} = -L_1^2 m_1 - L_2^2 m_2 - L_1'^2 m_2 - 2L_2 L_1' m_2 \cos\cos\theta_2 + I_1^{b3} + I_2^{c3} \tag{20}$$

$$M_{12} = M_{21} = -L_2^2 m_2 - L_2 L_1' m_2 \cos\cos\theta_2 + I_2^{c3} \tag{21}$$

$$M_{21} = -L_2^2 m_2 + I_2^{c3} \tag{22}$$

$$\begin{aligned} f_1 =& 2((L_1 m_1 + L_1' m_2)\cos\theta_1 + L_2 m_2 \cos(\theta_1 + \theta_2))\dot{L}\dot{\Omega} - (L_1^2 m_2 + L_2^2 m_2 + L_1'^2 m_2 \\ &- RL_1 m_1 \cos\theta_1 - RL_1' m_2 \cos\theta_1 + 2L_2 L_1' m_2 \cos\theta_2 - RL_2 m_2 \cos(\theta_1 + \theta_2))\ddot{\Omega} \\ &- (L_1 m_1 \sin\theta_1 + L_1' m_2 \sin\theta_1 + L_2 m_2 \sin(\theta_1 + \theta_2))\ddot{L} + (RL_1 m_1 \sin\theta_1 + RL_1' m_2 \sin\theta_1 \\ &+ RL_2 m_2 \sin(\theta_1 + \theta_2))\dot{\Omega}^2 - \frac{\mu R(L_1' m_2 \sin\theta_1 + L_2 m_2 \sin(\theta_1 + \theta_2)}{R_3^3} - \frac{\mu R(L_1 m_1)}{R_2^3} \\ &- m_2(L_1' + L_2 \cos\theta_2)\dot{L}_1'(\dot{\Omega} + \dot{\theta}_1) + 2L_2 L_1' m_2 \dot{\Omega} \sin\theta_2 \dot{\theta}_2 + 2L_2 L_1' m_2 \sin\theta_2 \dot{\theta}_1 \dot{\theta}_2 \\ &+ L_2 L_1' m_2 \sin\theta_2 \dot{\theta}_2^2 + I_1^{b3}\ddot{\Omega} + I_2^{c3} \end{aligned} \tag{23}$$

$$\begin{aligned} f_2 =& \frac{\mu L_2 m_2 (L_1' \sin\theta_2 - R\sin(\theta_1 + \theta_2))}{R_3^3} + I_2^{c3}\ddot{\Omega} - L_2 m_2(-2\cos(\theta_1 + \theta_2)\dot{L}\dot{\Omega} + L_2\ddot{\Omega} \\ &+ L_1' \cos\theta_2 \ddot{\Omega} - R\cos(\theta_1 + \theta_2)\ddot{\Omega} + L_1' \dot{\Omega}^2 Sin\theta_2) + \ddot{R}\sin(\theta_1 + \theta_2) - R\dot{\Omega}^2 \sin(\theta_1 + \theta_2) \\ &+ 2L_1' \dot{\Omega} \sin\theta_2 \dot{\theta}_1 + L_1' \sin\theta_2 \dot{\theta}_1^2 + \cos\theta_2 \dot{L}_1'(\dot{\Omega} + \dot{\theta}_1) \end{aligned} \tag{24}$$

2.3 Experimental Setup, Simplified Double Pendulum Model on Earth

The experimental setup, shown in Fig. 2, has been proposed to analyze the in-plane motion of the bodies, by recreating the orbital conditions that affect the satellite behavior in a ground-based environment.

The schematic representation and components used for this apparatus are shown in Fig. 3. As it was previously mentioned, it isn't possible to emulate, both the in-plane and out-plane conditions on the ground, therefore, this study will focus on the first ones. The proposed mechanism consists of an experimental platform with 3 degrees of freedom, composed by an AC motor connected to a pulley that rotates (1) a turntable, which at the same time supports an (2) inclined table. Finally, mounted parallel to the plate there's a pulley, connected to (3) a parallel slider which modifies the proximity of the analyzed bodies to the center of the apparatus. The rotation of the main base resembles the orbital motion, while the inclination and linear movements help to simulate more complex trajectories like elliptical ones by modifying the orbital radius, as well as modifying the gravitational effects on the bodies. In order to avoid twisting of electrical wires, all the motors are energized through the slip ring located lower part of the experimental setup and controlled via radio frequency by a pair of receiver-transmitter devices with a commercial microcontroller and to monitor the position of the pendulums a camera will be installed in the parallel slider (Fig. 4).

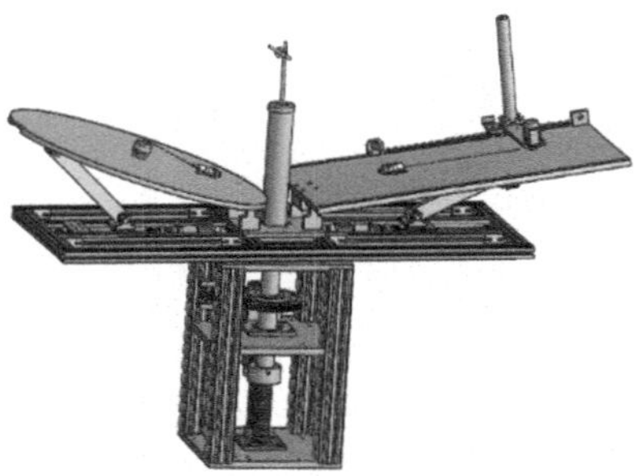

Fig. 2. Overview of the experimental apparatus.

One of the main differences between the ground-based experimental setup is gravity. In orbital conditions, gravity depends on the orbital radius, while in the turntable it is constant. These differences are reflected in the equations of motion for the experimental setup and those of the orbital system, in terms of gravity and angular velocity. The effects of gravity affecting the bodies in the inclined table are modified as follows

$$\boldsymbol{F}'_{g1} = -m_1 g(\sin\beta\,\boldsymbol{e}_2 + \cos\beta\,\boldsymbol{e}_3) \tag{25}$$

$$\boldsymbol{F}'_{g2} = -m_2 g(\sin\beta\,\boldsymbol{e}_2 + \cos\beta\,\boldsymbol{e}_3) \tag{26}$$

Where g is the gravitational acceleration constant on earth ($g = 9.8\frac{m}{s^2}$) and β is the inclination angle of the table. The angular velocity vector is now considered as

$$\boldsymbol{\omega} = \omega\,\sin\beta\,\boldsymbol{e}_2 + \omega\,\cos\beta\,\boldsymbol{e}_3 \tag{27}$$

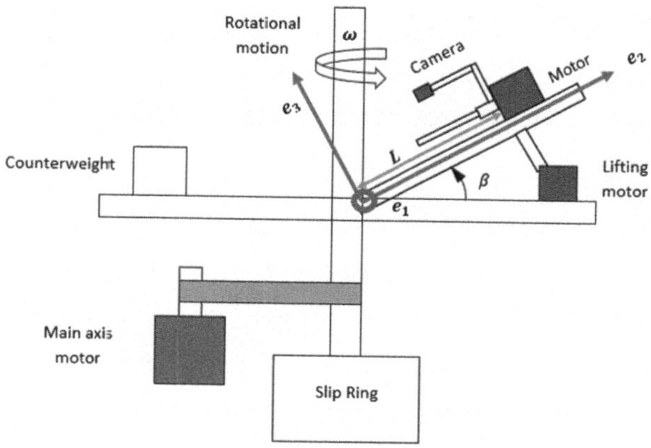

Fig. 3. Experimental platform

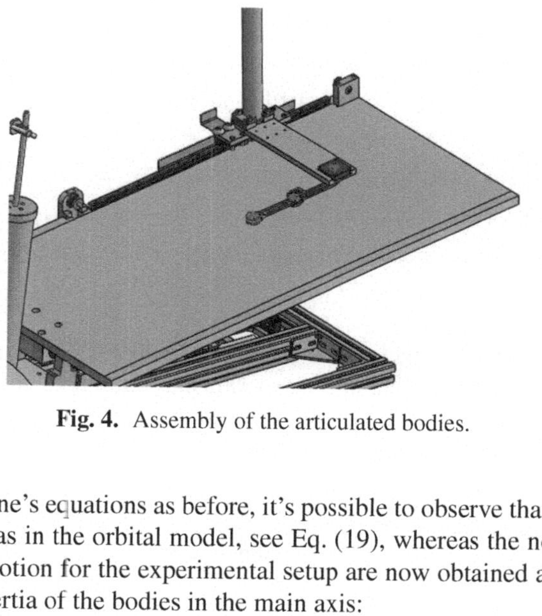

Fig. 4. Assembly of the articulated bodies.

Employing Kane's equations as before, it's possible to observe that the mass matrix remains the same as in the orbital model, see Eq. (19), whereas the nonlinear terms in the equations of motion for the experimental setup are now obtained as follows, where I_1^{b3}, I_2^{c3} are the inertia of the bodies in the main axis:

$$
\begin{aligned}
f_1' = & -g\ \sin\beta\left(\left(L_1 m_1 + L_1' m_2\right)\sin\theta_1 + L_2 m_2 \sin(\theta_1+\theta_2)\right) + I_1^{b3}\left(\ddot{\omega}\cos\beta - \dot{\omega}\dot{\beta}\ \sin\beta\right) \\
& - L_1 m_1\big(2\dot{\omega}\dot{R}\ \cos\beta\ \cos\theta_1 + \ddot{\omega}\cos\beta(L_1 - R\ \cos\theta_1) + R\dot{\omega}\dot{\beta}\ \sin\beta\ \cos\theta_1 \\
& - L_1\dot{\omega}\dot{\beta}\sin\beta\ \cos^2\theta_1 + \ddot{R}\sin\theta_1 - R\dot{\omega}^2\cos^2\beta\ \sin\theta_1 - L_1\dot{\omega}^2\cos\theta_1\sin\theta_1 \\
& + L_1\dot{\omega}^2\cos^2\beta\ \cos\theta_1\sin\theta_1 - L_1\dot{\omega}\dot{\beta}\ \sin\beta\ \sin^2\theta_1\big) + I_2^{c3}\left(\ddot{\omega}\ \cos\beta - \dot{\omega}\dot{\beta}\ \sin\beta\right) \\
& - m_2\left(L_1'\cos\theta_1 + L_2\cos(\theta_1+\theta_2)\right)\big(-2\dot{R}\dot{\omega}\cos\beta - R\ddot{\omega}\cos\beta + L_1'\ddot{\omega}\cos\beta\ \cos\theta_1 \\
& + L_2\ddot{\omega}\cos\beta\ \cos(\theta_1+\theta_2) + R\dot{\omega}\dot{\beta}\ \sin\beta - L_1'\dot{\beta}\dot{\omega}\ \sin\beta\ \cos\theta_1 \\
& - L_2\dot{\beta}\omega\ \sin\beta\ \cos(\theta_1+\theta_2) - L_1'\dot{\omega}^2\cos^2\beta\ \sin\theta_1 - L_1'\dot{\omega}\ \sin^2\beta\ \sin\theta_1
\end{aligned}
$$

$$
\begin{aligned}
&- L_2\dot{\omega}^2\cos^2\beta\ \sin(\theta_1+\theta_2) - L_2\dot{\omega}\sin^2\beta\ \sin(\theta_1+\theta_2)\\
&- 2L_1'\dot{\omega}\dot{\theta}_1\cos\beta\ \sin\theta_1 - 2L_2\dot{\omega}\dot{\theta}_1\cos\beta\ \sin(\theta_1+\theta_2) - L_1'\dot{\theta}_1^2\sin\theta_1\\
&- L_2\dot{\theta}_1^2\sin(\theta_1+\theta_2) + \dot{L}_1'\cos\theta_1(\dot{\omega}\cos\beta+\dot{\theta}_1) - 2\dot{\omega}\dot{\theta}_2 L_2\cos\beta\ \sin(\theta_1+\theta_2)\\
&- 2L_2\dot{\theta}_1\dot{\theta}_2\sin(\theta_1+\theta_2) - L_2\dot{\theta}_2^2\sin(\theta_1+\theta_2)\Big)\\
&+ \left(L_1'\sin\theta_1 + L_2\sin(\theta_1+\theta_2)\right)(\ddot{R} - \cos^2\beta(R - L_1'\cos\theta_1\\
&- L_2\dot{\omega}^2\cos(\theta_1+\theta_2) + L_1'\ddot{\omega}\cos\beta\ \sin\theta_1 + L_2\ddot{\omega}\cos\beta\ \sin(\theta_1+\theta_2) + \dot{L}_1'\dot{\theta}_1\sin\theta_1\\
&+ L_1'\dot{\theta}_1^2\cos\theta_1 + L_2\dot{\theta}_1^2\cos(\theta_1+\theta_2) + 2L_2\dot{\theta}_1\dot{\theta}_2\cos(\theta_1+\theta_2) + L_2\dot{\theta}_2^2\cos(\theta_1+\theta_2)\\
&+ \dot{\omega}\left(\dot{L}_1'\cos\beta\ \sin\theta_1 - \dot{\beta}\ \sin\beta\left(L_1'\sin\theta_1 + L_2\sin(\theta_1+\theta_2)\right)\right)\\
&+ 2\cos\beta((L_1'\cos\theta_1 + L_2\cos(\theta_1+\theta_2))\dot{\theta}_1 + L_2\dot{\theta}_2\cos(\theta_1+\theta_2)
\end{aligned}
\tag{28}
$$

$$
\begin{aligned}
f_2' = &- gL_2m_2\sin\beta\ \sin(\theta_1+\theta_2) + I_2^{c3}(\cos\beta\ddot{\omega} - \dot{\omega}\dot{\beta}\ \sin\beta)\\
&- \frac{1}{8}L_2m_2(-16\dot{R}\dot{\omega}\cos\beta\ \cos(\theta_1+\theta_2) + 8L_2\ddot{\omega}\cos\ \beta + 4L_1'\ddot{\omega}\cos(\beta-\theta_2)\\
&- 4R\ddot{\omega}\cos(\beta-\theta_1-\theta_2) + 4L_1'\ddot{\omega}\cos(\beta+\theta_2)\\
&- 4R\ddot{\omega}\cos(\beta+\theta_1+\theta_2) - 8L_2\dot{\omega}\dot{\beta}\ \sin\beta - 4L_1'\dot{\omega}\dot{\beta}\ \sin(\beta-\theta_2) - L_1'\dot{\omega}^2\sin(2\beta-\theta_2)\\
&- L_1'\dot{\omega}^2\sin(2\beta-2\theta_1-\theta_2) + 4R\dot{\omega}\dot{\beta}\ \sin(\beta-\theta_1-\theta_2) - L_2\dot{\omega}^2\sin(2(\beta-\theta_1-\theta_2))\\
&+ 2R\dot{\omega}^2\sin(2\beta-\theta_1-\theta_2) + 6L_1'\dot{\omega}^2\sin\theta_2 - 4L_1'\dot{\omega}\dot{\beta}\ \sin(\beta+\theta_2) + L_1'\dot{\omega}^2\sin(2\beta+\theta_2)\\
&+ 8\ddot{R}\sin(\theta_1+\theta_2) - 4R\dot{\omega}^2\sin(\theta_1+\theta_2) - 2L_2\omega^2\sin(2(\theta_1+\theta_2)) + 4R\dot{\omega}\dot{\beta}\ \sin(\beta+\theta_1+\theta_2)\\
&+ L_2\dot{\omega}^2\sin(2(\beta+\theta_1+\theta_2)) - 2R\dot{\omega}^2\sin(2\beta+\theta_1+\theta_2) - 2L_1'\dot{\omega}^2\sin(2\theta_1+\theta_2)\\
&+ L_1'\dot{\omega}^2\sin(2\beta+2\theta_1+\theta_2) - 8L_1'\dot{\omega}\dot{\theta}_1\sin(\beta-\theta_2) + 8L_1'\dot{\omega}\dot{\theta}_1\sin(\beta+\theta_2) + 8L_1'\dot{\theta}_1^2\sin\theta_2\\
&+ 8\dot{L}_1'\cos\theta_2(\dot{\omega}\cos\beta+\dot{\theta}_1))
\end{aligned}
\tag{29}
$$

One form to obtain the inclination angle of the turn table that will emulate the gravity effects of the orbit system in the experimental setup is by relating the gravidity gradient torque on the orbit with the one obtained in the ground-based platform [7], so the inclination angle will be obtained as follows:

$$\beta = \sin^{-1}\left\{\frac{-g + \sqrt{g^2 + 4\omega^4(R-L)(R+3L)}}{2(R-L)\omega^2}\right\} \tag{30}$$

In order to simplify the analysis for elliptical orbits, this angle can also be obtained through algebraic relations between the gravity force in orbit (Fg_e) and in the experimental setup (Fg_p,) as follows:

$$Fg_e = \frac{Gm_e}{r^2}m_s,\ Fg_p = g\sin\beta\, m_s \tag{31}$$

$$Fg_e = Fg_p \tag{32}$$

$$\beta = \sin^{-1}\left(\frac{Gm_e}{gr^2}\right) \tag{33}$$

Where G is the universal gravitational constant, m_e the Earth's mass, r the orbital radius to be emulated and $g = 9.8\frac{m}{s^2}$.

3 Results of Numerical Simulations

In order to validate the equations of motion obtained for the orbital system (Eqs. 20–24) and the experimental setup (Eq. 28–29), some numerical simulations where made. It is assumed that the orbital radius R of the mother ship and the semimajor axis for the case of elliptical orbits are 6800, with an eccentricity of $e = 0.025$ for the last case. The orbital period is then 5580 s (93 min).

As for the experimental setup, the parameters are set as follows. The weight of each pendulum is 20.55 g, the lengths, L_1 and L_2, are both 100 mm while L_1' is 120 mm. The distance R between the end of the reel system and the main shaft along the inclined table is 70 cm. In the case of circular orbits (e = 0) the angular rotation rate of the main shaft is set as 0.012, so the period evaluated is 50 s and the inclination is 0.011°. As for the case of the elliptic orbit, the inclination angle is evaluated as a function that depends on the orbital radius which is variable over time (Figs. 5 and 6).

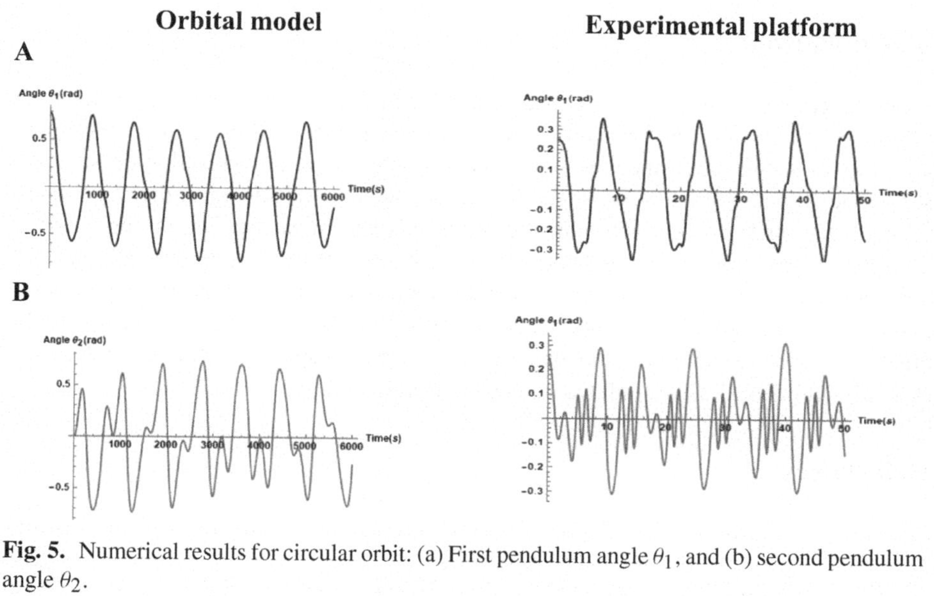

Fig. 5. Numerical results for circular orbit: (a) First pendulum angle θ_1, and (b) second pendulum angle θ_2.

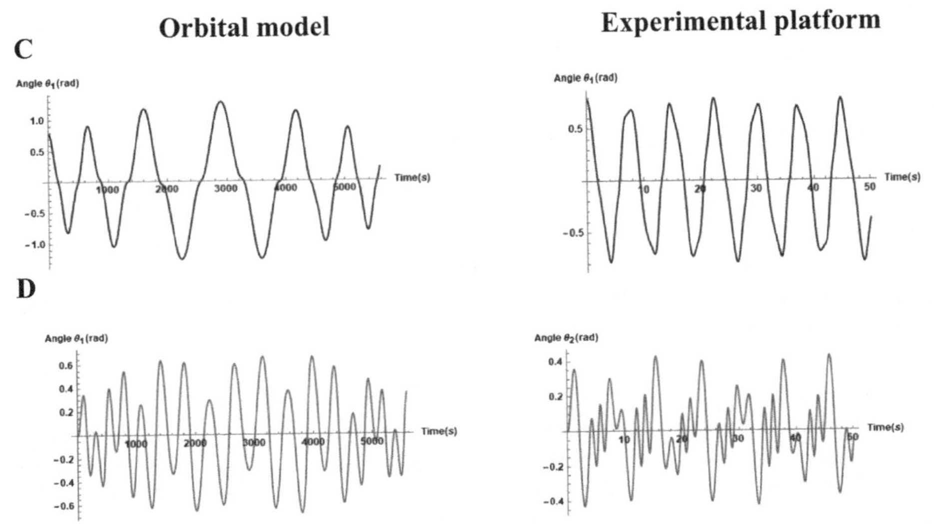

Fig. 6. Numerical results for circular orbit: (a) First pendulum angle θ_1, and (b) second pendulum angle θ_2.

4 Conclusion

From the graphics it is possible to observe that the mathematical model for the behavior of orbital system simplified as a group of pendulums and from the proposed experimental setup are similar, but since the system is very sensitive to any perturbation there are still some slight differences produced by the parameters considered in the experimental platform, such as the position of the center of mass in each pendulum and inertia. Therefore, some adjustments and considerations need to be done in order to increase the accuracy and precision of the experimental setup, as well as the real-life experiments and programming of all the components. Nevertheless, the simulations prove the mathematical model considered in this paper gives an optimal approximation to analyze the behavior and control of articulated bodies in orbital conditions.

References

1. Menon C, Bombardelli C (2007) Self-stabilising attitude control for spinning tethered formations. Acta Astronaut 60(10–11):828–833
2. NASA Orbital Debris Program Office (2013) Monthly number of objects in earth orbit by object type. Orbit Debris Q News 17(1):8
3. European Space Agency January 2019 «Space debris by the numbers» European Space Agency. https://www.esa.int/Safety_Security/Space_Debris/Space_debris_by_the_numbers. Accessed June 2019
4. Tyc G (1999) Dynamics and stability of spinning flexible space tether systems
5. Misra A, Xu D, Modi V (1986) On vibrations of orbiting tethers. Acta Astronaut 13(10):587–597
6. Fritzkowski P, Kaminski H (2008) Dynamics of a rope as a rigid multibody system. J Mech Mater Struct 3(6):1059–1075

7. Kaminski H, Fritzkowski P (2013) Application of the rigid finite element method to modeling ropes. Latin Am J Solids Struct 10(1):91–99
8. Kojima H, Iwasaki M, Fujii HA, Blanksby C, Trivalio P (2004) Nonlinear control of libration motion of tethered satellites in elliptics orbits. J Guid Control Dyn 27(2): 229–239
9. Kojima H, Sugimoto Y, Furukawa Y (2011) Experimental study on dynamics and control of tethered satellite systems with climber. Acta Astronaut (69): 98–108
10. Kane TR (2005) Dynamics: theory and applications. McGraw-Hill Series in Mechanical Engineering, New York

Design and Control of a Passive Solar Tracking System Using a Sky Imager

M. Angulo[1], A. Díaz-Ponce[1(✉)], L. Valentín[1], R. Valdivia[2], and S. Keshtkar[2]

[1] CONACYT-Centro de Investigaciones en Óptica, A.C., Aguascalientes, Mexico
{mangulo,adiaz,luismvc}@cio.mx
[2] Instituto Tecnológico y de Estudios Superiores de Monterrey, Tamaulipas, Mexico
{r.valdivia,skeshtkar}@tec.mx
https://www.cio.mx/investigacion/energia_solar/

Abstract. Solar energy is considered as one of the most important sources of renewable energy due to the possibility of being converted into electrical power, which could result in a feasible approach to global energy problems. Concentrated Solar Power (CSP) and High Concentration Photovoltaics (HCPV) systems present an alternative by the use of cost-efficient concentrating optics. However, these systems require a high-precision solar tracking system, allowing to stay perpendicular to the sun rays in order to achieve its correct functioning. In this work, a methodology for the detection and monitoring of the apparent movement of the sun through a passive image acquisition system is presented. The setup operates by processing sky dome images to find the solar-center. The development cost of this device is relatively low, it does not require articulated elements, can detect the presence of clouds, and could be compatible with multiple CSP systems. Preliminary results of a two-axis solar tracking by locating position of the sun in the sky with the proposed system are depicted.

Keywords: Concentrating solar power · Image processing · Sky-imaging system · Solar photovoltaic · Sun detection · Solar tracking system

1 Introduction

In the last two decades, renewable alternatives have been considered as an important source of energy. The axiomatic depletion of fossil fuels and global warming have encouraged the development and implementation of systems that generate energy from renewable sources.

Solar energy has protruded by a popular approach and is implemented in many nations of the world compared to conventional power sources. In addition, it is considered as an unlimited reserves, it has no cost, it is pollution-free, among other multiple benefits [1–3].

E. E. Hernandez et al. (Eds.): LASIRS 2019, MMS 86, pp. 170–178, 2020.
https://doi.org/10.1007/978-3-030-45402-9_17

Solar technologies transform solar radiation into another type of energy, which can be exploited by the user. Typically, this kind of technology has been implemented by the photovoltaic (PV) industry and the CSP industry in regions where there are scarce clouds, haze and, of course, where sun hours are abundant [4,5]. It is important to mention that these configurations achieve their optimum efficiency when the sun rays impact perpendicular to the surface of the converting part of the setup, such as thermal solar collectors, solar cells, among others [6–9]. This is possible due to a solar tracking system. Eventually, the power generation and performance of solar systems decrease dramatically if there is a significant solar orientation error, therefore, it is important to keep the error as low as possible.

There are several configurations of solar tracking systems in the literature, classified based on their degrees of freedom (DoFs) and/or their performance. The former can follow the apparent sun position just in one-axis and, for greater accuracy, there are two-axis solar tracking systems [10–12]. The two-axis solar tracking can be either: polar tracking (equatorial) or azimuth/elevation tracking (altitude-azimuth).

In terms of the performance, the tracking systems can be classified as passive or active. In the same way, active trackers can be classified into two types: open and closed loop [13]. A passive tracker system works without electronic sensors, actuators or electronic control systems. On the other hand open loop systems, works through mathematical model to predict the apparent sun position and use the time and geographical position in which the system is installed. Unfortunately, the angular position of the actuator that is being controlled can not be known, so it is impossible to re-adjust any positioning error caused by a disturbance [14–16]. The closed-loop systems employ, typically, the information of electro-optical sensors to orient the tracking mechanism and estimate the location of the sun. The most used are photoresistors or photodiodes arrangements, complementary metaloxide-semiconductor (CMOS), microelectro-mechanical system (MEMS), pixel sensor (image sensor), among others. These sensors have different advantages and disadvantages and their selection depends on solar technology, cost-effectiveness, the requirements of the system of interest and the strategies of commercialization [17]. The main disadvantages of this sensor are: its spectral response changes with respect to temperature or radiation variations, its response can presents areas of non-linearity, it has low repeatability, it can be easily damaged due to the humidity and dust and suffer degradation over time.

One of the most important parameters in solar sensors is the field of view (FOV) – a parameter which allows predicting the possible angle of sun position. Typically, the wider the field of view of a sensor, the lower the accuracy of the solar location (FOV_b in Fig. 1). This phenomenon can be explained by the fact that the sensor is susceptible to the measurement of noise caused by diffuse radiation in the environment. On the other hand, a narrow field of view (FOV_a in Fig. 1) allows to a achieve higher accuracy by capturing only direct solar radiation; however, when the sun is out of range, it can not be located without a pre-location, which requires a more complex control system.

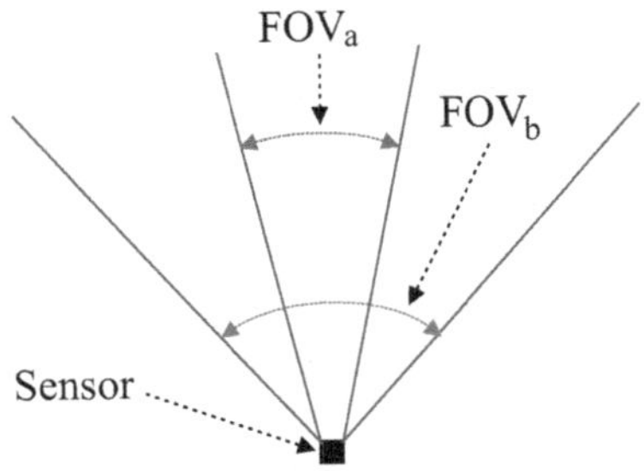

Fig. 1. FOV demonstration of two solar sensors.

In this work, a low-cost Passive Sun Tracking System (PSTS) based on image processing is proposed. The developed methodology depend on the implementation of several algorithms, which allow to identify and evaluate the apparent sun location with a FOV of 180°. The proposed system was tested in Aguascalientes city, Mexico (21°51'N 102°18'W).

2 Methodology

The experimental schematic representation of the system is shown in Fig. 2, this system is composed of two parts: the PSTS and the solar tracking system detailed in Sect. 2.1 and 2.3 respectively. The operation of the system begins with the capture of the solar rays reflected from the convex mirror towards the Raspberry-Pi camera of the PSTS.

In order to compute the apparent position of the sun, the image that was acquired is processed. Next, the Zenith and Azimuth angles are sent to the solar tracker controller. This procedure is done with a frequency set by the user. The control system of the solar tracking system takes as reference the received Zenith and Azimuth angles and, finally, the controller drives the actuators until they reach the desired position.

2.1 Passive Sun Tracking System

The PSTS developed consists of a low-cost sky imager device, image acquisition, processing algorithms and the support structure (See Table 1). The prototype is based on a Raspberry Pi Camera module V2 that uses a CMOS IMX219PQ sensor with 8-megapixel native resolution. The camera is protected by a case particularly designed for outdoor applications able to protect it from rain and dust. The convex mirror is mounted in a structure built by 20 mm structural aluminum using an upward facing configuration that achieves a FOV of 180° (See Fig. 3) , which allows to visualize the apparent sun position at any time of the day. Furthermore, in order to stabilize the camera, a set of clamping cables have been fastened from a point near the camera housing to the structure base.

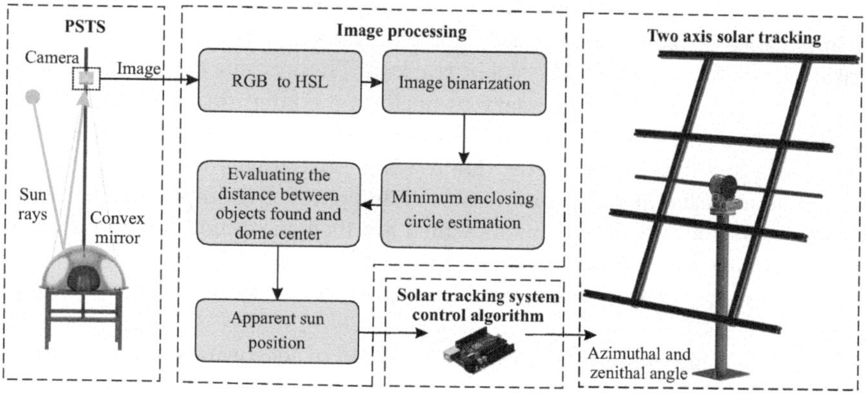

Fig. 2. Schematic of experimental setup.

Table 1. Cost of the materials used in the construction of the prototype.

Description	Quantity	Cost [$ USD]
Structural Al. 20 mm	6m	38.5
Convex mirror 18′	1pc	70.0
Camera board V2	1pc	23.0
Raspberry Pi 3 B	1pc	77.0
Miscellaneous materials		33.0
	Total	245.5

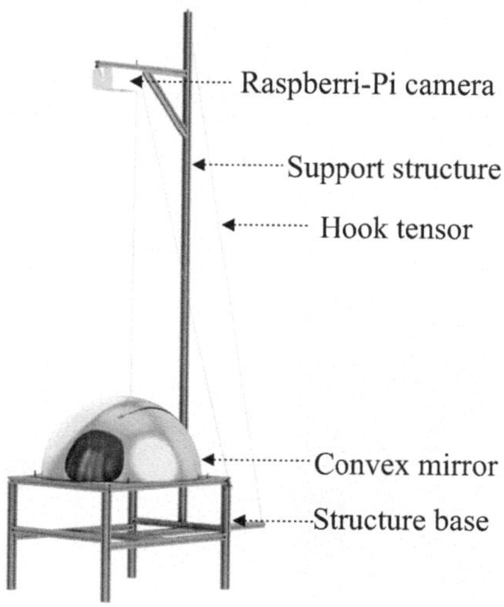

Fig. 3. PSTS designed.

Although the system needs to know neither its geographical location(longitude, latitude or altitude) nor the date and time, it requires a pre-orientation with respect to the solar tracker. The system can be installed on irregular surfaces, but it is necessary to perform a configuration routine in order to position the plane of the normal dome towards the zenith. In addition, the system has a low development cost due to some advantages such as the inclusion of economic material and the absence of moving parts. It is important to mention that the setup does not have sensors which are likely to suffer deterioration over time and change their linearity or their spectral response. On the other hand, a single prototype is able to command the position of several single or two-axis tracking systems.

One of the main advantages of using this kind of sensor in comparison of photodiodes or light-dependent resistors is its low sensitivity to weather conditions such as temperature, humidity and diffuse radiation from the environment. To achieve optimal solar tracking, It is possible to adjust the detection frequency of the sun position, and know the state of cloudiness in the day. The information is sent wirelessly to a user interface, where the apparent position of the sun, obtained with the developed prototype, can be displayed in real-time.

2.2 Sun Position Detection Process

In order to detect and monitor the apparent position of the sun, an image-based sun detection algorithm has been developed. The sun detection process begins with the acquisition of an image in an RGB color space, then a color space conversion from RGB to HSL color space is performed. HSL is another color space which represents the color on three channels: hue, saturation, and brightness (how bright the color is). Thus, objects with very large brightness values can be easily separated from the background in the image; it is assumed that the center of the brightest region in the sky image represents the position of the solar center. In this sense, to distinguish the sun from the rest of the elements in the image, a binarization of the image is performed. Figure 4 shows the resulting image after the binarization process, where the segmented objects are those with the highest brightness.

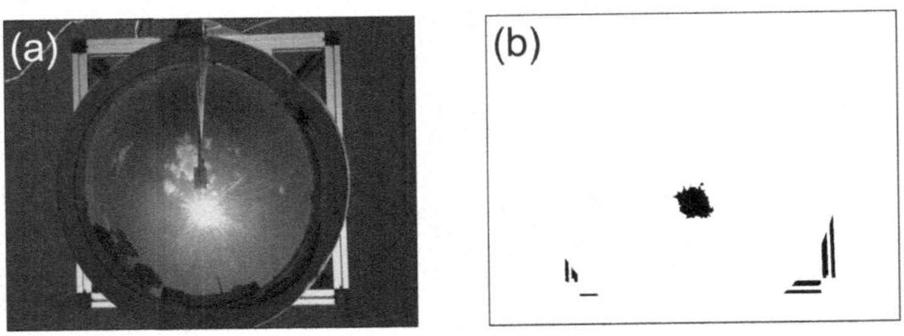

Fig. 4. Binarization of the image. Figure 4a shows the input color image. Figure 4b shows the result of the binarization.

From each object found, the minimum enclosing circle to get the center, as well as the radius, is computed. To discriminate the sun from all those objects found in this process, a distance to the dome center (which has also been detected) is calculated. If the distance of an element is lower than the dome radius then, the algorithm considers this element as the circle that represents the sun. A sun detection example of this process is shown in Fig. 5.

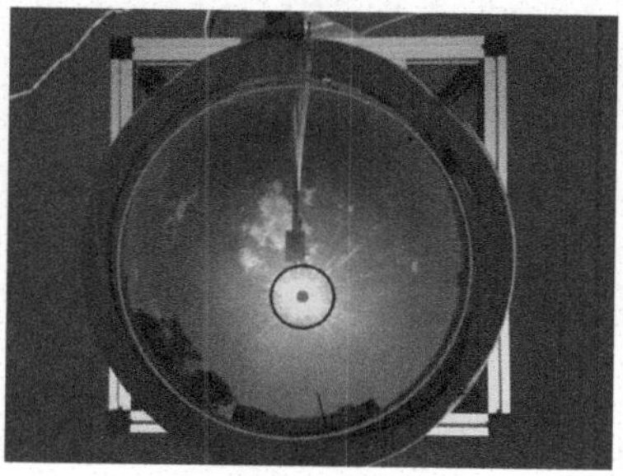

Fig. 5. Sun detection example.

2.3 Solar Tracker

To verify the proposed passive solar location strategy, a two-axis solar tracker is used. This mechanism operates with a slewing drive that provides torque, accuracy, precision, efficiency, and reliability in solar tracking (see Fig. 6).

Once the developed system locates the apparent position of the sun with respect to its reference frame, the azimuthal and zenithal angles are determined, which will then be sent to an Arduino controller. These angular values are interpreted by the solar tracking system control algorithm as input signals. In subsequent work, the characterization of solar tracking accuracy will be carried out.

3 Results and Discussion

A graphical user interface (GUI) was the developed in which the user can observe relevant information, such as the zenith and azimuth angles sent to the tracker system, the current image of the sky with the sun detected, among other information. Additionally, the user has the possibility of using the interface to make modifications of some of the parameters directly related to the detection of the sun position. Figure 7 shows the start window of the GUI; from this window the user is able to configure the initial system parameters.

In order to carry out the final solar tracking tests, it is necessary to instrument the PSTS with a magnetometer to geographically orientating the system. Once its geographical orientation is determinated, it will be sent wirelessly to the solar tracker controller, where a coordinate transformation is required to drive the actuators and align the system to the apparent position of the sun.

Fig. 6. Two-axis solar tracking system.

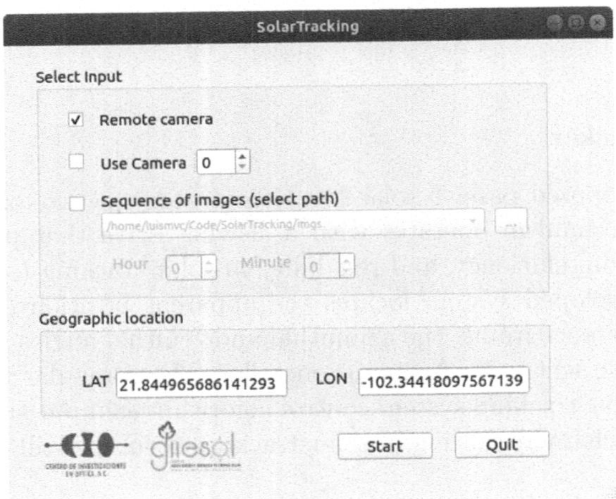

Fig. 7. Interface start window.

On the other hand, Fig. 8 shows the system GUI, where it is possible to observe the detection of the apparent position of the sun (performed by the proposed system), in addition to other relevant information such as geographical location, time, date, etc.

As mentioned in this document, the proposed PSTS has several advantages over the tracking sensors commonly used in the literature; however, it is necessary to measure the final tracking error to determine its viability in CSP and HCPV systems.

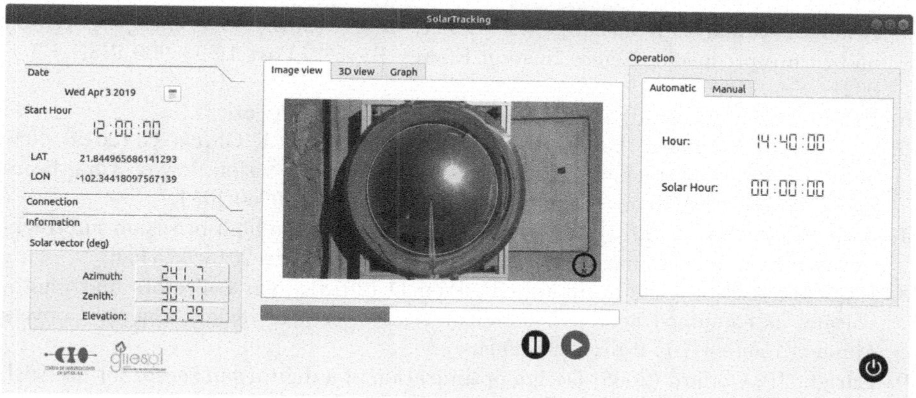

Fig. 8. System graphical user interface.

4 Conclusions

In this work, solar tracking algorithm was developed through the implementation of a novel passive solar location strategy. The estimated cost of the prototype, without considering the solar tracking system, was \$ 245.5 USD, which makes it economically viable to be implemented in low-cost solar applications that require solar tracking. One of the biggest advantages of the system developed is that it reaches a FOV of 180°; that is, it is possible to locate the position of the sun at any time of the day, requires low maintenance, has no moving parts and does not use sensors that deteriorate over time so the efficiency of the system does not decrease. In addition, it is important to mention that the PSTS is not affected by diffuse radiation; therefore, the accuracy of solar detection of the system does not decrease on sunny and cloudy days. However, in days with severe cloudiness the sun position can not be localized.

As future work, the characterization of the solar tracking system will be performed in order to know the tracking precision; nevertheless, it is essential to mention that there is no official procedure for the calibration of the tracking error measurement tool established by the International Electrotechnical Commission (IEC) or some other entity.

References

1. Sobri S, Koohi-Kamali S, Rahim N (2018) Solar photovoltaic generation forecasting methods: a review. Energy Convers Manage 156:459–497 ISSN 0196-8904
2. Herche W (2017) Solar energy strategies in the US utility market. Renew Sustain Energy Rev 77:590–595 ISSN 1364-0321
3. Alemán-Nava G, Casiano-Flores V, Cárdenas-Chávez D, Díaz-Chavez R, Scarlat N, Mahlknecht J, Dallemand J, Parra R (2014) Renewable energy research progress in Mexico: a review. Renew Sustain Energy Rev 32:140–153 ISSN 1364-0321

4. Kabir E, Kumar P, Kumar S, Adelodun A, Kim K (2018) Solar energy: potential and future prospects. Renew Sustain Energy Rev 82, Part 1:894–900 ISSN 1364-0321
5. Renewables 2013 (2013) Global Status Report 2013, Network R
6. Carballo J, Bonilla J, Berenguel M, Fernández-Reche J, García G (2019) New approach for solar tracking systems based on computer vision, low cost hardware and deep learning. Renew Energy 133:1158–1166 ISSN 0960-1481
7. Azizi K, Ghaffari A (2013) Design and manufacturing of a high-precision sun tracking system based on image processing. Int J Photoenergy 2013:7 754549
8. Kasaeian A, Nouri G, Ranjbaran P, Wen D (2018) Solar collectors and photovoltaics as combined heat and power systems: a critical review. Energy Convers Manage 156:688–705 ISSN 0196-8904
9. Enrigh, JP, Godard (2008) Design optimization of a digital sun sensor for use with parametric processing. IEEE Trans. Instrum. Meas. 57(10):2188–2195
10. Mousazadeh H, Keyhani A, Javadi A, Mobli H, Abrinia K, Sharifi A (2009) A review of principle and sun-tracking methods for maximizing solar systems output. Renew Sustain Energy Rev 13(8):1800–1818
11. Singh R, Kumar S, Gehlot A, Pachauri R (2018) An imperative role of sun trackers in photovoltaic technology: a review. Renew Sustain Energy Rev 82, Part 3:3263–3278 ISSN 1364-0321
12. AL-Rousan N, Mat Isa N, Mat Desa M (2018) Advances in solar photovoltaic tracking systems: a review. Renew Sustain Energy Rev 82, Part 3:2548–2569 ISSN 1364-0321
13. Huynh DC, Nguyen TM, Dunnigan MW, Mueller MA Comparison between open- and closed-loop trackers of a solar photovoltaic system. In: 2013 IEEE Conference on Clean Energy and Technology (CEAT), pp 128–133 IEEE
14. Nsengiyumva W, Chen S, Hu L, Chen X (2018) Recent advancements and challenges in Solar Tracking Systems (STS): a review. Renew Sustain Energy Rev 81, Part 1:250–279 ISSN 1364-0321
15. Chia-Yen L, Po-Cheng C, Che-Ming C, Chiu-Feng L Sun tracking systems: a review. Sensors (Basel) 9:3875–3890 https://doi.org/10.3390/s90503875
16. Lynch W, Salameh Z (1990) Simple electro-optically controlled dual-axis sun tracker. Sol Energy 45(2):65–9
17. Salgado-Conrado L (2018) A review on sun position sensors used in solar applications. Renew Sustain Energy Rev 82, Part 3:2128–2146 ISSN 1364-0321

Attenuation of the Amplitude Vibration Level by Using Nonlinear Active Control Techniques Applied to a Hybrid Short Journal Bearing Using a Particular Rotor-Bearing System Configuration

D.-E. Rivera-Arreola[1], J.-C. Gómez-Mancilla[2], T. Hernández-Cortés[1(✉)], and H. F. Abundis-Fong[3]

[1] Univ. Politéc. de Pach., Carretera Ciudad Sahagún-Pachuca Km. 20, 43830 Pachuca, Hidalgo, México
{eduardorivera,tonatiuh}@upp.edu.mx
[2] Inst. Politéc. Nal., Av. Luis Enrique Erro S/N, Unidad Profesional Adolfo López Mateos, Zacatenco Alcaldía Gustavo A. Madero, C.P. 07738 Ciudad de Mexico, Mexico
gomezmancilla@gmail.com
[3] TecNM/I.T. Pachuca, Felipe Angeles Km. 84.5, Venta Prieta, 42083 Pachuca de Soto, Hidalgo, Mexico
habundis@itpachuca.edu.mx

Abstract. Inherent to turbomachinery are some common operational problems, such as the critical amplitude of motion that the machine may experience when passing through one or more of its natural frequencies. High range of vibration levels can be catastrophic to the equipment and its attenuation is needed to keep the system working healthily. To that end, the present work proposes the use of nonlinear active control techniques applied to a single hybrid short journal bearing using a particular rotor-bearing system configuration, which not require the assumption of an exact symmetry in the operation of the two journal bearings located at both ends of the rigid shaft such as the classical Jeffcott rotor model. The analysis is focused by reducing the amplitude shaft vibration of the rigid rotor crossing its first resonance. The oil injection port is modeled by mean of the Dirac Delta Function.

Keywords: Hybrid-short journal bearing system · PID controller · Reynolds equation · State-space · Dirac delta function

1 Introduction

Turbomachinery is currently used across several industries because its technology can be applied in a wide range of rotating equipment, such as aircraft engines, wind turbines, and other forms of transfer energy between a rotor and a fluid.

E. E. Hernandez et al. (Eds.): LASIRS 2019, MMS 86, pp. 179–192, 2020.
https://doi.org/10.1007/978-3-030-45402-9_18

Thus, the health of rotating machinery is crucial. Some resonance might occur within the machine's range of operations. Therefore, it is of paramount importance to identify the critical speeds of the rotor, to prevent the turbo rotor from operating at such speeds. Hydrodynamic bearings are commonly used to attenuate high vibration levels due to because they behave with incomparable smoothness. However, the problem of self-excited vibration, called fluid-induced vibration, on the thin film in the rotating components may lead to large amplitudes of motion and possible damage to the rotor [1,2]. Externally pressurized bearings have been developed and successfully used in machinery that operates under high-speed and heavy-load conditions. Hybrid bearings combine the physical mechanisms of both hydrodynamic and hydrostatic bearings. Some advantages of hybrid bearings over purely hydrodynamic bearings are that wear and tear are avoided when the shaft starts and stops, eliminating metal to metal contact and allowing for very low starting torque. Attractive features include tolerance of wider variations in bearing clearance, good bearing film stiffness at zero rotating speed and the ability to employ higher viscosity oils at high speed. In this sense, hybrid bearings are designed to take maximum advantage of hydrodynamic and hydrostatic effects and, when designed appropriately, have superior performance at high speeds [3,4]. Donald-Bently Corp. USA [5] carried out experimental studies related to an externally pressurized fluid-film bearing mainly focused on the diagnosis of machinery malfunction. The most important conclusions can be summarized as follows:

- When externally pressurized, the rigidity of the bearings increases remarkably.
- Pressurization can produce dynamic instability.
- Under certain conditions the external pressurization might modify the resonance speeds.
- Pressure variation can allow adequate control of the rigidities in the bearing and can be implemented in more than one injection port.

The theory of fluid film lubrication is based mainly on the work of Osborne Reynolds (1842–1912). Under ideal conditions, the Reynold's equation is widely used to compute the film pressure for hydrodynamic journal bearings [6–12]. Over the last few decades, several numerical approaches have been used, such as the multigrid techniques [13], the finite element methods [14,15] or simpler, faster yet still accurate methods like the mobility method [16] or the impedance method [17,18]. In general, different strategies are employed to perform analysis of hydrodynamic journal bearings. Nowadays, with the fast development in technology, combining mechanics and mechatronics has given a great impulse to the development of the turbomachinery. This interdisciplinary knowledge has allowed the use of new electromagnetic devices as well as the application of control laws which can be useful and practical for the attenuation in highs levels of vibration. From this point of view Nicoletti et al. applied some linear and nonlinear control techniques to tilting-pad journal bearings. Theoretical results showed that the proper selection of the controllers can be used to control the oil supply to reduce the level of vibration amplitude of the rotor-bearing system [19]. Santos et al. derived theoretical expressions considering the $L/D = 1$ ratio for multirecess hybrid journal bearings with active lubrication. Analytical results of the bearing dynamic coefficients calculations were presented [20]. Alejandro et al. studied active lubrication applied to tilting-pad journal bearings. The equations governing the actively lubricated bearings was derived by including the pipelines dynamics effects into the mathematical models [21]. Arias-Montiel et al. applied a Linear Quadratic Regulator (LQR) to reduce the vibration amplitudes in a rotor system with active control scheme. Their results demonstrated significant reductions in the unbalance response [22]. Jorge et al. studied active tilting-pad journal bearings supported by flexible rotors. A reduction about 30% were achieved around the resonant zone with hybrid lubrication [23]. Leonardo Biagiotti et al. studied the reduction of a rotor vibration amplitude using PID tuning methods. Five tuning methods were used and compared for the PID. The vibration amplitude of the critical speed was reduced and the system damping was increased for every tuning method [24].

In the present work, the proposed configuration provides an analysis by using a single hybrid short journal bearing with four oil injection ports. Thus, does not require the assumption of an exact symmetry in the operation of the two journal bearings located at both ends of the rigid shaft. The hybrid bearing is modeled ad hoc by the termed closed solution to the Reynolds Lubrication Equation, and the Dirac Delta Function is introduced in order to model the oil injection port. Also, this work will incorporate a PID controller using the Ziegler-Nichols (ZN) tuning method applied to the nonlinear analytical expression to eliminate and can cope with unexpected operating conditions. To determine the stability of the nonlinear equation of motion and thus ensure the optimal functionality of the PID controller, the mathematical tools related to the theory of Lyapunov's stability criterion are used. Finally, the governing equations of motion corresponding to the proposed physical system are given to validate our experiments.

The remainder of the paper is organized as follows: a formulation and description of the mathematical models are presented in Sect. 2, the study an implementation of control laws are analyzed and studied in Sect. 3, the tests are presented in Sect. 4 and, the conclusion are given in Sect. 5.

2 Mathematical Modeling of the Rotor-Bearing System

The rotor-bearing configuration studied in the present work is a system based on a rigid shaft and two equidistant disks supported by only one hybrid short journal bearing located in the middle of the two disks. Figure 1 depicts a schematic diagram of the physical rotor-system configuration and the stiffness and damping coefficients. Where ω is the shaft speed considering a counterclockwise rotation, the force A_x is generated by the imbalance and weight of the disk 1, on X-axis and it is located through. Besides, A_y is the disk imbalance force in the Y-axis; forces on B_x and B_y are generated by oil pressure in the hybrid short journal bearing located through the segment. Similarly, force C_x is caused by the imbalance and weight of the disk 2, on the X-axis and C_y is the disk imbalance force in the second disk on the Y-axis, located at $\overline{OA} = l$. To simplify the mathematical expressions, some considerations must be assumed:

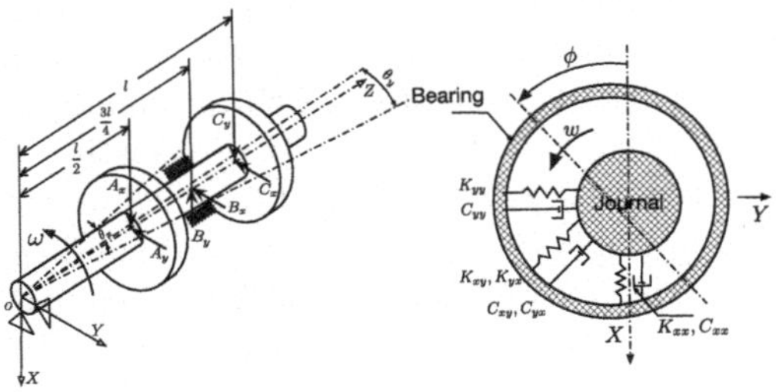

Fig. 1. Rigid shaft model and schematic representation of the cross section considering stiffness and damping coefficients.

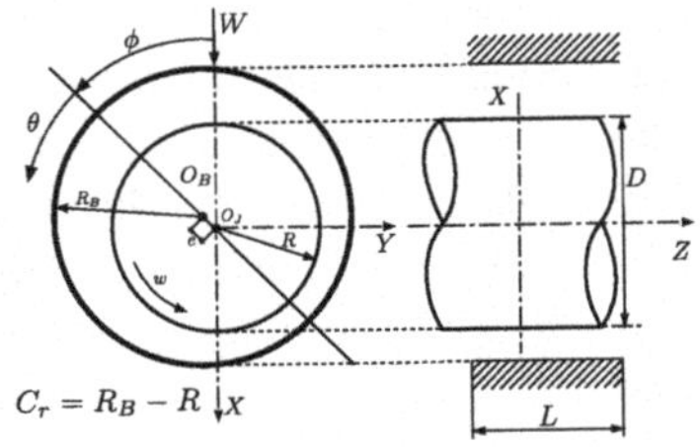

Fig. 2. Cross section for a journal bearing with nomenclature.

1. The shaft is Rigid.
2. The disks are equidistant to the bearing.
3. The moment of inertia of the shaft is ignored since it is much smaller than the moment of inertia of the disks.

The equations of motion presented below are derived by Newton-Euler formulation, where the angular displacements generated by the shaft are described by θ_x and θ_y respectively.

$$I_0\ddot{\theta}_x = \sum M_{xz}, \quad I_0\ddot{\theta}_y = \sum M_{yz} \tag{1}$$

Rewriting Eq. (1) and taking into account the equilibrium system based on [25], results.

$$\begin{aligned} I_0\ddot{\theta}_x + \left(\frac{9}{16}\right) l^2 \Big[& C_{XX}\dot{\theta}_x(1+\tan^2\theta_x) + C_{XY}\dot{\theta}_y(1+\tan^2\theta_y) \\ & + K_{XX}\tan\theta_x + K_{XY}\tan\theta_y \Big] = \\ \frac{l}{2} m_{d1}(\omega^2 e_1^x \cos(\omega t) & - \alpha e_1^y \sin(wt)) + l m_{d2}(\omega^2 e_2^x \cos(\omega t) - \alpha e_2^y \sin(wt)) \end{aligned} \tag{2}$$

$$\begin{aligned} I_0\ddot{\theta}_y + \left(\frac{9}{16}\right) l^2 \Big[& C_{YX}\dot{\theta}_x(1+\tan^2\theta_x) + C_{YY}\dot{\theta}_y(1+\tan^2\theta_y) \\ & + K_{YX}\tan\theta_x + K_{YY}\tan\theta_y \Big] = \\ \frac{l}{2} m_{d1}(\omega^2 e_1^x \sin(\omega t) & + \alpha e_1^y \cos(wt)) + l m_{d2}(\omega^2 e_2^x \cos(\omega t) + \alpha e_2^y \cos(wt)) \end{aligned} \tag{3}$$

where:

$$I_O = \left[\frac{1}{4} m_{d1} r_{d1}^2 + m_{d1}\left(\frac{l}{2}\right)^2\right] + \left[\frac{1}{4} m_{d2} r_{d2}^2 + m_{d2}\,(l)^2\right] \tag{4}$$

is the mass moment of inertia, $[kg \cdot m^2]$; α is the angular acceleration of the shaft, $[\frac{rad}{s^2}]$; l is the length of the shaft, $[m]$; m_{d1} is the disk's mass $[kg]$; $e_1^x = e_2^x = e_1^y = e_2^y$ are the imbalance distance of the disk masses in the x and y direction, $[m]$; $k_{ij} = \frac{W}{C_r} K_{ij}$ and $c_{ij} = \frac{W}{C_r \omega} C_{ij}$ are the dimension stiffness and damping coefficient $[\frac{N}{m}]$ and $[\frac{Ns}{m}]$ respectively in the direction i, j where, $W = m_d \cdot g$ is the weight of the system, $[N]$, C_r is the radial clearance, $[m]$. Finally, Φ is the equilibrium attitude angle $[rad]$, see Fig. 1 and 2. To compute the values of K_{ij} and C_{ij} the interested reader can consult references [26,27].

2.1 Reynolds Equation and Boundary Conditions

In lubrication theory, the Reynold's equation is a partial differential equation expressed in terms of pressure. Once the pressure profile is evaluated, all other bearing performance parameters can be determined. According to the movement of the shaft, the pressure field within a journal bearing is generated by the formation of a lubricant wedge, depicted in Fig. 2. Hence, it is of the utmost importance to determine the location of the maximum pressure and the regions where the lubricant oil film can produce cavitation [28,29]. The Reynolds theory of lubrication can be obtained based on the assumptions of [10,11]:

1. The fluid is assumed to be Newtonian.
2. Inertia and body forces terms are assumed to be negligible compared to the viscous terms.
3. Variation of pressure across the film is assumed to be negligible, i.e., $\frac{\partial P}{\partial z} = 0$, so $P = (x, y)$ only.
4. Flow is laminar.
5. Curvature effects are negligible. This implies that the thickness of the lubricant film is much smaller than the length or width of the bearing so that the physical domain of the flow can be unwrapped, allowing for the use of the cartesian coordinate system.

$$\frac{\partial}{\partial \theta}\left(h^3 \frac{\partial P}{\partial \theta}\right) + R^2 \frac{\partial}{\partial z}\left(h^3 \frac{\partial P}{\partial z}\right) = 12\pi \frac{\mu R^2}{C_r^3}\left[C_r \dot{\varepsilon} \cos\theta + C_r \varepsilon \left(\dot{\phi} - \frac{\omega}{2} \sin\theta\right)\right] \quad (5)$$

$$-\frac{L}{2} \leq z \leq \frac{L}{2}, \quad 0 \leq \theta \leq 2\pi, \quad h(\pi) = 1 + \varepsilon \cos\theta$$

$$P\left(\frac{L}{2}\right) = 0, \quad P\left(\frac{-L}{2}\right), \quad P(\theta + 2\pi) = P(\theta)$$

where, P is the pressure, $[Pa]$; θ is the circumferential coordinate, $[rad]$; z is the axial coordinate, $[m]$; ϕ is the equilibrium angle (attitude), $[rad]$; C_r is the radial clearance, $[m]$; e is the eccentricity, $[m]$; ε is the dimensionless eccentricity, $\varepsilon = \frac{e}{C_r}$; R is the journal radius, $[m]$; R_B is the bearing radius, $[m]$; μ is the dynamic viscosity, $[Pa \cdot s]$; H is the thickness of the oil film, $[m]$; h is dimensionless thickness of the oil film, $h = \frac{H}{C_r}$; w is the angular speed considering a counterclockwise rotation, $[rad/s]$; D is the journal bearing diameter, $[m]$; and L the journal bearing length, $[m]$. Also, this work is restricted to short length bearings, i.e. $L << D$. As a consequence, the pressure gradients, $\frac{\partial P}{\partial \theta} << \frac{\partial P}{\partial z}$, allow us to ignore the first term in Eq. (5). Further, a short bearing in steady-state ($\dot{\varepsilon} = \dot{\phi} = 0$), and by using the following dimensionless variables: $z = \frac{L}{2}$, $\overline{p} = \frac{p}{P_{\text{dim}}}$ and $P_{\text{dim}} = \mu N \left(\frac{R}{C_r}\right)^2 = \mu \left(\frac{\omega}{2\pi}\right)\left(\frac{R}{C_r}\right)^2$ the Reynolds equation for a hydrodynamic journal bearing expressed in dimensionless form provides:

$$\frac{\partial}{\partial \overline{z}}\left(h^3 \frac{\partial \overline{p}}{\partial \overline{z}}\right) = 12\pi \left(\frac{L}{D}\right)^2 (\varepsilon \sin \theta) \tag{6}$$

Now, by considering Half-Sommerfeld Boundary Conditions and solving Eq. (6) for $\overline{p}$ it is possible to compute the pressure field in dimensionless form. This expression is known as the Ocvirk solution and is defined by [30,31]:

$$\overline{P}_{ocv} = -6\pi \left(\frac{L}{D}\right)^2 \frac{\varepsilon \sin \theta}{1 + \varepsilon \cos \theta} \left(1 - \overline{z}^2\right) \tag{7}$$

On the other hand, the mathematical model for the pressurized case employs an oil injection port. Thusly, to simulate the effect of the external pressurization in the hydrodynamic journal bearing the generalized spatial Dirac delta function is used. Thus, $\delta = [\theta - \theta_{in}] \begin{cases} \infty & \theta - \theta_{in} \\ 0 & \theta \neq \theta_{in} \end{cases}; \delta = [z - a] \begin{cases} \infty & z - a \\ 0 & z \neq a \end{cases}$ where, $\theta_{in} = \pi + \beta - \phi$. The symbolic function $\delta(z - a)$ represents the pressure in a source point located in $z = a$, and the accumulated pressure is termed as the Heaviside function [32], in such a way $\Delta F_{pres} = \Delta p \cdot \Delta s$, where, ΔF_{pres} is the pressurization force, $[N]$; Δs is the area of the oil injection port, $[m^2]$, and Δp is the pressure of the oil injection port, $[Pa]$. Consequently, when $\Delta p \to 0$, then $\Delta s \to 0$ so that: $\Delta F_{pres} = \Delta p \cdot \Delta s = q = constant$. These conditions provide the mathematical model that represents a pressurization point in the dimensionless form described by the spatial Dirac delta function, which results in: $(\Delta \overline{p})_{prt} = \overline{q}_{prt} \delta(\overline{z} - a) \delta\left[\theta - (\pi + \beta - \phi)\right]$ [33]. Hence, $\overline{q}_{prt} = \frac{\delta F_{pres}}{DLP_{\text{dim}}} = \frac{\delta F_{pres}}{F_{\text{dim}}} = \frac{f_{prt}}{S}; P_{\text{dim}} = \mu N \left(\frac{R}{C_r}\right)$. Knowing that, $(\Delta \overline{p})_{prt}$ is the dimensionless pressure in the oil source port; $\overline{q}_{prt}$ is the dimensionless pressure in the point source regard to the equivalent characteristic force; f_{prt} is dimensionless pressurization force in the point source respect to the weight; S is the dimensionless Sommerfeld's number, $\delta(z)$ is the Dirac delta function; a is the arbitrary dimensionless axial position for the oil injection port; β defines the port position or angular coordinate of the oil injection port, $[rad]$; and F_{dim} is the equivalent characteristic force, $[N]$. Then, the solution of Eq. (6) for $\bar{p}_{pres}$ becomes:

$$\bar{p}_{pres}(\bar{z},\theta) = \left(\frac{L}{D}\right)^2 \bar{q}_{prt} \frac{\delta\left[\theta - (\pi + \beta - \phi)\right]}{2(1+\epsilon\cos\theta)^3} \left(1 - a\bar{z} - |(\bar{z} - a)|\right) \tag{8}$$

where, $\tilde{n}$ is the peak value of Dirac Delta approximation. Finally, the total pressure field for hybrid short journal bearing with a simple oil injection port is given by the addition of the unpressurized mathematical model (classic case) Eq. (7) plus the pressurized solution (pressurized case) Eq. (8), which can be written as:

$$\overline{p} = \overline{p}_{ocv} + \overline{p}_{pres} = -6\pi \left(\frac{L}{D}\right)^2 \frac{\varepsilon\sin\theta}{1+\varepsilon\cos\theta}\left(1 - \overline{z}^2\right) + \left(\frac{L}{D}\right)^2 \overline{q}_{prt} \frac{\delta\left[\theta - (\pi + \beta - \phi)\right]}{2(1+\varepsilon\cos\theta)^3} \left[1 - a\overline{z} - |\left(\overline{z} - a\right)|\right] \tag{9}$$

Two particular cases of study are analyzed: first, two horizontal oil injection ports located at $\beta = \pi/2$ and $\beta = 3\pi/2$, and second, at the top $\beta = \pi$ and bottom $\beta = 2\pi$, respectively. The peak value of the Dirac Delta approximation is proposed to simulate an adequate function of the equivalent oil injection port. The equilibrium eccentricity $\varepsilon = e/C_r$ is evaluated from $\varepsilon = 0.1$ to 0.9 [10] and is located in the middle of the journal bearing respect to the Z-axis to suppress conditions of imbalance forces. Plotting Eq. (9) and considering $\tilde{n} = 10$, $\varepsilon = e/C_r$ from 0.01 to 0.9, $a = 0$ and f_{prt} from 0 to 100 results:

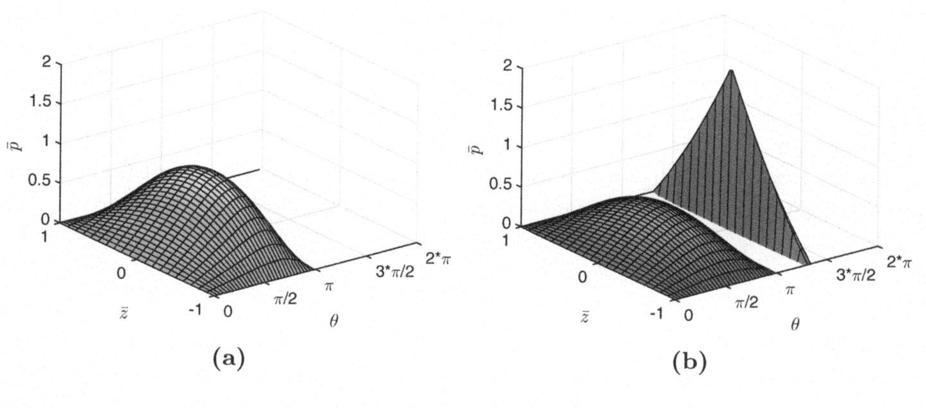

Fig. 3. Results for the computer simulation where: (a) $a = 0$, $f_{prt} = 20$, $\beta = 3\pi/2$, $\epsilon = 0.5$, $S = 1.6$, $\phi = 0.99$ (b) $a = 0$, $f_{prt} = 20$, $\beta = \pi/2$, $\epsilon = 0.5$, $S = 1.8$, $\phi = 0.56$.

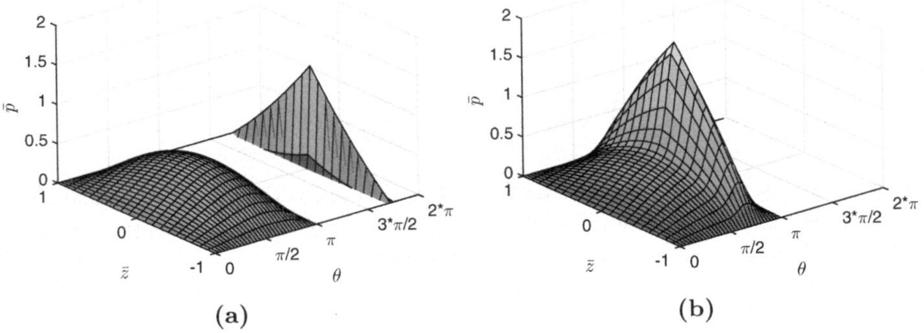

Fig. 4. Results for the computer simulation where: (a) $a = 0$, $f_{prt} = 20$, $\beta = \pi$, $\epsilon = 0.48$, $S = 2$, $\phi = 0.96$ (b) $a = 0$, $f_{prt} = 20$, $\beta = 2\pi$, $\epsilon = 0.39$, $S = 2$, $\phi = 1.07$.

Figs. 3 and 4 illustrate the hydrodynamic pressure in the oil film on the sliding surface of a hybrid short journal bearing by a specific port respectively.

3 PID Control Techniques

A nonlinear dynamic system can usually be represented by a set of nonlinear differential equations in the form: $\dot{x} = f(x, u, t)$ and $y = h(x, u, t)$. To obtain a full description of the system for further analysis, the following state variables are used: $x_1 = \theta_x$, $x_2 = \theta_y$, $x_3 = \dot{\theta}_x$, $x_4 = \dot{\theta}_y$. In such a way to determine the stability of the nonlinear system Eqs. (2) and (3) and thus ensure the optimal functionality of the PID controller, the Lyapunov's method for stability is used. First, the direct method of stability criterion states that: be the origin $x = 0$ in an equilibrium point of $\dot{x} = f(x)$ and be $D \subset \Re^n$ a domain that contains the origin. Be $V \rightarrow \Re$ a continuously differentiable function such that: $V(0) = 0$ and $V(x) \geq 0$ in $D - \{0\}$. Thus, $\dot{V}(x) \leq 0$. In D, then $x = 0$ the nonlinear rotor-system is asymptotically stable [34–36]. In this sense, the proposed Lyapunov's equation based on the mathematical model of the energy is:

$$V(x) = \frac{1}{2}k_t x_1^2 + \frac{1}{2}k_t x_2^2 + \frac{1}{2}I_O x_3^2 + \frac{1}{2}I_O x_4^2 \quad (10)$$

Satisfying that $V(x) \geq 0$ where: $k_t = \frac{GJ_P}{l}$ is the torsional stiffness, $[Nm]$; G is the shear modulus of the shaft, $[Pa]$; J_P is the polar moment of inertia $[m^4]$. Thus,

$$\dot{V}(x) = \frac{\partial V(x)}{\partial x_1}\dot{x}_1 + \frac{\partial V(x)}{\partial x_2}\dot{x}_2 + \frac{\partial V(x)}{\partial x_3}\dot{x}_3 + \frac{\partial V(x)}{\partial x_4}\dot{x}_4 \quad (11)$$

The computed results can define that the nonlinear rotor-system is stable in the sense of Lyapunov where, $0 \leq \|x_1\| \leq 71\mu\text{m}$, $0 \leq \|x_2\| \leq 71\mu\text{m}$, $0 \leq \|x_3\| \leq 6000\ rpm.$, $0 \leq \|x_4\| \leq 6000\ rpm.$. To validate the proposed Eq. (10) the indirect method of Lyapuno's stability was also used. This approach uses the linearization of the system by means of the Jacobian matrix for determining the local stability of the original system. Thus, the stability criterion states that: be the origin $x = 0$ an equilibrium point of the nonlinear system $\dot{x} = f(x)$ where $f : D \subset \Re^n$ is a continuously differentiable function and $D \subset \Re^n$ is an environment of the origin. Hence, consider the system: $\dot{x} = f(x,t)$ with $f = (0,t) = 0$ for all $t \geq 0$. Define by $A(t) = \left.\frac{\partial f(x,t)}{\partial x}\right|_{x=0}$. To be the Jacobian matrix of $f(x,t)$ regarding to x, evaluated. Now, considering a quadratic Lyapunov function candidate [37] $V(x) = x^T P x$, where P is a real symmetric positive definite matrix. The derivative of V along the trajectories of the linear system Eq. (12) is given by:

$$\dot{V}(x) = x^T P\dot{x} + \dot{x}^T P x = -x^T Q x \leq 0 \tag{12}$$

where Q is a symmetric matrix defined. If Q is positive definite, we can conclude that the origin is asymptotically stable; that is, $\text{Re}\lambda_i < 0$ for all eigenvalues of A, in other words must be positive $\det(P - \lambda I) \geq 0$. Currently, PID controllers are distinguished by their flexibility, reliability and computational advantage. From this point of view, active control was applied to the proposed rotor-system configuration to reduce the amplitude of the shaft vibration crossing its first resonance. Thus, if controlled actively, Eqs. (2) and (3) becomes [38]:

$$\begin{aligned}
\dot{x}_1 &= x_3 \\
\dot{x}_2 &= x_4 \\
\dot{x}_3 &= -\frac{9l^2 c_{xx} x_3 (1+tg^2 x_1)}{16 I_O} - \frac{9l^2 c_{xy} x_4 (1+tg^2 x_2)}{16 I_O} - \frac{9l^2 u k_{xx} \tan x_1}{16 I_O} \\
&\quad - \frac{9l^2 u k_{xy} \tan x_2}{16 I_O} + \frac{M_1}{I_O} + \frac{u}{I_O} \\
\dot{x}_4 &= -\frac{9l^2 c_{yx} x_3 (1+tg^2 x_1)}{16 I_O} - \frac{9l^2 c_{yy} x_4 (1+tg^2 x_2)}{16 I_O} - \frac{9l^2 u k_{yx} \tan x_1}{16 I_O} \\
&\quad - \frac{9l^2 u k_{yy} \tan x_2}{16 I_O} + \frac{M_2}{I_O} + \frac{u}{I_O}
\end{aligned} \tag{13}$$

where:

$$M_1 = \frac{l}{2} m_{d1} \left(\omega^2 e_1^x \cos\omega t - \alpha e_1^y \sin\omega t\right) + l m_{d2} \left(\omega^2 e_2^x \cos\omega t - \alpha e_2^y \sin\omega t\right) \tag{14}$$

$$M_2 = \frac{l}{2} m_{d1} \left(\omega^2 e_1^x \sin\omega t + \alpha e_1^y \cos\omega t\right) + l m_{d2} \left(\omega^2 e_2^x \sin\omega t + \alpha e_2^y \cos\omega t\right) \tag{15}$$

besides, $u(t)$ is the active control represented by the characteristic PID and can be expressed mathematically as: $u(t) = K_d \dot{e}(t) + K_i \int_0^t e(t)\,dt + K_p e(t)$, where: K_p, K_i, and K_d are the coefficients for the proportional, integral, and derivative terms respectively and denote tuning parameters. Also: $e = X_R - X_1$ is the error, where X_R is the setpoint and X_1 is the process variable. The classic tuning Ziegler-Nichols (ZN) method was selected due to its efficiency and fast response. Therefore, the tuning parameters were calculated as: $K_i = \frac{K_p}{T_i}$, $K_d = K_p T_d$ and $K_u = 4(h - (DB/2))/\pi d$, knowing that (ZN) $K_p = 0.60 * K_u$, $T_i = 0.50 * T_u$ and $T_D = 0.125 * T_u$ where, h is the amplitude of the relay, d is the amplitude output of the plant and T_u is the ultimate period of the closed-loop system in the steady-state of the test [24,39,40].

3.1 Simulation

To validate the numeric simulation, the following parameters presented in the Table 1 were taken from the experimental test. The simulation results were performed in Matlab-Simulink® using ODE45. Hence, by simulating Eqs. (2) and (3) the Fig. 5 depicts for (a), (b) and (c) the vibratory and orbit response for $\beta = \pi$ without control techniques and its correspondence with the experimental test in Fig. 7. However when the simulation is performed the Eq. (13) depicts the vibratory and orbit response for $\beta = \pi$ by using active control techniques presented in Fig. 5 for (d), (e) and (f). The behavior was smooth and the amplitude did not exceed the response at 5μm (peak).

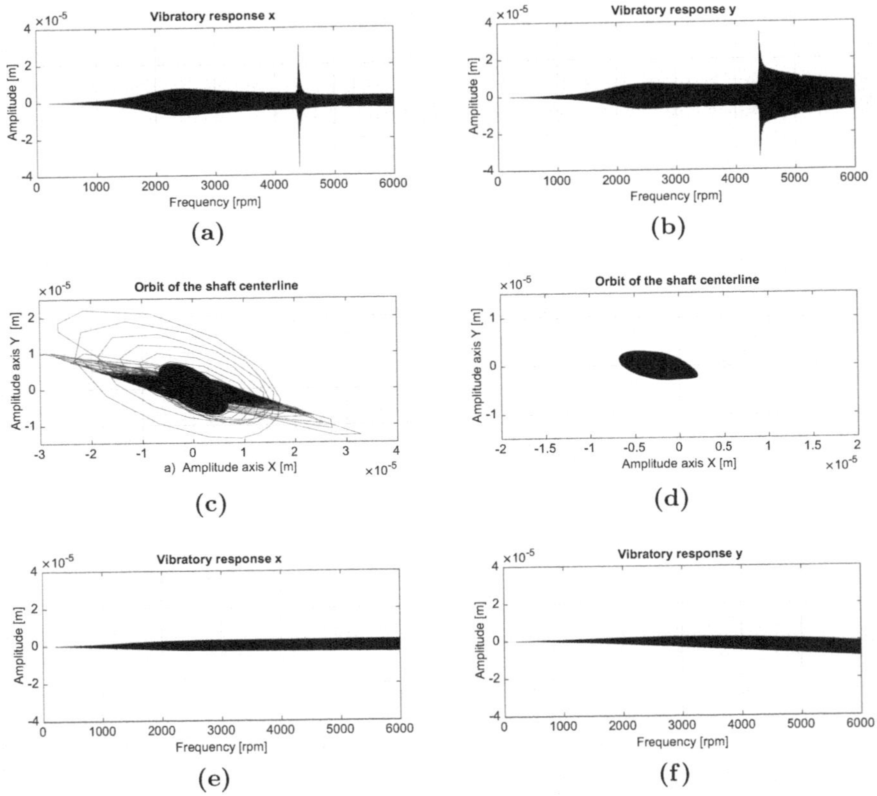

Fig. 5. System response as a function of the rotational frequency where: without law control (a), (b) and (c), and with active control techniques (d), (e) and (f) from $\omega = 200$ *to* 6000 *rpm.*, $\omega_n = 4550$ *rpm* and $\beta = \pi$.

4 Experimental Test

Fig. 6 illustrates the platform Bently Nevada® model RK4 rotor kit [41]. In the experimental setup "H" is the electrical motor a.c. that has a speed controller and acceleration ramp, "O" is a pivot self-aligning bronze bushing, "C" is the hydrodynamic journal bearing which can be pressurized by means of a pump; "A" and "B" are the positions of the equidistant disks in the rotor-system. The second disk is not visible since it is inside the supporting shell B. The parameters for the overall system are given in Table 1.

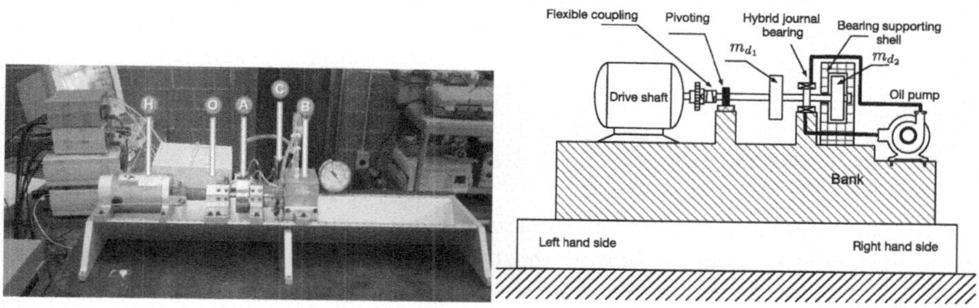

(a) (b)

Fig. 6. (a) Modified Rotor Kit type Bently Nevada® and (b) Graphical representation.

Table 1. System Parameters

Definition	Symbol	Value	SI units
Acceleration due to gravity	g	9.81	$[m/s^2]$
Angular speed considering counterclockwise rotation	ω	$0 - 6000$	$[rpm.]$
Angular acceleration	α	0, 1.2	$[rad/s^2]$
Damping factor	ζ	0.45	$[dimensionless]$
Diameter of the journal bearing	D	0.0254	[m]
Disk mass1	m_{d1}	0.8	[kg]
Disk mass2	m_{d2}	0.8	[kg]
Distance of the residual unbalance of mass	$e_1^x = e_2^x$	0.7×10^{-6}	[m]
Distance of the residual unbalance of mass	$e_1^y = e_2^y$	0.7×10^{-6}	[m]
Length of the journal bearing	L	0.0127	[m]
Length of the shaft	l	0.24	[m]
Lubricant viscosity	μ	0.0124	[Pa.s]
Polar moment of inertia	J_p	4.0863×10^{-8}	$[m^4]$
Radial clearance	C_r	71×10^{-6}	[m]
Radio of both Disks	$r_{d_1} = r_{d_2}$	0.0367	[m]
Shear modulus of the shaft	G	77.2×10^9	[Pa]
Weight of the system	$W = \Sigma m_d \cdot g$	15.6960	[N]

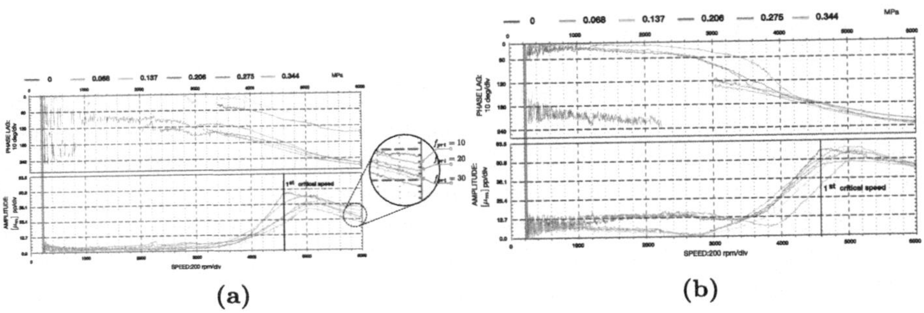

Fig. 7. Experimental test with external oil supply without law control considering $\beta = \pi$, $\omega_n = 4600\ rpm$ where: (a) Experimental bode diagram for X-axis and (b) Experimental bode diagram for Y-axis.

5 Conclusions

In this paper, a simple mathematical model for a rigid rotor-system with two disks supported by a single hybrid short journal bearing has been proposed and analyzed. The analytical expressions were derived by arranging four oil injection points orthogonally and modeled employing the Dirac Delta function. The equations yield results that support the experimental tests. These test successfully validate the applied lubricant pressurizing action; yet, currently the experimental set up lacks rapid active control capability. Also, to enhance the rotor-system stability with a wider and safer operational frequency range a PID active control technique was applied to the numerical method. To analyze the stability of the nonlinear rotor-system two different approaches based on the methodology of the direct and indirect Lyapunov's theory were used. These mathematical tools allowed to ensure the optimal functionality of the PID controller. The results have shown that the PID active control technique led to reducing significantly the amplitude of the shaft vibration crossing its first resonance. Moreover, the proposed experimental system shows that by increasing the bearing's pressure supply at $\beta = \pi$ and $\beta = 3\pi/2$ the amplitude vibration level can be attenuated. Finally, the results of the numerical solutions and tests show that the simplicity of the proposed rotor-system can be useful for designing and analyze new options of control laws.

References

1. Muszynsk A (1986) Whirl and whip-rotor bearing stability problems. J Sound Vib 111:443–462. https://doi.org/10.10007/1234567890
2. Tuma J, Bilos J (2007) Fluid induced instability of rotor systems with journal bearings. Eng Mech 14:69–80
3. Martinez E, Nossov VR, Gómez-Mancilla JC (2004) Validation and analysis of dynamic operation of a rotor-bearing system with hybrid lubrication. J Brazilian Soc Mech Sci Eng 7:83–90

4. Garg HC, Sharda HB, Kummar V (2006) On the design and development of hybrid journal berings: a review. Tribotest 12:1–19. https://doi.org/10.1002/tt.1
5. Donald BE (2002) The rotor system model. Fundamentals of rotating machinery diagnostics, pp 499–515
6. Fedor JV (1961) Journal Bearings With Arbitrary Position of Source. J Basic Eng ASME 83:572–578. https://doi.org/10.1115/1.3662269
7. Lund JW (1979) Evaluation of stiffness and damping coefficients for fluid-film bearings. Shock Vib Digest 11(1):5–10. https://doi.org/10.1177/058310247901100103
8. Childs DW (1993) Rotordynamic introduction to hydrodynamic bearing and squeeze-film Dampers. Turbomachinery Rotordynamics: Phenomena, Modeling and Analysis, pp 132–223
9. Norton RL (2006) Hydrodynamic lubrication theory. Machine Design: An Integrated Approach, USA, pp 653–664
10. Michael MK, Richard BE (2008) Reynolds Equation and Applications. Applied tribology: bearing design and lubrication, USA, pp 143–171
11. Andras ZS (2005) Thick-Film Lubrication. Fluid Film Lubrication, pp 88–146
12. Chasalebris A, Sfyris D (2013) Evaluation of the finite journal bearing characteristics, using the exact analytical solution of the Reynolds equation. Tribology Int 57:216–234. https://doi.org/10.1016/j.triboint.2012.08.011
13. Woods CM, Brewe DE (1989) The solution of the elrod algorithm for a dynamically loaded journal bearing Using multigrid techniques. Trans ASME J Tribol 111:302–308. https://doi.org/10.1115/1.3261914
14. Goenka PK (1984) Dynamically loaded journal bearings: finite element method analysis. Trans ASME J Tribol 106:429–439. https://doi.org/10.1115/1.3260954
15. Gómez-Mancilla JC, Nossov VR, Silva-Navarro G (2005) Rotor bearing system stability performance comparing hybrid versus conventional bearings. Int J Rot Mach 16–22:2005. https://doi.org/10.1155/IJRM.2005.16
16. Booker JF (1965) Dynamically loaded journal bearings: mobility method of solution. ASME J Basic Eng 3:537–546. https://doi.org/10.1115/1.3650602
17. Childs D, Moes H, Leeuwen H (1976) Journal Bearing Impedance Description for Rotordynamic Applications. Trans ASME J Lub Tech 1–24 https://doi.org/10.1115/1.3453021
18. Moes H, Sikkes EG, Bosma R (1986) Mobility and impedance tensor methods for full and partial-arc journal bearings. Trans ASME J Tribol 108:612–620. https://doi.org/10.1115/1.3261282
19. Nicoletti R, Santos IF (2002) Linear and non-linear control techniques applied to actively lubricated journal bearings. J Sound Vib 260:927–947. https://doi.org/10.1016/S0022-460X(02)00951-3
20. Santos IF, Watanabe FY (2003) Feasibility of influencing the dynamic fluid film coefficients of a multirecess journal bearing by means of active hybrid lubrication. J Brazil Soc Mech Sci Eng 25:154–163. https://doi.org/10.1590/S1678-58782003000200007
21. Alejandro CV, Santos IF (2014) Tilting-pad journal bearings with active lubrication applied as calibrated shakers: theory and experiment. J Vib Accoust 136:1–11. https://doi.org/10.1115/1.4028452
22. Arias-Montiel M, Silva-Navarro G, Antonio-Garcia A (2014) Active vibration control in a rotor system by an active suspension with linear actuators. J Appl Res Technol 12:898–907. https://doi.org/10.1016/S1665-6423(14)70596-6
23. Jorge GS, Santos IF (2017) Active tilting-pad journal bearings supporting flexible rotors: Part I –The hybrid lubrication. Tribo Int 107:94–105. https://doi.org/10.1016/j.triboint.2016.11.018

24. Leonardo BSM, Diogo SA, Ricardo UM, Katia LC (2019) Reduction of rotor vibration amplitude using PID tuning methods. IFToMM 2018. Mechanisms and Machine Science, vol 62, pp 74–88 https://doi.org/10.1007/978-3-319-99270-9_6
25. Gómez-Mancilla JC, Meda CJA, Torres CSG, Nossov VR (2005) Model derivation and numerical simulation for a pivoted rigid rotor-bearing system. Key Eng Mater 15:89–94. https://doi.org/10.4028/www.scientific.net/AMM.15.89 Trans Tech Publications ltd
26. Nosov VR, Gómez-Mancilla JC, Rivera-Arreola DE, Ramírez VI (2010) Stationary position for hybrid journal bearing with horizontal Point injection ports. Key Eng Mater 449:29–36. https://doi.org/10.4028/www.scientific.net/KEM.449.29 Trans Tech Publications ltd
27. Nosov VR, Rivera-Arreola DE, Gómez-Mancilla JC (2013) Stability threshold velocity for a hybrid short journal bearing with the injection ports in the symmetric points. J Sci Res Rep 2:74–92. https://doi.org/10.9734/JSRR/2013/2142
28. Nosov VR, Gómez-Mancilla JC (2010) On the appearance of lubricant film rupture in cylindrical journal bearings. Tribology Trans 47:233–238. https://doi.org/10.1080/05698190490439120
29. Nosov VR, Gómez-Mancilla JC, Meda CJA (2011) Occurrence of hydrodynamic cavitation. Water Sci Technol 64:595–601. https://doi.org/10.2166/wst.2011.608
30. Ocvirk FV (1952) Short-bearing approximation for full journal bearings. NASA Tech Rep Serv (NTRS) 2808:1–61
31. Stachowiak GW, Andrew AW (1993) In hydrodynamic lubrication. Engineering tribology, pp 121–206
32. Kanwal RP (1998) The Dirac delta function and delta sequences. Generalized Functions: Theory and Applications, pp 1–14
33. George A, Hans W, Frank HE (2012) Dirac delta function. Mathematical Methods for Physicists, pp 83–95
34. Lyapunov AM (1992) The general problem of the stability of motion, pp 19–247
35. Afanasev VN, Kolmanovskii VB, Nosov VR (1995) Mathematical theory of control system design. 3:–651
36. Jean-Jacques E, Slotine E, Weiping L (1991) Applied nonlinear control, pp 1–433
37. Khalil HK (1996) Lyapunov Stability. Nonlinear Systems, 2nd edn, pp 120–128
38. Tammi k (2007) Active Control of Radial Rotor Vibrations: Identification, Feedback, Feedforward, and Repetitive Control Methods. Doctoral thesis: Department of Automation and Systems Technology at the Helsinki University of Technology, Espoo, Finland, pp 1–165
39. Ziegler JG, Nichols NB (1943) Optimum settings for automatic controllers. Trans ASME 65:433–444
40. Parinya A, Zongli L (2013) An experimental study on PID tuning methods for active magnetic bearing systems. Int J Adv Mech syst 5(2):146–154. https://doi.org/10.1504/IJAMECHS.2013.055991
41. SEPILAB Instituto Politécnico Nacional, Laboratorio de vibraciones y rotodinámica. Ciudad de México, México

Transient Vibrations Suppression in Parametrically Excited Resonators

Miguel Ramírez-Barrios[1(✉)], Fadi Dohnal[2], and Joaquín Collado[3]

[1] Professional Interdisciplinary Unit of Biotechnology, National Polytechnic Institute, Mexico City, Mexico
mlramirezb@ipn.mx

[2] Division for Mechatronics Lienz, UMIT - Private University for Health Sciences, Medical Informatics and Technology, Lienz, Austria
fadi.dohnal@umit.at

[3] Automatic Control Department, Cinvestav, Mexico City, Mexico
jcollado@ctrl.cinvestav.mx

Abstract. Micromechanical resonators with a high-quality factor have been studied in recent years due to their excellent performance and potential applications. Some of these micromechanical resonators describe a long settling time, which can be disabled for fast switching applications. A novel design reduces the settling time by coupling two micromechanical resonators with parametric excitation; the parametric excitation tuned at the anti-resonant frequency increases the damping into the system; thus, the responses vibrations are attenuated, and the settling time is minimized. The scheme can be classified as open-loop control and belongs to the vibrational control field. Through a mathematical approach, this article describes this phenomenon in detail; by the averaging technique, the parametric excitation effects are analyzed. Moreover, we give new analytic conditions and formulas to minimize the settling time; these formulas might be helpful for future experimentation. Finally, using the current analysis, this contribution proposes a methodology to suppress the transient vibrations in this class of systems.

Keywords: Parametric excitation · Vibrational control · Micromechanical resonators

1 Introduction

Due to their high-quality factor ($Q = \omega_i/\gamma_i$) and accurate resonance spectra, high-Q mechanical resonators have been studied and used in many applications, such as filters, high precision sensors, amplifiers, and logic circuits [10]. However,

This project was partially funded by the Austrian IWB-project:LaZu-COLD_ IWB_ TIROL_POST_005 (Campus Technik Lienz).

E. E. Hernandez et al. (Eds.): LASIRS 2019, MMS 86, pp. 193–205, 2020.
https://doi.org/10.1007/978-3-030-45402-9_19

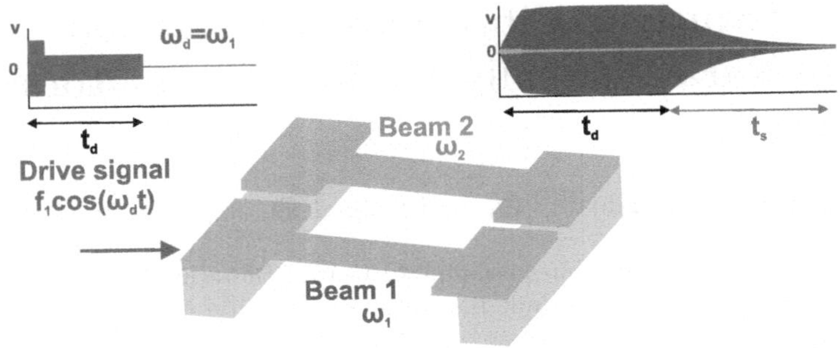

Fig. 1. Weakly coupled resonators scheme. The time response of beam one describes a long settling time t_s.

high-Q resonators have a long settling time, i.e., its transition time to steady-state is slow, which is challenging for rapid repetitive operations in this class of devices. Recently, an experimental implementation demonstrated for the settling time reduction using a piezoelectrically induced parametric in paired micromechanical resonators [10,11]. These paired micromechanical resonators can be represented in a scheme, as it is shown in Fig. 1, Fig. 2 and Fig. 3. The experimentation, reported in [10,11], proves that a periodic signal adequately tuned, can be used as an open-loop control to minimize the response's vibrations in one mode of the system. This experiment describes how to use the parametric excitation to fast transfer energy between the two beams of the system. With this idea, the parametric excitation is activating during a short time window, called a *pulse*, to transfer the response's energy in the first mode into the second mode. Thus, the second mode dissipates the energy of the system, wherewith, the transient vibrations on the first mode are suppressed; therefore, the settling time in the first mode is minimized. The above experiment can be classified as a particular case of the **anti-resonance** phenomenon [5]. In the current contribution, we explain in a mathematical point of view these experiments; then, we provide an analytical formula to compute the length of the *pulse.*

The dynamics of motion of two micromechanical resonators coupled by parametric excitation can be expressed by two coupled Mathieu equations: [11]:

$$\begin{aligned}\ddot{x}_1 + \gamma_1\dot{x}_1 + \omega_1^2 x_1 + \beta\cos(\omega_p t)\, x_2 &= f_1\cos(\omega_d t)\\ \ddot{x}_2 + \gamma_2\dot{x}_2 + \omega_2^2 x_2 + \beta\cos(\omega_p t)\, x_1 &= f_2\cos(\omega_d t)\end{aligned} \qquad (1)$$

where γ_i is the constant damping factor, β is the amplitude of the parametric excitation signal $p(t) = \beta cos(\omega_p t)$, and $f_i cos(\omega_d t)$ is the drive signal (or forced signal). Many publications deal with single or coupled differential equations with periodic coefficients, where it is possible to find analytic and numerical methods to describe the dynamics behavior. The main focus of these works is the destabilizing effect of parametric excitation [2,3,19]. Other publications [1,16] deal with the stabilizing effect of the parametric excitation.

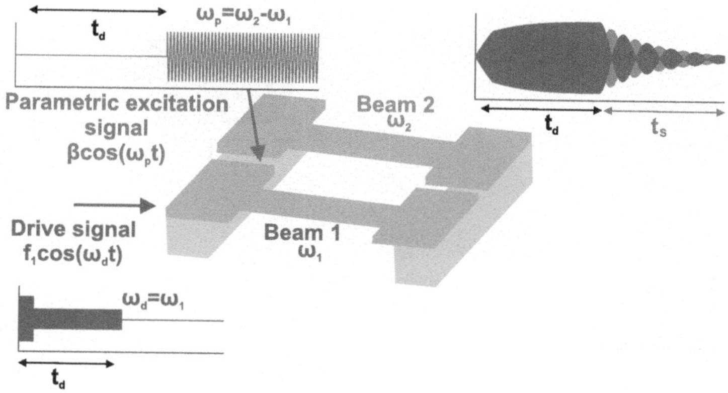

Fig. 2. Strongly coupled resonators scheme, the coupling term is the parametric excitation tuned at the anti-resonance frequency $\omega_p = \omega_2 - \omega_1$.

On the other hand, the suppression of vibrations by parametric excitation [5,7,12,13] arises from the experiments reported in [17]. This experiment establishes that in the system of Eq. (1), the parametric excitation effect increases the damping in the system when the parametric excitation is tuned at the specific frequencies: $\omega_p = \frac{|\omega_2 - \omega_1|}{n}$ with $n \in \mathbb{N}^+$, these are called as **anti-resonance** frequencies. By singular perturbations and averaging techniques the treatment [6] studies and describes the **anti-resonance** phenomena in detail.

Since the linear time-varying system, Eq. (1), cannot be solved analytically, asymptotic methods become powerful analytical tools. Well-known techniques such as the Poincaré Lindstedt method, harmonic balance, averaging method, multiple scales method, and successive approximations assume that β is a small parameter; thus the above techniques provide valid approximations on the time-scale $1/\beta$. Mathematicians, physicists, and engineers have used the techniques listed above because they allow computing analytic formulas that describe the system dynamics [8,9]. A numerical method based on the Lyapunov-Floquet theory can be found in [3,14,15], which allows arbitrary values of β. Therefore, to describe the experiments presented in [11], this article deals with the averaging method developed in [6,8]. We analyzed the dynamics behavior when the system is parametrically excited at anti-resonance frequency $\omega_p = \omega_2 - \omega_1$. Averaging the Eq. (1), the linear time-variant system is transformed into an approximated linear time-invariant represented by the amplitude-phase variables. With this theory, we obtain an analytic condition to guarantee the transfer of the energy between the two modes of the system. Then we provide a formula to compute the length of the *pulse* of parametric excitation. This analysis establishes an algorithm to suppress transient vibrations on micromechanical resonators.

The paper is organized as follows: The next section establishes the system description and the problem to solve. Then, Sect. 3 briefly describes the averaging method for the Eq. (1). The following section analyzes the effect of the

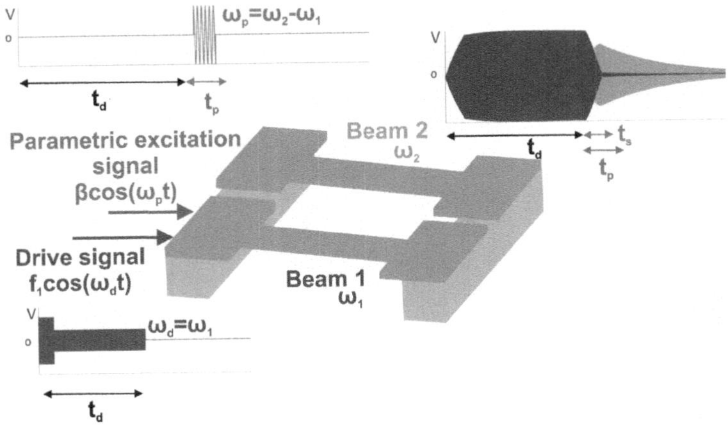

Fig. 3. Strongly coupled resonators scheme, the coupling term is a pulse of parametric excitation. The settling time of beam one t_s is minimized.

anti-parametric resonance and gives the formula to compute the length of the *pulse* which guarantees the suppression of vibration in the first beam of the system. In the last section, we reproduce the experiments of [11] to confirm the successful prediction of the formulas.

2 Problem Statement

The Fig. 1 depicts a schematic representation of two coupled microelectromechanical resonators, where it is assumed that each beam resonates at ω_1 and ω_2 respectively. Same Fig. 1 shows the time response of each beam, defined x_1 and x_2 in Eq. (1). Because $\omega_d = \omega_1$, the beam one resonates[1] with drive signal $f_1 cos(\omega_d t)$ during the time t_d. The free response of the system starts when $f_1 = 0$, this reveals settling time t_s in the first beam. Thus, this class of devices demands the waiting for at least t_s to initialize again, which disabling them to quickly operations.

If the two mechanical resonators are coupled by parametric excitation properly tuned, the vibration energy in mode one (beam 1) can be quickly transferred to the second mode (beam 2), as shown in the Fig. 2. Here, the envelope amplitudes of the responses describe an anti-phase behavior. Then, if the parametric excitation is turning off at the time t_p, the vibration energy in the first mode is fast transferred to the second mode during t_p, after the time t_p the second mode dissipates the energy vibration of the system, leading to the significant reduction of the settling time t_s in the first mode, see the Fig. 3. We named

[1] This fact can be understood, when the micromechanical resonator is used as a sensor.

the parametric excitation signal active in the time t_p as a *pulse* of parametric excitation. Namely, with a *pulse* of parametric excitation, the transient vibrations of the beam one are suppressed. The present contribution fully describes this phenomenon in a mathematical sense, which leads to obtaining a formula to compute t_p.

3 Transformation and Averaging

The experiments with new models of high-Q micromechanical resonators [11] can be dynamically described by the system Eq. (1). For the sake of simplicity, we applied the averaging technique to the homogeneous part of the system only. Then, the system can be rewritten as:

$$\ddot{x} + \Gamma\dot{x} + (\Omega + Bp(t))\, x = \mathbf{0}, \tag{2}$$

with $x = \begin{bmatrix} x_1 \; x_2 \end{bmatrix}^T$, where $p(t) = \beta \cos(\omega_p t)$ is the parametric excitation signal, being the matrices

$$\Gamma = \begin{bmatrix} \gamma_1 & 0 \\ 0 & \gamma_2 \end{bmatrix}, \; \Omega = \begin{bmatrix} \omega_1^2 & 0 \\ 0 & \omega_2^2 \end{bmatrix}, \; B = \begin{bmatrix} b_{11} & b_{12} \\ b_{21} & b_{22} \end{bmatrix}.$$

Rewritten in the comprehensive index notation (Einstein summation, repeated indices are implicitly summed over):

$$\ddot{x}_i + \omega_i^2 x_i = -\beta \left\{ \gamma_i \dot{x}_j - b_{ij} x_j \cos(\omega_p t) \right\} \tag{3}$$

with $i, j = 1, 2$ and applying the time transformation

$$\tau = \omega_p t \tag{4}$$

to normalize the frequency, the Eq. (3) turns to

$$x_i'' + \nu_i^2 x_i = -\frac{\beta}{\omega_p^2} \left\{ \omega_p \gamma_i x_j' + b_{ij} x_j \cos(\tau) \right\} \tag{5}$$

where $\frac{d}{d\tau}() = ()'$ and $\nu_i = \frac{\omega_i}{\omega_p}$. To study the behavior of (5) one assumes that the solutions can be written into the amplitude $A_i(\tau)$ and phase $\phi_i(\tau)$ terms [18]:

$$x_i(\tau) = A_i(\tau)\, c_i + \phi_i(\tau)\, s_i \tag{6}$$

and

$$x_i'(\tau) = -A_i(\tau)\, \nu_i s_i + \phi_i(\tau)\, \nu_i c_i \tag{7}$$

to simplify the process. we omitted the time dependence $A = A(\tau)$, $\phi = \phi(\tau)$ and with the abbreviations $s_i = \sin(\nu_i \tau)$, $c_i = \cos(\nu_i \tau)$, we performed the coordinate transformation (6)–(7) into (5):

$$A_i' c_i + \phi_i' s_i = 0 \tag{8}$$

by differentiating (7) the next relation holds

$$x_i'' = -A_i'\nu_i s_i - A_i\nu_i^2 c_i + \phi_i'\nu_i c_i - \phi_i\nu_i^2 s_i$$

and substituting into (5) we obtain

$$-A_i'\nu_i s_i + \phi_i'\nu_i c_i = \frac{\beta}{\omega_p^2}\tilde{F}_i \tag{9}$$

where $\tilde{F}_i = \tilde{F}_i\left(A_1, A_2, \phi_1, \phi_2, \tau\right)$ is given by:

$$\tilde{F}_i = -\left\{\omega_p \sum_j \gamma_i \left[\left(-A_j\nu_j\right)s_j + \phi_j\nu_j c_j\right] + \sum_j b_{ij}\left(A_j c_j + \phi_j s_j\right)\cos\left(\tau\right)\right\}$$

if we define the state vector $u = \left[A_1\; A_2\; \phi_1\; \phi_2\right]^T$ and insert the relation (8) in (9) the equations of motion finally become:

$$\begin{aligned} A_i' &= -\tfrac{\beta}{\omega_p\omega_i}\tilde{F}_i\left(u,t\right)s_i = \beta\tilde{G}_i\left(u,\tau\right) \\ \phi_i' &= \tfrac{\beta}{\omega_p\omega_i}\tilde{F}_i\left(u,t\right)c_i = \beta\tilde{H}_i\left(u,\tau\right) \end{aligned} \tag{10}$$

with $i = 1, 2$. The functions $\tilde{G}_i\left(u,\tau\right)$ and $\tilde{H}_i\left(u,\tau\right)$ on the right-hand side of these system of equations are not periodic, they are quasi-periodic. However, they can be split into a finite sum of different periodic terms:

$$\tilde{G}_i\left(u,\tau\right) = \sum_{k=1}^{N} \tilde{G}_{i,k}\left(u,\tau\right) \text{ and } \tilde{H}_i\left(u,\tau\right) = \sum_{k=1}^{N} \tilde{H}_{i,k}\left(u,\tau\right)$$

where N is fixed, $\tilde{G}_{i,k}\left(u,\tau\right)$ and $\tilde{H}_{i,k}\left(u,\tau\right)$ are T_k periodic functions in τ. In this case, the *averaging in the general case* method can be applied [4,18]:

$$\begin{aligned} \hat{A}_i' &= \beta \sum_{k=1}^{N} \tfrac{1}{T_k}\int_0^{T_k} \tilde{G}_{i,k}\left(u,\tau\right)d\tau \\ \hat{\phi}_i' &= \beta \sum_{k=1}^{N} \tfrac{1}{T_k}\int_0^{T_k} \tilde{H}_{i,k}\left(u,\tau\right)d\tau \end{aligned} \tag{11}$$

where $(\hat{\ })$ indicates time-averages variables. Hence, for averaging the system, the periods of the right hand sides of Eq. (10) must be determined. Using the trigonometric identities, products of the trigonometric terms on the right hand sides of Eq. (10) can be rewritten as a sum of basic trigonometric periodic terms. Thus, the integrals in Eq. (11) over a basic trigonometric terms yield always zero, with the exception of the frequencies $\omega_p = \frac{2\omega_i}{n}$ and $\omega_p = \frac{|\omega_2 \pm \omega_1|}{n}$ with $n \in \mathbb{N}^+$, namely **resonant frequencies** [18,19].

4 Parametric Anti-resonance Frequency

Averaging the Eq. (10) at the anti-resonance frequency: $\omega_p = \omega_2 - \omega_1$ we obtain the averaged system:

$$\hat{u}' = \mathbf{A}\hat{u} \tag{12}$$

with

$$\mathbf{A} = \frac{\beta}{\omega_p}\begin{bmatrix} -\frac{1}{2}\gamma_1 & 0 & 0 & \frac{1}{4\omega_1}b_{12} \\ 0 & -\frac{1}{2}\gamma_1 & -\frac{1}{4\omega_1}b_{12} & 0 \\ 0 & \frac{1}{4\omega_2}b_{21} & -\frac{1}{2}\gamma_2 & 0 \\ -\frac{1}{4\omega_2}b_{21} & 0 & 0 & -\frac{1}{2}\gamma_2 \end{bmatrix}$$

a linear time invariant system, where the difference between the original $u = \left[A_1\, A_2\, \phi_1\, \phi_2\right]^T$ and the averaged solutions $\hat{u} = \left[\hat{A}_1\, \hat{A}_2\, \hat{\phi}_1\, \hat{\phi}_2\right]^T$ is close to β order, $\hat{u}(\tau) - u(\tau) = \mathcal{O}(\beta)$ on the time scale $1/\beta$ [18].

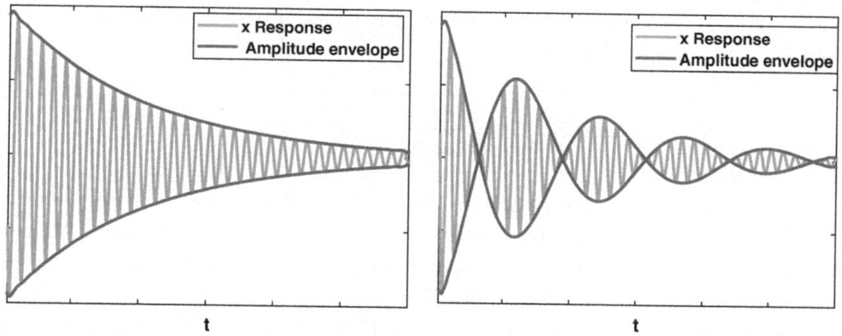

(a) $(\gamma_1 - \gamma_2)^2 > \frac{b_{12}b_{21}}{\omega_1\omega_2} > 0$ (b) $\frac{b_{12}b_{21}}{\omega_1\omega_2} > (\gamma_1 - \gamma_2)^2 \geq 0$

Fig. 4. Amplitude envelope conditions. a) Exponential envelope time-response b) Beating envelope time-response.

The motion system (12) does not depend of terms b_{11} and b_{22}, i.e., the coupling terms b_{12}, b_{21} fully describe the effect of the parametric excitation at $\omega_p = \omega_2 - \omega_1$. Then computing the eigenvalues of the matrix $\mathbf{A}$, we obtain

$$\lambda_{12} = -\frac{\beta}{4\omega_p}\left((\gamma_1 + \gamma_2) \pm \sqrt{(\gamma_1 - \gamma_2)^2 - \frac{b_{12}b_{21}}{\omega_1\omega_2}}\right) \tag{13}$$

(the matrix has two repeated eigenvalues: $\lambda_1 = \lambda_3$ and $\lambda_2 = \lambda_4$ given by relation (13)). The transformation of the system to the phase amplitude coordinates provides the 'slow motion' which is represented by the amplitude envelope time evolution response, see Fig. 4. Then the response envelope can be described by the eigenvalues λ_i of the matrix $\mathbf{A}$. If the condition

$$(\gamma_1 - \gamma_2)^2 > \frac{b_{12}b_{21}}{\omega_1\omega_2} \geq 0 \tag{14}$$

is satisfied the eigenvalues are real. Therefore, the envelope decreases or increase exponentially as is shown in the Fig. 4(a).

Alternatively, when the matrix $\mathbf{A}$ has complex eigenvalues the parametric excitation term introduces a modulation in the slow motion, thus if the inequality

$$\frac{b_{12}b_{21}}{\omega_1\omega_2} > (\gamma_1 - \gamma_2)^2 \geq 0 \tag{15}$$

is fulfilled the parametric excitation leads to a modulated responses signals, sometimes called as a 'beating signal'. The imaginary part of λ_i can be interpreted as: the **frequency of the slow motion**. Thus, the frequency of the envelope amplitudes is given by:

$$\hat{\omega}_A = \frac{\beta}{4\omega_p}\sqrt{\frac{b_{12}b_{21}}{\omega_1\omega_2} - (\gamma_1 - \gamma_2)^2} \tag{16}$$

thus, the envelope amplitude responses of Eq. (12) describe a 'beating' behavior, see the Fig. 4. Furthermore, there are two amplitude responses, since there are two frequencies $\pm\hat{\omega}_A$ the two envelope responses are in anti-phase when $\omega = \omega_2 - \omega_1$ and Eq. (15) is fulfilled. Therefore, the envelope amplitude in the time response in Eq. (1) can be approximated with the averaged system in Eq. (12).

Rescaling the time $\tau \rightarrow t$ by the Eq. (4) the period T_A of envelope amplitude of the responses in Eq. (1), can be computed as:

$$T_A = \frac{8\pi}{\beta\sqrt{\frac{b_{12}b_{21}}{\omega_1\omega_2} - (\gamma_1 - \gamma_2)^2}} \tag{17}$$

5 Simulations and Results

Table 1. Simulations parameters obtained from the experiments in [11].

Natural frequencies	**Parametric Anti-Resonance**
$\omega_1 = 291.51 \times 2\pi \times 10^3$ rad/s $\omega_2 = 291.95 \times 2\pi \times 10^3$ rad/s	$\omega_p = \omega_2 - \omega_1 = 440 \times 2\pi$ rad/s $\beta = 1$ V
Coupling Coefficients	**Dissipation Coefficients**
$b = b_{12} = b_{21} = 1.2 \times 10^9$ and $b_{11} = b_{22} = 0$	$\gamma_1 = \gamma_2 = 39$
Drive signal	**Quality factor**
$\omega_d = \omega_1,\ f_1 = f_2 = \begin{cases} 0.9\text{ V} & 0 \leq t \leq 14\text{ ms} \\ 0.4\text{ V} & 14\text{ ms} < t \leq 0.1\text{ s} \\ 0 & t > 0.1\text{ s} \end{cases}$	$Q = \frac{\omega_1}{\gamma_1} = 46964$

Simulating the system (1) for the experiment parameters in [11], which are shown in the Table. 1. The Fig. 5, Fig. 6 and Fig. 7 show the expected behavior in the averaged method. The three figures depict the time responses for beam one and two, in blue and red color, respectively. The drive signal is active during the

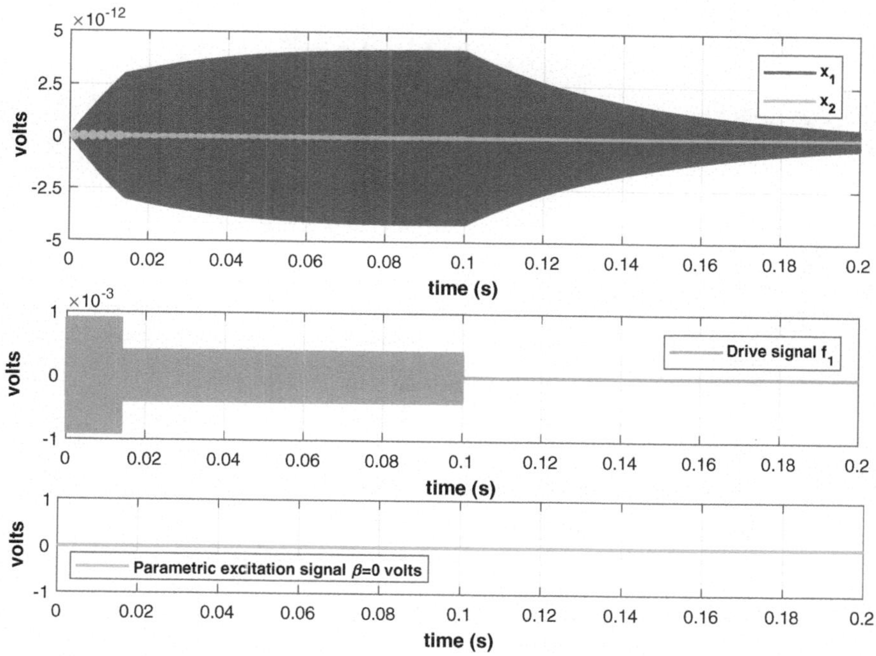

Fig. 5. Simulation for parameters in Table 1. No parametric excitation. The time response in beam one exhibits a long settling time.

time $t_d = 0.1$ s, in green color. And the parametric excitation signal in brown color. The Fig. 5 shows the natural response of the system without parametric excitation, after the time $t_p = 0.1$, the time response of beam one exhibits a long settling time approximately $t_s = 0.1$ s.

Since the mathematical model proposed in [11] considers: $b_{11} = b_{22} = 0$, $b = b_{12} = b_{21}$ and $\gamma = \gamma_1 = \gamma_2$, the condition (15) is always fulfilled and the envelope amplitude period is given by the formula (17), which is simplified to

$$T_A = \frac{8\pi\sqrt{\omega_1\omega_2}}{\beta b}. \tag{18}$$

Then, in Fig. 6 and also in Fig. 8 can be observed the expected anti-phase behavior in the envelope amplitude time-responses, then using the Eq. (18) the period of the envelope amplitude responses is $T_A = 64$ ms. It can be observed that the maximum energy transfer between the two modes occurs at the time $\frac{T_A}{4} = 16$ ms. At this time, the x_1 state achieves zero, x_2 has a maximum amplitude and vice versa. Accordingly, this fact suggests switch-off the parametric

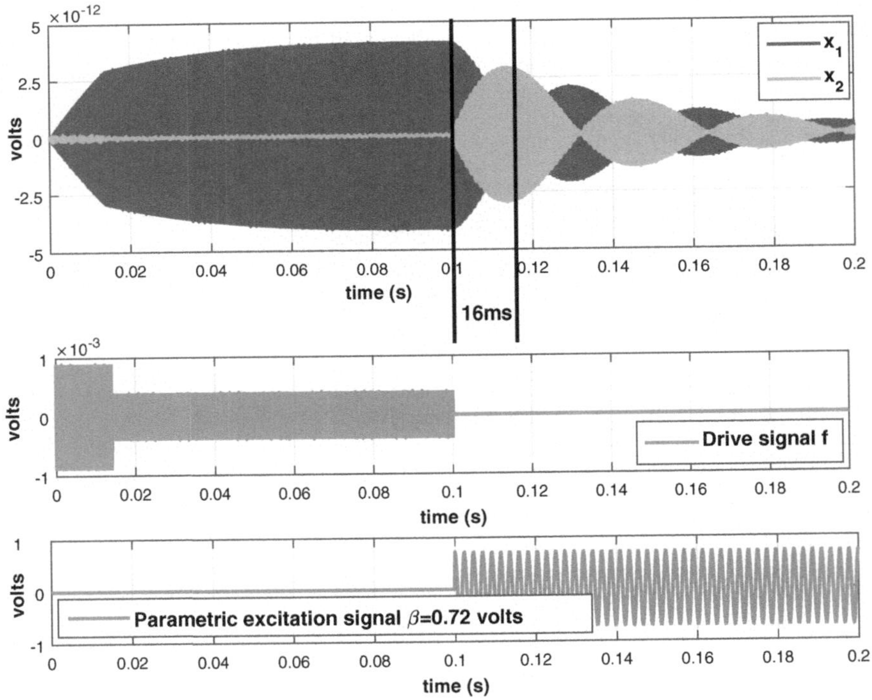

Fig. 6. Simulation for parameters in Table 1. Parametric excitation signal tuned at $\omega_p = \omega_2 - \omega_1$. The maximum transfer of energy occurs at the time $\frac{T_A}{4} = 16$ ms.

excitation at $\frac{T_A}{4}$. Therefore, the time length of the *pulse* of parametric excitation must be $t_p = \frac{T_A}{4}$ hence:

$$t_p = \frac{2\pi\sqrt{\omega_1\omega_2}}{\beta b} \tag{19}$$

In this case, ($\gamma = \gamma_2 = \gamma_1$) the slow-motion frequency, does not depend on γ, the damping term, γ is related to exponential behavior only. Thus, the length of the *pulse* t_p does not depend on γ either. From the formula (19) it can be concluded that if the amplitude of parametric excitation increases, t_p is reduced. Therefore, to minimize the settling time in the first mode, the parameter βb must be as large as possible. Thus, Fig. 7 and Fig. 8 show the time-response when the system is coupled with a *pulse* of parametric excitation. Here, the settling time of beam one was reduced from $t_s = 0.1$s to $t_s = 16$ ms, which means that the settling time was minimized by 84%.

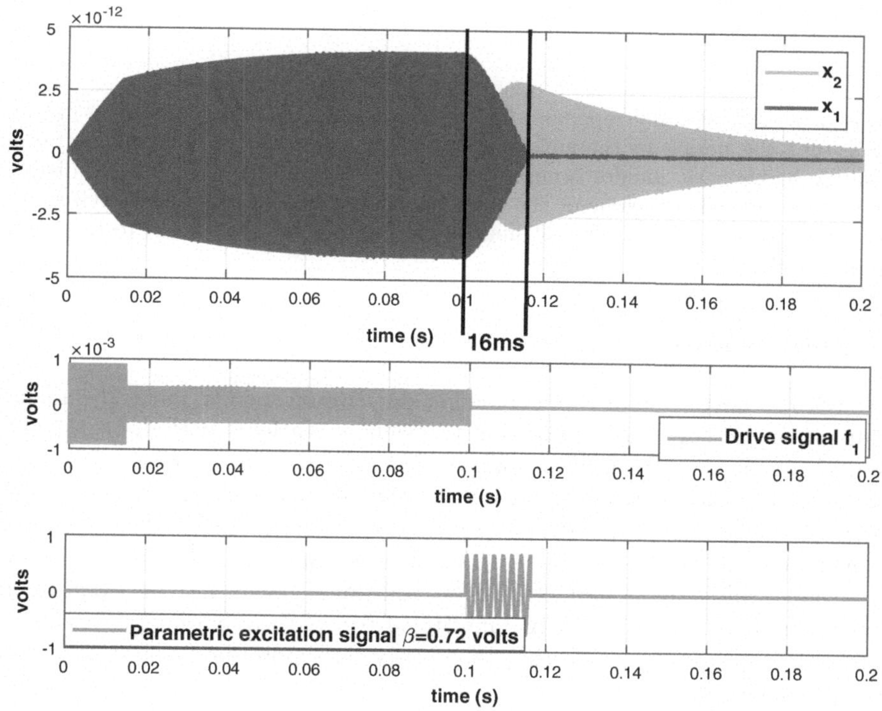

Fig. 7. Simulation for parameters in Table 1. Parametric excitation signal tuned at $\omega_p = \omega_2 - \omega_1$ with the length of the *pulse* $t_p = 16$ ms. The settling time of beam one is minimized to $t_s = 16$ ms approximately.

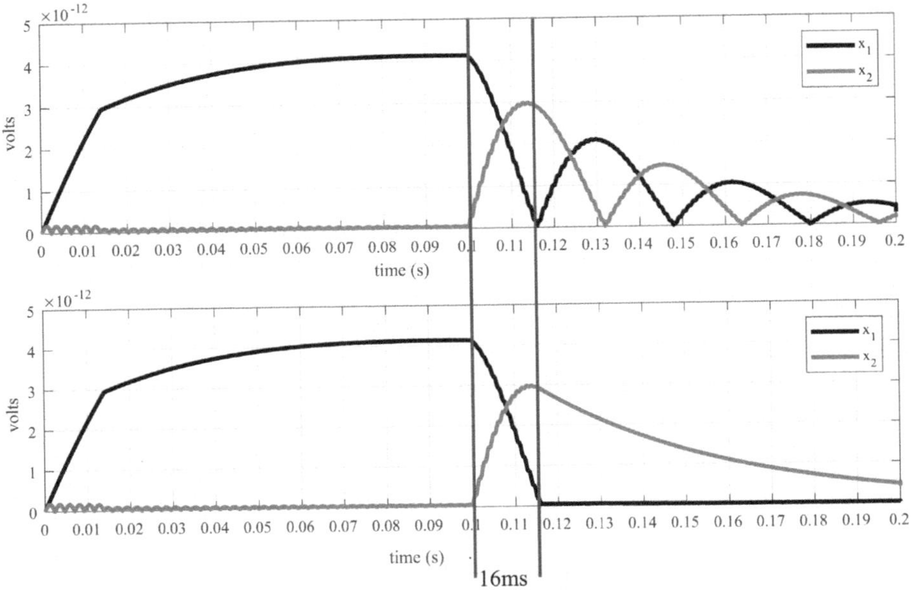

Fig. 8. Envelope of time-responses, for the parameters in Table 1. The top diagram shows, that the period of the envelope amplitude is T_A and the maximum transfer of energy between the modes occurs at the time $\frac{T_A}{4} = 16$ ms. The inferior diagram shows the amplitude envelope of the time responses when the system is excited by a *pulse* of length $t_p = 16$ ms. The settling time of beam one is minimized to $t_s = 16$ ms approximately.

6 Conclusions

Through the anti-resonance phenomenon description, a new method and procedure were presented to suppressing transient vibrations in High-Q micromechanical resonators. We obtain analytic formulas to tune properly a pulse of parametric excitation signal to quickly transfer the response energy between the modes of the system, herewith the responses vibrations in the first mode are suppressed significantly, i.e., the settling time is minimized. The micromechanical schemes allow for manufacturing an additional beam, i.e., an extra mode. More research is needed to clarify if multiple frequencies of parametric excitation signals are capable of transferring the energy into more modes and improving the current results.

References

1. Blekhman I (2004) Selected Topics in Vibrational Mechanics, vol 11. World Scientific, Singapore
2. Champneys A (2009) Dynamics of parametric excitation. In: Meyers R (ed) Encyclopedia of Complexity and Systems Science. Springer, New York, pp 2323–2345. https://doi.org/10.1007/978-3-642-27737-5

3. Collado J (2018) Hill equation: from 1 to 2 degrees of freedom. In: Clempner J, Yu W (eds) New Perspectives and Applications of Modern Control Theory. Springer, Cham, pp 43–71
4. Dohnal F (2008) General parametric stiffness excitation-anti-resonance frequency and symmetry. Acta Mechanica 196(1–2):15–31
5. Dohnal F (2008) Damping by parametric stiffness excitation: resonance and anti-resonance. J. Vibr. Control 14(5):669–688
6. Dohnal F (2009) Damping of Mechanical Vibrations by Parametric Excitation: Parametric Resonance and Anti Excitation. Sudwestdeutscher Verlag fur Hochschulschriften
7. Dohnal F, Ecker H, Springer H (2008) Enhanced damping of a cantilever beam by axial parametric excitation. Archive Appl. Mechan. 78(12):935–947
8. Dohnal F, Verhulst F (2008) Averaging in vibration suppression by parametric stiffness excitation. Nonlinear Dyn. 54(3):231–248
9. Nayfeh AH (2008) Perturbation Methods. Wiley, Hoboken
10. Okamoto H, Gourgout A, Chang CY, Onomitsu K, Mahboob I, Chang EY, Yamaguchi H (2013) Coherent phonon manipulation in coupled mechanical resonators. Nat. Phys. 9(8):480
11. Okamoto H, Mahboob I, Onomitsu K, Yamaguchi H (2014) Rapid switching in high-Q mechanical resonators. Appl. Phys. Lett. 105(8):083114
12. Ramírez M (2017) Attenuation of vibrations by parametric excitation (In Spanish). Ph.D. thesis, Cinvestav
13. Ramírez M, Collado J (2016) Attenuation vibration by parametric excitation using the meissner equation. In: 2016 13th International Conference on Electrical Engineering, Computing Science and Automatic Control (CCE), pp. 1–6. IEEE
14. Ramírez M, Collado J, Dohnal F (2019) Coupled mathieu equations: γ-hamiltonian and μ-symplectic. In: Awrejcewicz J (ed) Dynamical Systems Theory. IntechOpen
15. Ramírez Barrios M, Collado J, Dohnal F (2020) Stability of coupled and damped mathieu equations utilizing symplectic properties. In: Lacarbonara W, Balachandran B, Ma J, Tenreiro Machado J, Stepan G (eds) Nonlinear Dynamics of Structures, Systems and Devices. Springer, Cham. https://doi.org/10.1007/978-3-030-34713-0_14
16. Thomsen JJ (2013) Vibrations and Stability: Advanced Theory, Analysis, and Tools. Springer, Heidelberg
17. Tondl A (1998) To the problem of quenching self-excited vibrations. Acta technica ČSAV 43(1):109–116
18. Verhulst F (2006) Nonlinear Differential Equations and Dynamical Systems. Springer, Heidelberg. https://doi.org/10.1007/978-3-642-61453-8
19. Yakubovich V, Starzhinskii V (1975) Linear Differential Equations With Periodic Coefficients, vol 1,2. Wiley, Hoboken

Robust State Estimation for Underactuated Systems Using Sliding Modes and Attractive Ellipsoid Method

Bonifacio Sanchez[1(✉)], Patricio Ordaz[2], Ernesto Flores[3], and Alexander Poznyak[1]

[1] CINVESTAV-IPN, 07360 Mexico City, Mexico
{bsanchez,apoznyak}@ctrl.cinvestav.mx

[2] CITIS-UAEH, 42184 Pachuca Hidalgo, Mexico
jesus_ordaz@uaeh.edu.mx

[3] ESTi-UAEH, 43800 Tizayuca Hidalgo, Mexico
efloresg@uaeh.edu.mx

Abstract. The control of underactuated systems is a challenging and interesting topic in applied engineering mainly due to their high-range of applicability. This condition arises because they have less independent control inputs than configuration variables. In real-time experimental environments, the presence of uncertainties, disturbances and the unavailability of the full state vector complicate the control objectives. Furthermore, when the number of passive joints is greater than the number of active ones, the stabilizability radius is too small, and the control techniques based on the linear approximation are not robust enough to perform stabilization in practice. Under these conditions, the control problem when the full state-vector is not available becomes challenging. State-estimation may lead to a well-know peaking phenomenon. In this paper, the Attractive Ellipsoid Method and Sliding Modes are used to estimate the unavailable states. These give the possibility to reach a stability-zone in finite time, while uncertain and the peaking phenomenon effects are reduced, in order to provide sufficient conditions for its applicability in observer-based control. Experimental results are presented for the electromechanical triple-link inverted pendulum.

Keywords: Underactuated · Observer · Robust · Attractive Ellipsoid Method · Sliding Mode

1 Introduction

The control design for underactuated systems is still an active field in engineering, mainly due to its high-range of applicability in complex systems. Walking robots, mobile and aerial vehicles, elastic and flexible robots, pendular-type systems, *Human-Robot interaction* systems, mobile, aerial, water and underwater

E. E. Hernandez et al. (Eds.): LASIRS 2019, MMS 86, pp. 206–217, 2020.
https://doi.org/10.1007/978-3-030-45402-9_20

robots, are common examples [1–8]. This class of systems has more configuration variables than independent control inputs and put many challenging situations together. They are highly nonlinear systems and result to be controllable/stabilizable only in some regions of the state-space. For this reason, only stabilization tasks are considered,such as practical stability [10, 11]. This problem has been solved using many different techniques, but when the order of the system increases, the problem may become too difficult in practice [9, 16, 17]. Furthermore, on a real-time experimental environment, model and parameters uncertainties, matched and unmatched disturbances, the unavailability of some of the internal variables, or the drawbacks of approximating them using traditional techniques such as filters, differentiators and dirty derivative [12, 13] may appear. In this sense, the estimator must respond faster than the natural frequency of the plant, because, once the systems is initialized in some vicinity of the considered solution, system may leave the attractive set before the control acts, and the system may become unstabilizable. The problem turns to be more difficult when uncertainties and disturbance effects are latent. This require of robust estimation techniques, which leads to the *peaking phenomenon* [20]. High undershoots/undershots on the estimated signals occur and lead to instability when applied to an observer-based controller. A robust technique which allows to deal with uncertainties/disturbances and this phenomenon at the same time is thus required. The Attractive Ellipsoid Method is a robust technique which concludes with Ultimately Uniformly Bounded stability [10, 14]. As mentioned in [10, 11, 15], this type of stability establishes that system's trajectories reach an *invariant set* in finite time, if they start from an interior point of a bounded set. In this work, some robust Luemberguer-type observer structures are presented for pendular-type underactuated mechanical systems driven by *Permanent Magnet DC* (**PMDC**) motors on a real-time experimental environment. The **AEM** and *Sliding Modes* (**SM**) techniques are used for this aim. Results are presented for the *Triple-Link Inverted Pendulum* (**TLIP**) system. It is important to remark that this work has been partially submitted to the *IEEE Transactions on Industrial Electronics* journal recently. See the reference [18] for details. Thus, the main contribution of the work are some techniques to perform state-observation for electromechanical underactuated systems driven by **PMDC** motors when disturbances and uncertain effects are latent using the **AEM** and **SM** robust techniques, while a reduction of the peaking phenomenon is achieved, in order to provide the conditions for the observer-based stabilization in unstable configurations under experimental environments. The following section presents the problem statement in a formal manner, and some of the most important theoretic preliminaries The state-space description of underactuated mechanical systems driven by **PMDC** motors is presented in the *quasi-linear* format [9, 14, 18]. The Sect. 3 presents the propositions referred to the observers design. Some remarks about its applicability are also presented. Finally, Sect. 4 contain the experimental results for the electromechanical **TLIP** system, and last section describe the concluding remarks.

2 Preliminaries and Problem Statement

Consider an $n-$DoF Lagrangian system, coupled to the model of a set of m independent Permanent Magnet DC (**PMDC**) motors driving its joints directly:

$$\begin{aligned}\mathbf{D}(q)\ddot{q} + \mathbf{C}(q,\dot{q})\dot{q} + G(q) &= \mathbf{W}\mathbf{K}_\tau I_a - \tau_L, \\ \mathbf{L_a}\dot{I}_a + \mathbf{Ra}I_a + \mathbf{K_e}\mathbf{W}^\intercal\dot{q} &= v_a, \ \|\tau_L\| \leq \delta_0,\end{aligned} \tag{1}$$

where $q \in \mathbb{R}^n$ is the generalised coordinates vector, $I_a \in \mathbb{R}^m$ the vector of armature's currents, $v_a \in \mathbb{R}^m$ the armature's voltages, $0 < \mathbf{D}(q) \in \mathbb{R}^{n\times n}$ the inertia matrix, $\mathbf{C}(q,\dot{q}) \in \mathbb{R}^{n\times n}$ the centrifugal and Coriolis forces matrix and $G(q) \in \mathbb{R}^n$ the gravitational torques and forces vector. Furthermore, $\mathbf{L_a} \in \mathbb{R}^{m\times m}$ is a diagonal matrix containing the armature's inductances of the actuators, $\mathbf{R_a} \in \mathbb{R}^{m\times m}$ the armature's resistances $\mathbf{K_e} \in \mathbb{R}^{m\times m}$ and $\mathbf{K}_\tau \in \mathbb{R}^{m\times m}$ the back-emf and torque constants, respectively. The following properties are fulfilled for the mechanical part, as exposed in [4,19]:

1. *Inertia matrix is bounded in the following form:*

$$\lambda_m \mathbf{I_n} \leq \mathbf{D}(q) \leq \lambda_M \mathbf{I_n}, \ 0 < \lambda_m < \lambda_M, \ \lambda_m, \lambda_M \in \mathbb{R}. \tag{2}$$

2. *Consider two vectors γ, $\chi \in \mathbb{R}^p$. The Coriolis and centripetal forces vector is admitted to be radially unbounded with respect to χ as:*

$$\|\mathbf{C}(q,\gamma)\chi\| \leq \delta_C \|\gamma\|\|\chi\|, \ 0 < \delta_C, \ \delta_C \in \mathbb{R}. \tag{3}$$

3. *The gravitational torques and forces vector is bounded as*

$$\|G(q)\| \leq \delta_G, \ 0 < \delta_G, \ \delta_G \in \mathbb{R}. \tag{4}$$

Let select the change of variables as $x_1 = I_a$, $x_2 = q$, $x_3 = \dot{q}$, such that $x = [x_1^\intercal, x_2^\intercal, x_3^\intercal]^\intercal \in \mathcal{D} \subseteq \mathbb{R}^r$, $r = 2n + m$ is the state vector and $u_t = v_a \in \mathcal{U} \subseteq \mathbb{R}^m$ the control input. The nonlinear control affine representation results in the form $\dot{x}_t = f(x_t,t) + \mathbf{B}u_t + \vartheta(x_t,t)$ arises. Consider however the following remark about the state-space quasi-linear representation.

Remark 1 (***The quasi-linear representation***). *The linear approximation around the equilibrium solution $x_{eq} \in \mathcal{D}$ is found for the nominal system, i.e., when $\tau_L = 0$, using Taylor's series expansion as $\mathbf{A} = \left.\frac{\partial f(x_t)}{\partial x_t}\right|_{x_t = x_{eq}} \in \mathbb{R}^{r\times r}$. The so called quasi-linear format is obtained as*

$$\dot{x}_t = \mathbf{A}x_t + \mathbf{B}u_t + \xi(x_t,t), \ x(0) = x_0. \tag{5}$$

*For the particular case of electromechanical systems driven by **PMDC** motor:*

$$\begin{aligned}\mathbf{A} &= \begin{bmatrix} -\mathbf{L_a}^{-1}\mathbf{R_a} & \mathbf{0_{m\times n}} & -\mathbf{L_a}^{-1}\mathbf{K_e}\mathbf{W}^\intercal \\ \mathbf{0_{n\times m}} & \mathbf{0_{n\times n}} & \mathbf{I_n} \\ \mathbf{A_{31}} & \mathbf{A_{32}} & \mathbf{A_{33}} \end{bmatrix}, \ \mathbf{B} = \begin{bmatrix} \mathbf{L_a}^{-1} \\ \mathbf{0_{n\times m}} \\ \mathbf{0_{n\times m}} \end{bmatrix}, \ \xi(x_t,t) = \begin{bmatrix} 0 \\ 0 \\ o(x_t) + \mathbf{D}^{-1}(x_2)\tau_L(t) \end{bmatrix}, \\ o(\cdot) &= \mathbf{D}^{-1}(\cdot)\left\{\mathbf{W}\mathbf{K}_\tau x_1 - \mathbf{C}(\cdot)x_3 - G(\cdot)\right\} - \left[\mathbf{A_{31}}, \mathbf{A_{32}}, \mathbf{A_{33}}\right] x_t,\end{aligned} \tag{6}$$

where $o(x_t)$ is given and contains the nonlinear terms of the nominal model, and $\mathbf{A_{31}} \in \mathbb{R}^{n\times m}$, $\mathbf{A_{32}} \in \mathbb{R}^{n\times n}$ and $\mathbf{A_{33}} \in \mathbb{R}^{n\times m}$ come from the linear approximation of the mechanical part.

For this representation, it results that:

Lemma 1 (***On the boundedness of*** $\xi(\cdot)$). *Let define* $\mathbf{A_3} = [\mathbf{A_{31}}, \mathbf{A_{33}}, \mathbf{A_{32}}]$. *The uncertainties, disturbances and nonlinear terms vector* $\xi(x_t,t) \in \mathbb{R}^r$ *is quasi-Lipschitz, i.e., it is bounded as*

$$\|\xi(x_t,t)\|^2 \leq \gamma_0 + \gamma_1 \|x_t\|^2, \; \gamma_0 = 2\left(\frac{\delta_0}{\lambda_m}\right)^2, \; \gamma_1 = 2\left(\frac{\delta_f \sqrt{r}}{\lambda_m} + \sqrt{\lambda_{max}(\mathbf{A_3^\intercal A_3})}\right)^2 \tag{7}$$

for the scalar constant $\max\{\max(k_{\tau_j}), \delta_G, \delta_c\} \leq \delta_f$.

Proof. Refer to [18].

Also consider the quasi-linear representation in the regular observable form:

Remark 2 (***The regular observable form***). *Consider the non-singular coordinate transformation*

$$z_t = \begin{bmatrix} z_1 \\ z_2 \end{bmatrix} = \mathbf{\Gamma} x_t = \begin{bmatrix} \mathbf{N_c^\intercal} \\ \mathcal{C} \end{bmatrix} x_t,$$

where $\mathbf{N_c} \in \mathbb{R}^{(r-p)\times r}$ *is the null space of the output matrix. The alternate representation is found to be* $\dot{z}_t = \bar{\mathbf{A}} z_t + \bar{\mathbf{B}} u_t + \bar{\xi}(z_t,t)$, *as*

$$\begin{bmatrix} \dot{z}_1 \\ \dot{z}_2 \end{bmatrix} = \begin{bmatrix} \bar{\mathbf{A}}_{11} & \bar{\mathbf{A}}_{12} \\ \bar{\mathbf{A}}_{21} & \bar{\mathbf{A}}_{22} \end{bmatrix} \begin{bmatrix} z_1 \\ z_2 \end{bmatrix} + \begin{bmatrix} \bar{\mathbf{B}}_1 \\ \bar{\mathbf{B}}_2 \end{bmatrix} u_t + \begin{bmatrix} \bar{\xi}_1 \\ \bar{\xi}_2 \end{bmatrix}, \quad y_t = z_2, \tag{8}$$

where $\bar{\mathbf{A}} = \mathbf{\Gamma A \Gamma^{-1}}$, $\bar{\mathbf{B}} = \mathbf{\Gamma B}$ *and* $\bar{\xi}(z_t,t) = \mathbf{\Gamma}\xi(x_t,t)$. *Henceforth,* $\bar{\mathbf{A}}_{11} \in \mathbb{R}^{(r-p)\times(r-p)}$, $\bar{\mathbf{A}}_{12} \in \mathbb{R}^{(r-p)\times p}$, $\bar{\mathbf{A}}_{21} \in \mathbb{R}^{p\times(r-p)}$, $\bar{\mathbf{A}}_{11} \in \mathbb{R}^{p\times p}$, $\bar{\mathbf{B}}_1 \in \mathbb{R}^{(r-p)\times m}$, $\bar{\mathbf{B}}_2 \in \mathbb{R}^{p\times m}$, $\bar{\xi}_1 \in \mathbb{R}^{r-p}$ *and* $\bar{\xi}_2 \in \mathbb{R}^p$. *Here* $z_1 \in \mathbb{R}^{r-p}$ *and* $z_2 \in \mathbb{R}^p$ *contain the unavailable and measurable state-variables, respectively.*

This lead to the following lemma.

Lemma 2 (***On the boundedness of*** $\bar{\xi}(\cdot)$). *The disturbances, uncertainties and nonlinear terms vector* $\xi(x_t,t)$ *is quasi-Lipschitz, as seen in Lemma 2.1. This condition may also be expressed in the new coordinates frame* z_t *as:*

$$\|\bar{\xi}(\cdot)\|^2 \leq \kappa_0 + \kappa_1 \|z_t\|^2, \; \kappa_0 = \eta_0 \lambda_{\max}^2(\mathbf{\Gamma}), \; \kappa_1 = \eta_1 \left(\frac{\lambda_{\max}(\mathbf{\Gamma})}{\lambda_{\min}(\mathbf{\Gamma})}\right)^2 \tag{9}$$

Proof. Refer to [18].

The problem is now ready to be formulated.

Consider the case of a $n-$***DoF*** *mechanical system, driven by a set of* $m-$*independent dc motors* (1), *in the quasi-linear format* (5)–(6), *or* (8). *The problem is to estimate the unavailable state sub-vector* z_1, *using full or reduced order observers, while disturbances and uncertain effects are latent, and providing a way for the reduction of the peaking phenomenon.*

Consider the following theoretical preliminaries.

Definition 1 (UUB stability *[10]*). *The nonlinear dynamical system $\dot{x}(t) = f(t, x(t))$, $x(t_0) = x_0$, $t \geq t_0$, is Ultimately Uniformly Bounded with bound b, if there exists $a > 0$ such that, for every $\delta \in (0, a)$ there exists $T = T(\delta, b) > t_0$ such that :*

$$\|x(t_0)\| \leq \delta, \implies \|x(t)\| \leq b, \ \forall t \geq t_0 + T. \tag{10}$$

The following corollary [10], gives sufficient conditions to conclude with this.

Corollary 1. *Consider the nonlinear dynamical system $\dot{x}(t) = f(t, x(t))$, $x(t_0) = x_0$, $t \geq t_0$. Assume that there exist a proper function $V : \mathbb{R}^r \to \mathbb{R}$ (i.e., continuously differentiable, positive definite and radially unbounded, [14]), and class $\mathcal{K}$ functions (strictly increasing, [10]) $\alpha(\cdot)$ and $\beta(\cdot)$ such that:*

$$\alpha(\|x_t\|) \leq V(x) \leq \beta(\|x_t\|), x \in \mathcal{D}, \ V'(x)f(x_t) \leq 0, x \in \mathcal{D}, \ \|x\| \geq \mu \tag{11}$$

where $\mu > 0$ is such that $\mathcal{B}_{\alpha^{-1}(\eta)}(0) \subset \mathcal{D}$ for some $\eta \geq \beta(\mu)$. Then the nonlinear dynamical system is bounded. If last condition is now given as $V'(x)f(x_t) < 0, x \in \mathcal{D}$, $\|x\| > \mu$, where $\mu > 0$ is such that $\mathcal{B}_{\alpha^{-1}(\eta)}(0) \subset \mathcal{D}$ for some $\eta > \beta(\mu)$. Then the nonlinear dynamical system is ultimately bounded with bound $\varepsilon = \alpha^{-1}(\eta)$. Furthermore, $\lim\ \sup_{t \to \infty} \|x(t)\| \leq \alpha^{-1}(\beta(\mu))$.

The **AEM** is closely related to this definition (see [14] for details). In order to elucidate this method, consider that the nonlinear system is given in the quasi-linear format (5). The **AEM** consists in designing a state-feedback controller $u_t = \mathbf{K}x_t$, $\mathbf{K} \in \mathbb{R}^{m \times r}$, such that **UUB** stability is concluded, for the Lyapunov Candidate Function in the quadratic classic form $V(x_t) = x_t^{\mathsf{T}}\mathbf{P}x_t$, $0 < \mathbf{P} = \mathbf{P}^{\mathsf{T}} \in \mathbb{R}^{r \times r}$. In this sense, the ultimate attractive set has the particular form

$$\mathcal{E} := \{x \in \mathbb{R}^r : x^{\mathsf{T}}\mathbf{P_{att}}x_t \leq 1\}, \ \mathbf{P_{att}} = \mu\mathbf{P}.$$

where $0 < \mathbf{P_{att}} \in \mathbb{R}^{r \times r}$ is the *shape matrix* of the ellipsoid. The length of the semi-axes of the ellipsoid are backward proportional to the eigenvalues of $\mathbf{P_{att}}$. For this reason, the **AEM** is related to the solution of the following *constrained minimization problem.*

$$\textit{minimize } \tfrac{1}{\mu}\mathbf{tr}\left(\mathbf{P}^{-1}\right) \textit{ by } \{\mathbf{P}, \mathbf{K}\} \in \mathbf{\Gamma}, \ \textit{subject to } 0 < \mathbf{P} = \mathbf{P}^{\mathsf{T}}, 0 < \mu. \tag{12}$$

where $\mathbf{\Gamma}$ is the set of restrictions that determine a class of admissible matrices $\mathbf{P}$, $\mathbf{K}$.

3 Robust Observation

The following propositions are done to solve the stated problem. The first one uses a full-order observer, with **AEM**.

Proposition 1. *For the nonlinear system in the form (5)–(6), if there exist a solution set $(\mathbf{P}, \mathbf{L}|\alpha, \varepsilon)$ for the matrix inequality:*

$$\mathbf{\Theta} = \begin{bmatrix} \mathbf{PLC} + \mathbf{C}^{\mathsf{T}}\mathbf{L}^{\mathsf{T}}\mathbf{P} - \mathbf{PA} - \mathbf{A}^{\mathsf{T}}\mathbf{P} - \alpha\mathbf{P} - \varepsilon\eta_1\mathbf{I_r} \ , & -\mathbf{P} \\ -\mathbf{P} \ , & \varepsilon\mathbf{I_r} \end{bmatrix} > 0, \tag{13}$$

with matrices $0 < \mathbf{P} = \mathbf{P}^\intercal \in \mathbb{R}^{r \times r}$, $\mathbf{L} \in \mathbb{R}^{r \times p}$, *and the scalars* $0 < \alpha, \varepsilon$, *then the estimated state* $\hat{x}_t$, *from*

$$\begin{aligned} \dot{\hat{x}}_t &= \mathbf{A}\hat{x}_t + \mathbf{B}u_t + \mathbf{L}(y_t - \hat{y}_t), \ \hat{x}(0) = \hat{x}_0, \\ \hat{y}_t &= \boldsymbol{\mathcal{C}}\hat{x}_t, \end{aligned} \tag{14}$$

converges to a invariant set about the real one x_t, *given by the ellipsoid:*

$$\mathcal{E} = \{\tilde{x}_t \in \mathbb{R}^r \ : \ \tilde{x}_t^\intercal \mathbf{P}_{\mathbf{att}} \tilde{x}_t \leq 1\}, \ \mathbf{P}_{\mathbf{att}} = (\alpha/\beta)\mathbf{P},$$

and the so-called Ultimate Uniformly Bounded stability is concluded, with:

$$b = \sqrt{\tfrac{\beta}{\alpha}\lambda_{min}\left(\mathbf{P}^{-1}\right)}, \ T = \tfrac{1}{\alpha}\ln\left\{\tfrac{\alpha V(\tilde{x}_0) - \beta}{\alpha\kappa}\right\},$$

for small enough $0 < \kappa << 1$.

In the second one a reduced order observer was designed.

Proposition 2. *Consider the electromechanical system in the quasi linear format* (8)*. If there exists a solution set* $(\alpha, \varepsilon | \mathbf{P}, \mathbf{L})$ *for the matrix inequality:*

$$\boldsymbol{\Theta} = \begin{bmatrix} \left\{ \begin{matrix} \mathbf{P}\bar{\mathbf{A}}_{11} + \bar{\mathbf{A}}_{11}^\intercal \mathbf{P} - \mathbf{P}\mathbf{L}\bar{\mathbf{A}}_{21} \\ -\bar{\mathbf{A}}_{21}^\intercal \mathbf{L}^\intercal \mathbf{P} + \alpha\mathbf{P} + \varepsilon\eta_1 \mathbf{I}_{n-p} \end{matrix} \right\} & , \ \mathbf{P}\mathbf{N}_{\mathbf{c}}^\intercal - \mathbf{P}\mathbf{L}\boldsymbol{\mathcal{C}} \\ \mathbf{N}_{\mathbf{c}}\mathbf{P} - \boldsymbol{\mathcal{C}}^\intercal \mathbf{L}^\intercal \mathbf{P} & , \ -\varepsilon\mathbf{I}_{n-p} \end{bmatrix} < 0 \tag{15}$$

then, the unavailable state sub-vector $z_1 \in \mathbb{R}^{r-p}$ *is approximated by the solution of the Reduced Order Observer:*

$$\begin{aligned} &\dot{\varrho}_t = \boldsymbol{\Theta}_1 \varrho_t + \boldsymbol{\Theta}_2 y_t + \left(\bar{\mathbf{B}}_1 - \mathbf{L}\bar{\mathbf{B}}_2\right) u_t, \ \varrho(0) = \varrho_0, \\ &\hat{z}_1 = \varrho_t + \mathbf{L}y_t, \ \boldsymbol{\Theta}_1 = \bar{\mathbf{A}}_{11} - \mathbf{L}\bar{\mathbf{A}}_{21}, \ \boldsymbol{\Theta}_2 = \bar{\mathbf{A}}_{11}\mathbf{L} + \bar{\mathbf{A}}_{12} - \mathbf{L}\bar{\mathbf{A}}_{21}\mathbf{L} - \mathbf{L}\bar{\mathbf{A}}_{22} \end{aligned}$$

without need of reconstructing the available one $z_2 \in \mathbb{R}^p$*. Here,* $0 < \mathbf{P} \in \mathbb{R}^{(r-p)\times(r-p)}$ *and* $\mathbf{L} \in \mathbb{R}^{(r-p)\times p}$ *are constant matrices, while* $\alpha, \varepsilon \in \mathbb{R}$ *are positive scalars. Then, the estimated state* $\hat{z}_1$, *converges to the ellipsoid:*

$$\mathcal{E} = \left\{\tilde{z}_1 \in \mathbb{R}^{r-p} \ : \ \tilde{z}_1^\intercal \mathbf{P}_{\mathbf{att}} \tilde{z}_1 \leq 1\right\}, \ \mathbf{P}_{\mathbf{att}} = (\alpha/\beta)\mathbf{P}.$$

Under this conditions, the so-called Ultimate Uniformly Bounded stability is concluded, as in previous result.

The third approach joins AEM with Sliding Mode Observation.

Proposition 3. *Let suggest the state observer [18]:*

$$\begin{aligned} \dot{\hat{z}}_t &= \bar{\mathbf{A}}\hat{z}_t + \bar{\mathbf{B}}u_t + \mathbf{L}\phi(z_t), \ \hat{z}(0) = \hat{z}_0, \\ \hat{y}_t &= \bar{\boldsymbol{\mathcal{C}}}\hat{z}_t, \end{aligned} \tag{16}$$

where $\mathbf{L} \in \mathbb{R}^{r \times p}$ *is a constant matrix gain having the form* $\mathbf{L} = \left[\mathbf{L_1}^\intercal, \mathbf{I}_p\right]^\intercal$. *If the function* $\phi(\cdot)$ *has the following structure:*

$$\begin{aligned} &\phi(\sigma_t) = \bar{\mathbf{A}}_{22}\sigma_t + \rho_{\mathbf{0}} Sign(\sigma_t) + \rho_1, \\ &0 < \rho_{\mathbf{0}} \in \mathbb{R}^{p \times p}, \rho_1 = \eta + \delta_4\sqrt{\lambda_{\max}(\bar{\mathbf{A}}_{12}^\intercal \bar{\mathbf{A}}_{12})}, \end{aligned} \tag{17}$$

then, in the region $\|e_1\| \leq \delta_4$*, the estimated state variable* $\hat{z}_2$ *converges to real state variable* z_2 *in finite time, given as:*

$$t_r = \text{trace}(\rho_0)^{-1}\sqrt{2}V_1^{\frac{1}{2}}(\sigma(t_0)) + t_0, \tag{18}$$

where storage function $V_1(\sigma_t)$ *is given in the quadratic form* $V_1(\sigma_t) = \frac{1}{2}\sigma_t^{\intercal}\sigma_t$.

Now, consider the following storage function:

$$V_2(e_1) = e_1^{\intercal}\mathbf{P}e_1,\ 0 < \mathbf{P} \in \mathbb{R}^{(r-p)\times(r-p)}. \tag{19}$$

Proposition 4. *If for positive scalars* α *and* β*, there exists a solution of LMI:*

$$\boldsymbol{\Theta}=\begin{pmatrix} \mathbf{P}\mathcal{A}_{11}+\mathcal{A}_{11}^{\intercal}\mathbf{P}-\mathbf{Y}\mathcal{A}_{21}-\mathcal{A}_{21}^{\intercal}\mathbf{Y}^{\intercal}+\alpha\mathbf{P}, & \mathbf{P}, & -\mathbf{Y} \\ \mathbf{P} & -\varepsilon\mathbf{I}_p, & \mathbf{0} \\ -\mathbf{Y}^{\intercal} & \mathbf{0}, & -\varepsilon\mathbf{I}_p \end{pmatrix}<0,\ \mathbf{P}>0,\ \mathbf{Y}=\mathbf{P}\mathbf{L_1}, \tag{20}$$

is fulfilled, then for all $t \geq t_r$ *the storage function* $V_2(e_1)$ *satisfies:*

$$V_2(e_1(t))\leq\frac{\beta}{\alpha}+\left\{V_2(e_1(t_r))-\frac{\beta}{\alpha}\right\}e^{-\alpha(t-t_r)} \tag{21}$$

where $\beta = \varepsilon\eta$*, and the Ultimate Uniform Bounded Stability is concluded, for:*

$$b=\sqrt{\frac{\beta}{\alpha}\lambda_{\max}(\mathbf{P^{-1}})},\ T=\frac{1}{\alpha}\ln\left\{\frac{\alpha V_2(t_r)-\beta}{\alpha\zeta}\right\}+\frac{\sqrt{2}}{\text{trace}(\rho_0)}V_1^{\frac{1}{2}}(\sigma_0)+t_0, \tag{22}$$

for small-enough constant $\zeta \in \mathbb{R}^+$.

Next section presents experimental results for the **TLIP** system.

4 Experimental Results

Experimental results were obtained for the *Triple Link Inverted Pendulum* system of the *PendCon Company*, driven by a *Maxon DC motor 218009* at first joint. The instrumentation diagram is depicted in Fig. 1. The motor's driver *4-Q-DC Servo-control LSC 30/2* and *DAQ NI USB-6363* are used. System's numerical parameters are given in Table 1. Centre of masses were considered at the centre of the link on each case. The dynamic model considers the mass of the links and joints separately. With notation and parametric equations in Table 2, inertia matrix is given as:

$$\mathbf{D}(q)=\begin{bmatrix} \mathbf{D}_{11}, & \mathbf{D}_{12}, & m_3l_{c_3}(l_1C_{31}+l_2C_{321}) \\ \mathbf{D_{12}}, & I_{\omega_2}+(M_2+m_3)l_2^2, & m_3l_2l_{c_3}C_{321} \\ m_3l_{c_3}(l_1C_{31}+l_2C_{321}), & m_3l_2l_{c_3}C_{321}, & I_{\omega_3} \end{bmatrix}$$

where:

$$\mathbf{D}_{11}= I_{\omega_1}+J_{m_1}+(m_1+M_1+m_2+M_2+m_3)l_1^2+(M_2+m_3)l_2^2+2(M_2+m_3)l_1l_2C_2,$$
$$\mathbf{D}_{12}= m_2l_1l_{c_2}C_{21}+l_2(M_2+m_3)(l_2+l_1C_2).$$

Also, the centripetal-Coriolis forces matrix is found to be:

$$\mathbf{C}(q,\dot{q})=\begin{bmatrix} -2(M_2+m_3)l_1l_2S_2\dot{q}_2, & -\theta_1\dot{q}_2, & -(\theta_2+\theta_3)\dot{q}_3 \\ \theta_1\dot{q}_1, & 0, & -\theta_2\dot{q}_3 \\ (\theta_2+\theta_3)\dot{q}_1+\theta_2\dot{q}_2, & \theta_2(\dot{q}_1+\dot{q}_2), & 0 \end{bmatrix},$$

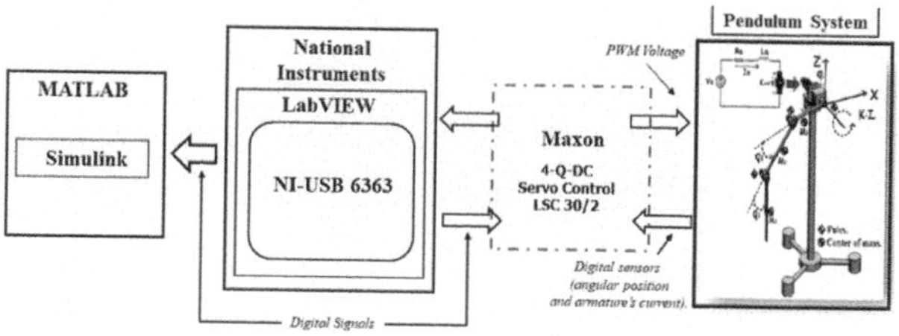

Fig. 1. Experimental system block diagram.

Table 1. Numerical parameters of the Triple-Link Inverted Pendulum.

Parameter	Value	Parameter	Value
Mass of the 1st link, m_1	$0.03\,[kg]$	Length of the third link, l_3	$0.29\,[m]$
Mass of the 2nd link, m_2	$0.035\,[kg]$	Armature's resistance, R_a	$2.44\,[\Omega]$
Mass of the 3d link, m_3	$0.035\,[kg]$	Armature's inductance, L_a	$612\,[\mu H]$
Mass of the 2nd joint, M_1	$0.1\,[kg]$	Torque constant, K_τ	$82.2\,\left[\frac{mNm}{A}\right]$
Mass of the 3d joint, M_2	$0.1\,[kg]$	Back-emf constant, K_e	$116\,\left[\frac{rpm}{V}\right]$
Length of the 1st link, l_1	$0.19\,[m]$	Inertia moment of the shaft, J_m	$116\,[gcm^2]$
Length of the 2nd link, l_2	$0.26\,[m]$		

and the gravitational torques and forces vector:

$$G(q)=\begin{bmatrix} -m_1gl_{c_1}S_1-M_1gl_1S_1-m_2g\theta_4-M_2g\theta_5-m_3g\theta_6 \\ -m_2gl_{c_2}S_{12}-M_2gl_2S_{12}-m_3g\left(l_2S_{12}+l_{c_3}S_{123}\right) \\ -m_3gl_{c_3}S_{123} \end{bmatrix}.$$

From [19], centripetal-Coriolis forces matrix may be decomposed into the following sub matrices:

$$\mathbf{C}_1=\begin{bmatrix} 0 & -(M_2+m_3)l_1l_2S_2 & 0 \\ -(M_2+m_3)l_1l_2S_2 & -\theta_1 & 0 \\ 0 & 0 & -\theta_2-\theta_3, \end{bmatrix},\mathbf{C}_2=\begin{bmatrix} \theta_1 & 0 & 0 \\ 0 & 0 & 0 \\ 0 & 0 & -\theta_2 \end{bmatrix},\ \mathbf{C}_3=\begin{bmatrix} \theta_2+\theta_3 & \theta_2 & 0 \\ \theta_2 & \theta_2 & 0 \\ 0 & 0 & 0 \end{bmatrix}$$

It follows that:

$$\delta_c=3^2\max\left\{(M_2+m_3)l_1l_2+m_2l_1l_{c_2},m_3l_{c_3}(l_1+l_2)\right\} \tag{23}$$

Table 2. Notation and parametric equations.

Notation	Parametric equations
$S_j=\sin(q_j)$, $C_j=\cos(q_j)$	$\theta_1=(m_2l_1l_{c_2}S_{21}+(M_2+m_3)\,l_1l_2S_2)$
$S_{jk}=\sin(q_j+q_k)$, $C_{jk}=\cos(q_j+q_k)$	$\theta_2=m_3l_2l_{c3}S_{321}$, $\theta_3=m_3l_1l_{c_3}S_{31}$
$S_{kj}=\sin(q_k-q_j)$, $C_{kj}=\sin(q_k-q_j)$	$\theta_4=l_1S_1+l_{c_2}S_{12}$, $\theta_5=l_1S_1+l_2S_{12}$
	$\theta_6=\theta_5+l_{c_3}S_{123}$

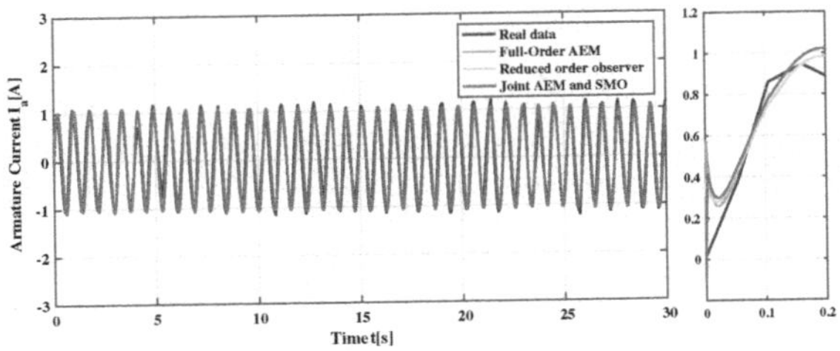

Fig. 2. Estimation of the armature's current.

On the other hand, with the Jacobian matrix of this vector respect to the generalized coordinate, it results evident that, with $\theta_7 = m_2 g l_{c_2} + M_2 g l_2$:

$$\delta_G = 3\theta_7 \max\{m_1 g l_{c_1} + (M_1 + m_2 + M_2) g l_1, m_3 g (l_2 + l_{c_3})\}, \tag{24}$$

Results are presented under the knowledge of all the links positions. See Figs. 2, 3, 4, 5, 6, 7 and 8. The gains are obtained using *Matlab 2015* and *cvx 2.1-toolbox*, in Table 3.

Table 3. Numerical gains for the proposed observers.

	α	β	**L**
Proposition 1	45	0.004	$\begin{bmatrix} -35.244 & -0.443 & -0.337 \\ 12.64 & -0.04 & 0.037 \\ -0.04 & 12.72 & -0.008 \\ 0.037 & -0.008 & 12.84 \\ 1109.5 & 22.06 & 4.55 \\ 21.87 & 1147.5 & -0.71 \\ 4.6 & -0.77 & 1158.5 \end{bmatrix} \times 10^3$
Proposition 2	45	0.004	$\begin{bmatrix} 2.00 & 0.034 & 0.0017 \\ 58.94 & 1.004 & 0.05 \\ 1.00 & 60.14 & 0.001 \\ 0.05 & 0.001 & 60.125 \end{bmatrix}$
Proposition 3	65	0.004	$\begin{bmatrix} 2.15 & 0.0034 & 0.0017 \\ 63.72 & 1.00 & 0.05 \\ 1.02 & 65.07 & 0.0008 \\ 0.05 & 0.0008 & 65.06 \end{bmatrix}$

The comparative survey is summarized in Table 4, where the parameters t_r and $|MEV|$ are the rise-time (in milliseconds) and the magnitude of the *mean-*

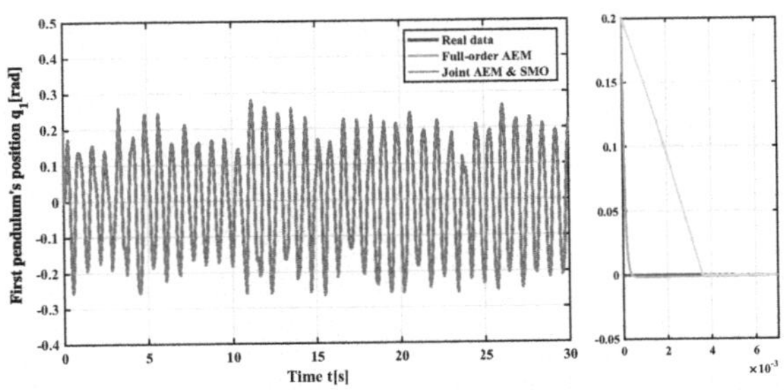

Fig. 3. Estimation of the first pendulum's position.

Table 4. Numerical results for the electromechanical **TLIP** system.

State-variable	$t_r[ms]$			$\lvert MEV \rvert \times 10^{-3}$		
$I_a(t)$	53.87	53.7	38.57	15	14.3	13.03
$q_1(t)$	0.173	—	2.85	0.008	—	0.007
$q_2(t)$	0.171	—	4.05	0.013	—	0.025
$q_3(t)$	0.169	—	2.84	0.008	—	0.008
$\dot{q}_1(t)$	24	34.27	32.6	11	12.57	20.7
$\dot{q}_2(t)$	25	42.34	39.8	1.33	4.12	11.64
$\dot{q}_3(t)$	23	38.71	35.04	5.4	3.98	13.56
	P1	**P2**	**P3**	**P1**	**P2**	**P3**

value (in the appropriate units) of the error's signal, respectively. The tags **P1**, **P2** and **P3** refer to the observers of propositions 1, 2 and 3 respectively.

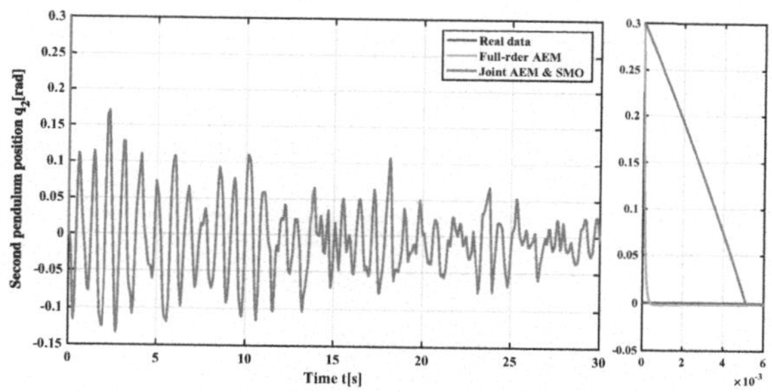

Fig. 4. Estimation of the second pendulum's position.

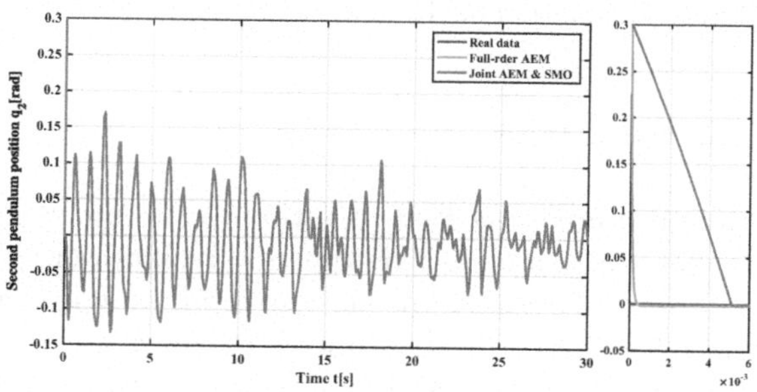

Fig. 5. Estimation of the third pendulum's position.

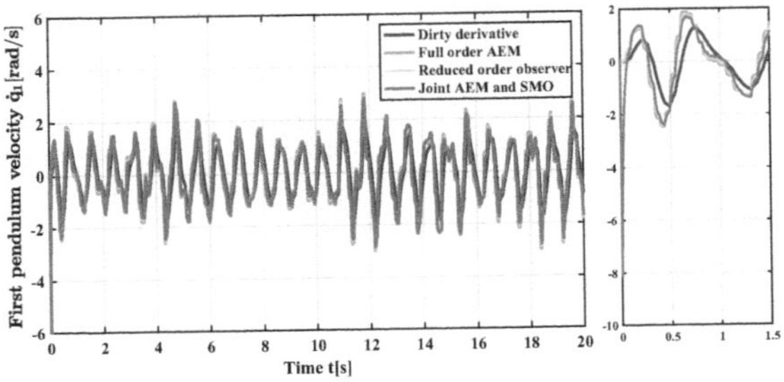

Fig. 6. Estimation of the first pendulum's velocity.

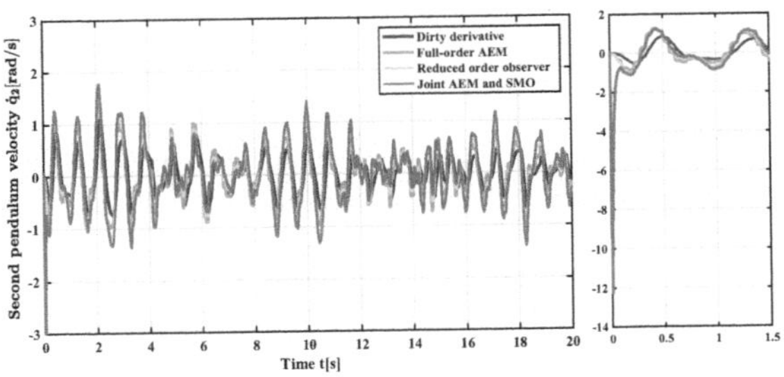

Fig. 7. Estimation of the second pendulum's velocity.

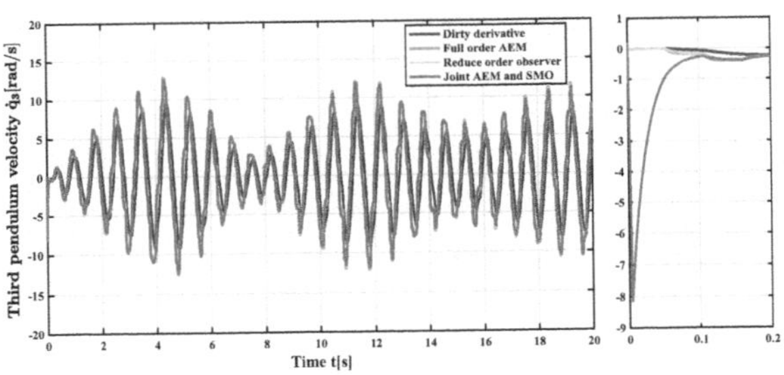

Fig. 8. Estimation of the third pendulum's velocity.

5 Concluding Remarks

Three robust techniques are presented to estimate the unavailable state variables for electromechanical systems. Experimental results were obtained for the **TLIP**. When the observed signals are used to achieve state-feedback control the problem becomes hard. For the experiments of this work, uncertainties were introduced in system's parameters and signals were estimated well. Finally it is worth to mention that, observers tuning have been done with care of the overshoot/undershoot effects in the transient response, in order to reduce the dangerous peaking phenomenon. In future, this techniques will be tested on real-time stabilization.

References

1. Olfati-Saber R (2002) Normal forms for underactuated mechanical systems with symmetry. IEEE Trans. Autom. Control 47(2):305–308
2. Xin X, Liu Y (2014) Control Design and Analysis for Underactuated Robotic Systems. Springer, Heidelberg
3. Fantoni I, Lozano R (2001) Non-linear Control for Underactuated Mechanical Systems. Springer, Heidelberg
4. Spong MW (1987) Modeling and control of elastic joint robots. J. Dyn. Syst. Meas. Control 109(4):310–318
5. Ortega R, Spong MW, Gómez-Estern F, Blankenstein G (2002) Stabilization of a class of underactuated mechanical systems via interconnection and damping assignment. IEEE Trans. Autom. Control 47(8):1218–1233
6. Birglen L, LalibertÃ T, Gosselin CM (2007) Underactuated Robotic Hands, vol 40. Springer, Heidelberg
7. Lefeber E, Pettersen KY, Nijmeijer H (2003) Tracking control of an underactuated ship. IEEE Trans. Control Syst. Technol. 11(1):52–61
8. Reyhanoglu M, van der Schaft A, McClamroch NH, Kolmanovsky I (1999) Dynamics and control of a class of underactuated mechanical systems. IEEE Trans. Autom. Control 44(9):1663–1671
9. Bonifacio SR, Patricio OO, Alexander PG (2018) Robust stabilising control for the electromechanical triple-link inverted pendulum system. IFAC-PapersOnLine 51(13):314–319
10. Haddad WM, Chellaboina V (2011) Nonlinear Dynamical Systems and Control: A Lyapunov-based Approach. Princeton University Press, Princeton
11. Khalil HK (2014) Nonlinear Control. Pearson Higher Education, London
12. Ramirez-Neria M, Sira-Ramirez H, Garrido-Moctezuma R, Luviano-Juarez A (2014) Linear active disturbance rejection control of underactuated systems: the case of the Furuta pendulum. ISA Trans. 53(4):920–928
13. Alonge F, D'ippolito F, Raimondi FM, December 2001 Trajectory tracking of underactuated underwater vehicles. In: Proceedings of the 40th IEEE Conference on Decision and Control (Cat. No. 01CH37228), vol. 5, pp. 4421-4426. IEEE
14. Poznyak A, Polyakov A, Azhmyakov V (2014) Attractive Ellipsoids in Robust Control. Birkhauser, Boston
15. Azhmyakov V, December 2011 On the geometric aspects of the invariant ellipsoid method: application to the robust control design. In: 2011 50th IEEE Conference on Decision and Control and European Control Conference, pp. 1353–1358. IEEE
16. Spong MW, September 1994 Partial feedback linearization of underactuated mechanical systems. In: Proceedings of IEEE/RSJ International Conference on Intelligent Robots and Systems (IROS 1994), vol. 1, pp. 314-321. IEEE
17. Lam S, Davison EJ, June 2006 The real stabilizability radius of the multi-link inverted pendulum. In: 2006 American Control Conference, p. 6. IEEE
18. Sanchez B, Cuvas C, Ordaz P, Santos-Sanchez OJ, Poznyak A (2019) Full-Order observer for a class of nonlinear systems with unmatched uncertainties: joint attractive ellipsoid and sliding mode concepts. IEEE Trans. Ind. Electron. 67:5677–5686
19. Kelly R, Davila VS, Perez JAL (2006) Control of Robot Manipulators in Joint Space. Springer, Heidelberg
20. Khalil HK (2017) High-Gain Observers in Nonlinear Feedback Control, Society for Industrial and Applied Mathematics (SIAM), Philadelphia

Modelling and Control of 5 DOF Serial Robot for Object Capturing and Manipulation Using a Semi-analytical Approach

Bolaños Adrian(✉), Sánchez Kevin, Gabino Juan, Gutiérrez-Carmona Irandi, and Martínez Manuel

Instituto Tecnológico de Monterrey Campus Tampico, Tamaulipas, Mexico
A01411983@itesm.mx, juan.diaz@tec.mx

Abstract. This paper details the kinematic analysis for a 5 degrees-of-freedom serial robot, the Scorbot-ER III. Based on the Denavit-Hartenberg convention is presented a semi-analytical relation between the end effector pose, and the joints angles. The computational time required to calculate the robot trajectories is reduced significantly, making this a suitable algorithm for real time applications; specially for those where the robot is intended to move in a continuously changing environment.

Keywords: Serial robot · Denavit-Hartenberg · Inverse kinematics · Forward kinematics · Trajectories

1 Introduction

The capturing of space objects is an area with room for improvement; two applications of this are the removal of space debris and on-orbit servicing (OOS). The former aims to remove defunct artificially created free-flying objects that threats any manned or unmanned spacecraft orbiting around Earth, while OOS missions are needed to repair and give maintenance. In most cases the robot must interact with non-cooperative objects, meaning that a constant analyzing of the motion of the environment is required in order to calculate dynamical trajectories and identifying grappling points in the bodies [1].

The unpredictability and inconvenience of non-cooperative targets has led scientists to propose different capturing methods using rigid or flexible connections [2]. One flexible capture mechanism consists of a spacecraft equipped with an ejector that applies acceleration to the masses on a net, afterwards the net expands and envelops the objective allowing a secure capture [3]. Harpoons are also flexible capture mechanism that instead of deploying a net, fires a harpoon that drills into the objective. Though both methods have been studied and simulated, the dynamics on them are quite complicated and the outcomes difficult to predict; furthermore, if the target is intended to be recovered without damage, these previous methods cannot be implemented.

As an alternative, a rigid capture mechanism is considered [2]. Due to their versatility, they are apt for capturing non-cooperative objects [4]. However, for the manipulator to

E. E. Hernandez et al. (Eds.): LASIRS 2019, MMS 86, pp. 218–227, 2020.
https://doi.org/10.1007/978-3-030-45402-9_21

be considered a suitable option, a strategy to efficiently construct trajectories in real time applications must be considered [5].

The problem of determining the actuators angle given the final position and orientation of the end-effector is known as the inverse kinematic problem. The inverse kinematic problem may be solved by numerical methods such as the Newton-Raphson method. Unfortunately, the time required to find a feasible solution is not known due to its dependency on arbitrarily set of initial conditions, and if a real-time trajectory is to be determined using this method, the correct operation of the robot may be compromised [6].

To circumvent this problem, artificial intelligence methods such as neural networks are combined with other numerical methods to have more efficient and faster calculations [6]. Although artificial intelligence methods are the right tools when the model is completely unknown or there are many uncertainties, one should resort to analytical results if available.

In this work, we show that using the Denavit-Hartenberg (D-H) convention [7, p. 68], a semi-analytical method can be easily found to solve the inverse kinematics problem in a fixed time. The semi-analytical nature of our solution comes from the Scorbot-ER III configuration, and the solution in a fixed time comes from the fact that a finite and known number of iterations is needed to find a feasible solution.

This paper is organized as follows: In Sect. 2 is formulated the Scorbot-ER III problem and defined the Denavit-Hartenberg matrices. In Sect. 3 the problem is solved using a semi-analytical approach. In Sect. 4 is shown the numerical validation. And finally, in Sect. 5 are shown the conclusions and future work is in Sect. 6.

2 Problem Formulation

The Scorbot-ER III robot is a versatile robot with 5 DOF and a working radius of 610 mm [8], designed to perform human tasks with basic actions that consist of gripper open-close, body, shoulder and elbow rotations, and wrist roll and pitch movements [4].

Capturing space debris requires that, given a position in space, our robot calculates in the smallest possible time a feasible trajectory to catch the flying object. Using numerical methods such as Newton-Raphson methods, although effective and well-studied, does not guarantee that the inverse kinematic problem will be solved fast enough. Hence, the design of a simple algorithm that guarantees fixed convergence time for the inverse kinematic problem is needed.

2.1 Denavit-Hartenberg Convention

The kinematic analysis of the Scorbot-ER III can be extremely complex, and the Denavit-Hartenberg convention was introduced to provide a systematic procedure for performing the analysis of the relationship between the individual joints of the robot manipulator and the position and orientation of end-effector, simplifying the analysis considerably.

In Fig. 1 is shown the Scorbot-ER III initial configuration, and the corresponding frames fixed to each of the articulations.

Fig. 1. Scorbot-ER III configuration and coordinate frames

$$
{}^{i-1}A_i = \begin{bmatrix} c_{\theta_i} & -c_{\alpha_i}s_{\theta_i} & s_{\alpha_i}s_{\theta_i} & a_i c_{\theta_i} \\ s_{\theta_i} & c_{\alpha_i}c_{\theta_i} & -s_{\alpha_i}c_{\theta_i} & a_i s_{\theta_i} \\ 0 & s_{\alpha_i} & c_{\alpha_i} & d_i \\ 0 & 0 & 0 & 1 \end{bmatrix} \tag{1}
$$

In the Denavit-Hartenberg convention [7, 9], a homogeneous transformation matrix ${}^{i-1}A_i \in \mathbb{R}^{4\times4}$ is used to describe the relation between the i-th and the $(i-1)$-th articulation,

where $c_{\xi_i} = cos(\xi_i)$, and $s_{\xi_i} = sin(\xi_i)$. Parameters α_i, θ_i, a_i, and d_i are described in Table 1 [10].

Table 1. D-H parameters

System parameters	
$\{X_i, Y_i, Z_i\}$	i-th coordinate frame
Q_i	Origin of the i-th coordinate frame
θ_i	**Joint angle**, The rotation angle of each actuator
a_i	**Link length**, distance from Z_{i-1} to Z_i with direction X_i^+
d_i	**Link offset**, distance from Q_{i-1} to X_i with direction Z_{i-1}^+
α_i	**Link twist**, rotation from Z_{i-1} to Z_i with respect to the X_i-axis

The homogenous matrix ${}^{i-1}A_i$ results from the multiplication of rotational and translational matrices that are found analyzing the robot configuration and the actuators positions, this is:

$$
{}^{i-1}A_i = R_{z,\theta_i} Trans_{z,d_i} Trans_{x,a_i} R_{x,\alpha_i} \tag{2}
$$

$$
R_{x,\alpha_i} = \begin{bmatrix} 1 & 0 & 0 & 0 \\ 0 & c_{\alpha_i} & -s_{\alpha_i} & 0 \\ 0 & s_{\alpha_i} & c_{\alpha_i} & 0 \\ 0 & 0 & 0 & 1 \end{bmatrix} \tag{3}
$$

$$Trans_{x,a_i} = \begin{bmatrix} 1 & 0 & 0 & a_i \\ 0 & 1 & 0 & 0 \\ 0 & 0 & 1 & 0 \\ 0 & 0 & 0 & 1 \end{bmatrix} \tag{4}$$

$$Trans_{z,d_i} = \begin{bmatrix} 1 & 0 & 0 & 0 \\ 0 & 1 & 0 & 0 \\ 0 & 0 & 1 & d_i \\ 0 & 0 & 0 & 1 \end{bmatrix} \tag{5}$$

$$R_{z,\theta_i} = \begin{bmatrix} c_{\theta_i} & -s_{\theta_i} & 0 & 0 \\ -s_{\theta_i} & c_{\theta_i} & 0 & 0 \\ 0 & 0 & 1 & 0 \\ 0 & 0 & 0 & 1 \end{bmatrix} \tag{6}$$

The Denavit-Hartenberg parameters for the Scorbot-ER III are shown in Table 2.

Table 2. D-H parameters for the Scorbot-ER III

Coordinate frame	a_i (cm)	d_i (cm)	α_i (degrees)	θ_i (degrees)
1	2.5	14	90	0
2	22	0	0	135
3	22	0	0	−135
4	0	0	90	0
5	0	14	0	0

2.2 Orientation and Position of End-Effector

The orientation of the end-effector can be described by defining the unit vectors of $\{X_5, Y_5, Z_5\}$ in terms of the inertial frame $\{X_0 Y_0 Z_0\}$ by the transformation matrix, ${}^0T_5 \in \mathbb{R}^{4\times4}$ (Fig. 2),

$${}^0T_5 = \begin{bmatrix} {}^0R_5 & \overline{p_5} \\ 0_{1\times3} & 1 \end{bmatrix} \tag{7}$$

where ${}^0R_5 \in \mathbb{R}^{3\times3}$, defines the orientation of the end-effector, this is

$$R_5^0 = \begin{bmatrix} c_\phi c_\gamma & c_\phi s_\alpha s_\gamma - s_\phi c_\alpha & s_\phi s_\alpha + c_\phi c_\alpha s_\gamma \\ s_\phi c_\gamma & c_\phi c_\alpha + s_\phi s_\alpha s_\gamma & s_\gamma c_\alpha s_\gamma - c_\phi s_\alpha \\ -s_\gamma & c_\gamma s_\alpha & c_\alpha c_\gamma \end{bmatrix} \tag{8}$$

Fig. 2. Unit vectors of end-effector in relation to the inertial frame

For simplicity, the elements of R_5^0 may be rewritten as

$$R_5^0 = \begin{bmatrix} n_x & o_x & a_x \\ n_y & o_y & a_y \\ n_z & o_z & a_z \end{bmatrix} = \begin{bmatrix} n & o & a \end{bmatrix} \tag{9}$$

where $n, o, a \in \mathbb{R}^3$ are defined by the direction cosines of the unit vectors $\{X_5, Y_5, Z_5\}$, in terms of the inertial frame $\{X_0 Y_0 Z_0\}$. Finally, the vector $\overline{p_5} \in \mathbb{R}^3$ defines the position of the end-effector, this is

$$\overline{p_5} = \begin{bmatrix} p_x \\ p_y \\ p_z \end{bmatrix} \tag{10}$$

Matrix ${}^0T_5$ can also be defined as the transformation of the end-effector pose through each coordinate frame using Eq. (1)

$${}^0T_5 = {}^0A_1 {}^1A_2 {}^2A_3 {}^3A_4 {}^4A_5 \tag{11}$$

3 Main Results

By equating Eq. (9) with Eq. (11),

$$\begin{bmatrix} {}^0R_5 & \overline{p_5} \\ 0_{1\times 3} & 1 \end{bmatrix} = {}^0A_1 {}^1A_2 {}^2A_3 {}^3A_4 {}^4A_5 \tag{12}$$

a set of nonlinear equations for the joints angles $\{\theta_1, \theta_2, \theta_3, \theta_4, \theta_5\}$ is found. In general, due to the complexity of the equations, numerical approximations are required to determine the angles that satisfy Eq. (12). However, in the next sub-section, it is shown that it may be found simple trigonometric relations to derive the θ_i angles.

3.1 Inverse Kinematics

In its construction, the Scorbot-ER III has 3 consecutive joints with parallel axes of rotation. This results in the manipulator having one redundant angle. To avoid this redundancy, θ_2 was left as a free variable calculated by cycling through every physically possible value and solving for each of the other angles using Eq. (12).

Pre-multiplying Eq. (12) by A_1^{-1}, the following trigonometric relations were found:

$$p_y c_{\theta_1} = p_y s_{\theta_1} \tag{13}$$

$$a_1 + a_3 c_{\{\theta_2+\theta_3\}} + a_2 c_{\theta_2} + d_5 \left(a_x c_{\theta_1} + a_y s_{\theta_1}\right) = p_x c_{\theta_1} + p_y s_{\theta_1} \tag{14}$$

$$d_1 + a_3 s_{\{\theta_2+\theta_3\}} + a_2 s_{\theta_2} = p_z - d_5 a_z \tag{15}$$

$$s_{\theta_5} + n_y c_{\theta_1} = n_x s_{\theta_1} \tag{16}$$

$$c_{\theta_5} + o_y c_{\theta_1} = o_x s_{\theta_1} \tag{17}$$

Pre-multiplying by $A_2^{-1} \cdot A_1^{-1}$ and post-multiplying by A_5^{-1} the following simultaneous equations were found:

$$s_{\{\theta_3+\theta_4\}} = a_z s_{\theta_2} + a_x c_{\theta_1} c_{\theta_2} + a_y s_{\theta_1} c_{\theta_2} \tag{18}$$

$$c_{\{\theta_3+\theta_4\}} = s_{\theta_2}\left(a_x c_{\theta_1} + a_y s_{\theta_1}\right) - a_z c_{\theta_2} \tag{19}$$

Using Eq. (13):

$$\theta_1 = atan2\left(p_y, p_x\right) \tag{20}$$

From Eq. (14), (15):

$$\begin{aligned}\theta_3 = atan2(p_z - d_5 a_z - d_1 - a_2 s_2, p_x c_{\theta_1} + p_y s_{\theta_1} - a_1 - a_2 c_{\theta_2} \\ -d_5(a_x c_{\theta_1} + a_y s_{\theta_1})) - \theta_2\end{aligned} \tag{21}$$

Using Eq. (16), (17):

$$\theta_5 = atan2\left(n_x s_{\theta_1} - n_y c_{\theta_1}, o_x s_{\theta_1} - o_y c_{\theta_1}\right) \tag{22}$$

With Eq. (18), (19):

$$\theta_4 = atan2\left(c_{\theta_2}\left(a_x c_{\theta_1} + a_y s_{\theta_1}\right) + a_z s_{\theta_2}, s_{\theta_2}\left(a_x c_{\theta_1} + a_y s_{\theta_1} - a_z c_{\theta_2}\right)\right) - \theta_3 \tag{23}$$

Where $atan2(y, x)$ returns the arc tangent of the two variable x and y, which is between $-\pi$ and π. Note that all angles are calculated using the inverse tangent, this is because inverse sine and inverse cosine are prone to get indeterminations more often due to their very limited domain and range.

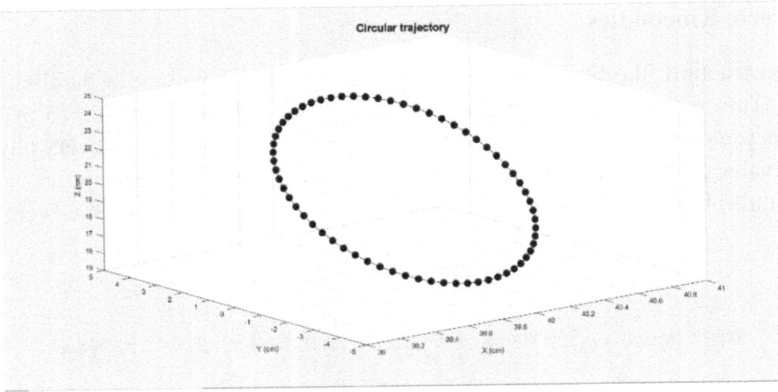

Fig. 3. Points defined to create a circular trajectory

3.2 Trajectories

Trajectories were constructed by a set of points defining the position and orientation of the final effector; this is, by construction the left-hand side of Eq. (12). Now, using Eqs. (20), (21), (22) and (23) the set of the joints angles may be determined. Thus, the Scorbot-ER III can move along a specific path maintaining the end effector on a variable or fixed orientation (Fig. 3).

4 Numerical Validation

To prove the correctness of the inverse kinematics equations, the error norm

$$\left\| \begin{bmatrix} {}^{0}R_5 & \overline{p_5} \\ 0_{1\times 3} & 1 \end{bmatrix} = -{}^{0}A_1{}^{1}A_2{}^{2}A_3{}^{3}A_4{}^{4}A_5 \right\|_2 \quad (24)$$

was observed for each point in the trajectory. The error norm was of the order of 10^{-2} units; this is, a highly accurate solution was found.

To evaluate the feasibility of the method based on its computational time, a circular trajectory with center {40, 0, 20} cm and a 5 cm radius was computed (Fig. 4). For this path, a total of 65 poses for the robot were calculated. The solution, shown in Table 3, is found with a total computational time of 700 ms.

Table 3. Sets of angles for 5 points of the circle

θ_1 (degrees)	θ_2 (degrees)	θ_3 (degrees)	θ_4 (degrees)	θ_5 (degrees)
7.125	70	−111.839	131.839	0
7.091	71	−111.532	130.532	0
6.989	72	−111.232	129.232	0
6.821	73	−110.948	127.948	0
6.588	74	−110.689	126.689	0

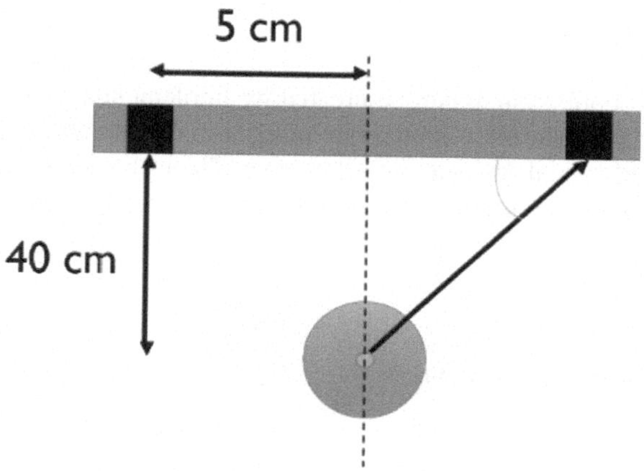

Fig. 4. Trajectory of a circle (top view)

It was found that the maximum variation between expected and calculated coordinates was of 0.18 cm (Table 4).

Table 4. Comparison between approximated and real points

Approximated coordinates (cm)			Real coordinates (cm)		
X	Y	Z	X	Y	Z
40.1028	5.0128	19.9984	40	5	20
40.0746	4.9852	20.5042	40	4.9759	20.4901
40.0396	4.9084	21.0091	40	4.9039	20.9755
39.9956	4.7841	21.5099	40	4.7847	21.4514
39.9401	4.6128	22.0034	40	4.6194	21.9134

5 Conclusions

The Denavit-Hartenberg formalism allowed a simple analysis to determine the joints angles in the Scorbot-ER III. Based on the transformation matrices and equations derived from this convention, the time required to compute an approximated solution based on a semi-analytical approach has proven to be considerably faster than numerical methods, making it a feasible solution to create trajectories with arbitrary paths. However, physical limitations as the workspace and finite orientations of the end-effector must be considered.

6 Further Projects

An autonomous robotic system can be created by implementing a set of camera and distance sensors. This allows a continuous recognition of the target´s pose, creating a ${}^{0}T_5$ matrix based in real on-time measurements. Then, the semi-analytical solution proposed in this paper can be applied to find the solution in this changing environment.

Appendix

Scorbot-ER III transformation matrices.

$$
\begin{aligned}
{}^{0}A_1 &= \begin{bmatrix} \cos(\theta_1) & 0 & sin(\theta_1) & a_1\cos(\theta_1) \\ sin(\theta_1) & 0 & -\cos(\theta_1) & a_1 sin(\theta_1) \\ 0 & 1 & 0 & d_1 \\ 0 & 0 & 0 & 1 \end{bmatrix} \\
{}^{1}A_2 &= \begin{bmatrix} \cos(\theta_2) & -sin(\theta_2) & 0 & a_2\cos(\theta_2) \\ sin(\theta_2) & cos(\theta_2) & 0 & a_2 sin(\theta_2) \\ 0 & 0 & 1 & 0 \\ 0 & 0 & 0 & 1 \end{bmatrix} \\
{}^{2}A_3 &= \begin{bmatrix} \cos(\theta_3) & -sin(\theta_3) & 0 & a_3\cos(\theta_3) \\ sin(\theta_3) & cos(\theta_3) & 0 & a_3 sin(\theta_3) \\ 0 & 0 & 1 & 0 \\ 0 & 0 & 0 & 1 \end{bmatrix} \\
{}^{3}A_4 &= \begin{bmatrix} \cos(\theta_4) & 0 & sin(\theta_4) & 0 \\ sin(\theta_4) & 0 & -\cos(\theta_4) & 0 \\ 0 & 1 & cos(\alpha_4) & 0 \\ 0 & 0 & 0 & 1 \end{bmatrix} \\
{}^{4}A_5 &= \begin{bmatrix} \cos(\theta_5) & -sin(\theta_5) & 0 & 0 \\ sin(\theta_5) & cos(\theta_5) & 0 & 0 \\ 0 & 0 & 1 & d_5 \\ 0 & 0 & 0 & 1 \end{bmatrix}
\end{aligned} \tag{A1}
$$

References

1. Xu W, Liang B, Li C, Xu Y (2010) Autonomous rendezvous and robotic capturing of non-cooperative target. Robotica 28:705–718
2. Hausmann G, Wieser M, Haarmann R, Brito A, Meyer J-C., Jäkel S, Lavagna M, Jakobsson B, Biesbroek R (2015) e. Deorbit Mission: OHB debris mission removal concepts. In: ASTRA 2015
3. Thomsen B, Sharf I (2016) Experiments on Tether-Net capture and net closing mechanism of space debris. In: 67th International astronautical congress, Guadalajara, México
4. Ali MH, Aizat K, Yerkhan K, Zhandos T, Anuar O (2018) Vision-based robot manipulator for industrial applications. Procedia Comput Sci 133:205–212

5. Basu R, Padage S (2017) Development of 5 DOF robot arm-gripper for sorting and investigating RTM concepts. Mater Tod Proc 4(2):1634–1643
6. Rios A, Hernandez E, Moreno J, Keshtkar S, De la Garza R (2018) Kinematics analysis of a new 3DOF parallel manipulator as walking rehabilitation device. In: 15th International conference on electrical engineering, computing science and automatic control (CCE), pp 1–6
7. Zhu Q, Zhang Z (2017) An efficient numerical method for forward kinematics of parallel robots. IEEE Access 99:1
8. Spong MW, Vidyasagar M (2008) Robot dynamics and control. Wiley, New York, p 68
9. Amatya S, Petchartee S (2015): Real time kinect based robotic arm manipulation with five degree of freedom. In: The 2015 Asian conference on defence technology
10. Rosales EM, Gan Q (2002): Forward and inverse kinematics model for a 5-dof pioneer 2 robot arm. Essex

Development of a Vision System for a Mobile Manipulator to Solid Residue Collection in the Food Sector

Jhoset Méndez Amador, P. A. Niño-Suárez(✉), and Raúl Rivera-Blas

Instituto Politécnico Nacional, ESIME Unidad Azcapotzalco, Ciudad de México, Mexico
joset_753@hotmail.com, {pninos,rriverab}@ipn.mx

Abstract. Due to its versatility, artificial vision has become very useful in different areas, like medicine, where robotic arms and cameras have been incorporated to carry out surgeries that involve a high level of precision. The combination between the robotics and the artificial vision has managed to optimize the performance of robotics arms in the object's subjection. In this paper, the methodology "V Model" is used for designing of the mobile manipulator, aided by algorithms of artificial vision, to carry out the recognition of objects. These algorithms are implemented on a FIT-PC4 embedded system. The object recognition algorithms can calculate the position of the object in the Euclidean space and through arm kinematics it can collect the objects and deposit them in a container inside the mobile.

Keywords: V model · Artificial vision · Robotics · Inverse kinematics

1 Introduction

In food industry, equipment and facilities must be designed and built to ensure that the safety of food can be guaranteed and prevent its contamination. It is essential that the workspaces are always free of residues that can cause incidents in order to satisfy with the highest quality parameters in the products. The food industry usually has a highly competitive production, but with a relatively small robotic participation compared to the manufacturing or automotive industries due to the products are highly variable in shape, size and structure [1]. A viable alternative for incorporating the automation in cleaning processes in industries (food sector), is the use of vision systems to support human on-site supervision in the production cycle, in such a way that the information sent to the robotic system be acquired without having any contact with the food that is being processed or transported. Artificial vision can be defined as a field of artificial intelligence that, by using the appropriate techniques, allows the collection, processing and analysis of any type of special information obtained through digital images; being one of the fields of research that has generated a lot of interest in recent decades. The inclusion of artificial vision in the field of robot-assisted production has increased in recent years, with cameras being the most suitable sensors for obtaining information about the robot's environment,

E. E. Hernandez et al. (Eds.): LASIRS 2019, MMS 86, pp. 228–239, 2020.
https://doi.org/10.1007/978-3-030-45402-9_22

due to its wide field of panoramic view [2]. In the same way, they can also obtain information about objectives, obstacles and even data that provide direct information on the situation being displayed. The cameras offer the possibility of supporting mobile robots in navigation tasks without the need for an operator [3–5].

The problem solved with the technological development presented, is common in bakery companies; this occurs in production process, specifically when the bread goes to the conveyor belt to continue the packaging process, there are times when the bread falls to the ground and someone must be picking it up. There are multiple jobs where artificial vision systems are used to support production processes; as presented by [6], where an artificial vision system implemented in an embedded system for controlling the operation of different types of wheeled mobile manipulators.

The solution devised supports medium and large bakery industries in cleaning tasks at the production level, specifically, collecting solid waste on the soil and deposit them in a predetermined place.

The vision algorithm identifies brown color object's (hamburger bread), the object is in the Cartesian coordinates of the mobile manipulator, and are sent to the platform and robotic arm through the control algorithm; that way, the platform is moved to a nearby point and then the arm is placed on the bread, so that it is recovered from the soil and placed in the adequate container. Moreover, the incorporation of this kind of technology in the production plants, will allow to carry out some of the routine and heavy work previously destined for cleaning personnel, or as an additional task for the plant operators, it will provide security in the production process because it does not have human intervention in the food handling tasks.

In Sect. 2 the design of the vision system concept and operation control of the mobile manipulator is presented, Sect. 3 presents the development of the vision and control algorithms for operating the mobile platform, Sect. 4 presents some results of laboratory experiments. Finally, there are the conclusions of the work carried out.

2 Design of the Vision System Concept

The system design was made by using the V-Model defined in [7], this tool presents a general information flow for the product development process, since the idea conception to its final design. It begins with the identification of user requirements, once these have been taken into account a translation of the requirements is made into engineering terms that will provide solution to needs; at the end, the V-Model ends with a system of user validation, each stage of the product must be tested, in order to avoid incidents with the customer, as mentioned in [8]. The initial part of the analysis, needs and customer requirements, was carried out using the QFD Methodology, presented in [9], from these results the design goals were obtained and from them a conceptual design was made, the hardware architecture was choose and the control algorithms were developed by mean of direct kinematic models of the manipulator and the mobile platform. These models allow the mobile could move along a predefined path assigned by the user and stop automatically when the camera detects the object of interest, then with the direct and inverse kinematics of the robotic arm, the position to be reached is calculated.

2.1 Design Goals

1. The vision system must be implemented in a mobile manipulator, according with the following features:

    a. The mobile platform must have wheeled locomotion.
    b. The arm must be articulated and have a gripper.
    c. The maximum load capacity of the mobile manipulator on its gripper, must be in accordance with the task to perform.
    d. The robot should move at a low speed (3 up to 5 m/s).
    e. The platform dimensions must be: 500 mm wide, 600 mm long and 782 mm high, or according to the workspace.
    f. The minimum (arm retracted) and maximum (arm fully extended) robotic arm reach and maximum must be in accordance with the task.

2. The vision and control system must provide the ability to acquire and process images at the mobile manipulator, for making decisions during the trajectory and could be autonomous in their movement.
3. Vision system must be able to operate in places with variable lighting.

2.2 Concept Generation

Based on the goals generated from the customer's requirements, the function tree that the vision and control system must possess for operating the mobile manipulator is defined (see Fig. 1). The system detects the object to be collected (hamburger bread), indicates to the mobile how far must move (to a point near the object), and the mobile manipulator arm will perform the collection of the found object. The concept idea is to use the system in general way to identify different types of objects and control different types of mobile manipulators, by easily modifying some modules of the program.

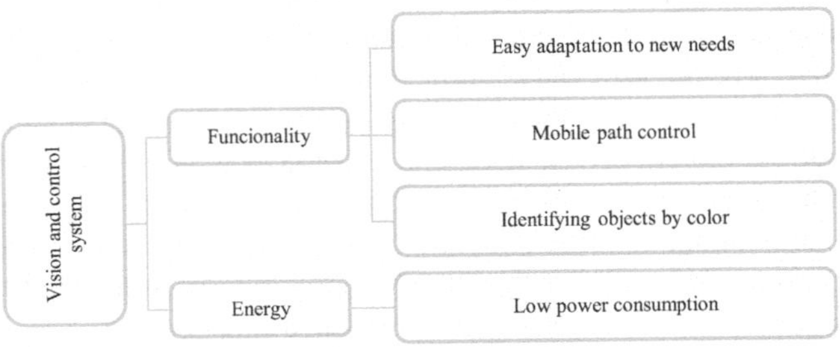

Fig. 1. Vision system function tree.

In Fig. 2a. the modules that make up the functional structure of the vision system and operation control of the mobile manipulator are observed. The main module is the digital processing system, in which the vision algorithms are implemented in order to give autonomy to the mobile manipulator. Similarly, a fundamental part of the vision system is the selection of the camera for capturing the image to the digital processing system. For the integral design of the vision system it is necessary to integrate the type of actuators that generate the locomotion of the mobile manipulator and the robotic arm.

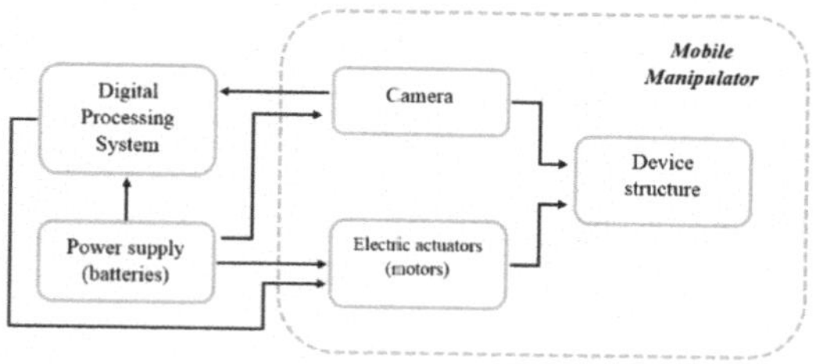

a. Hardware architecture design proposal

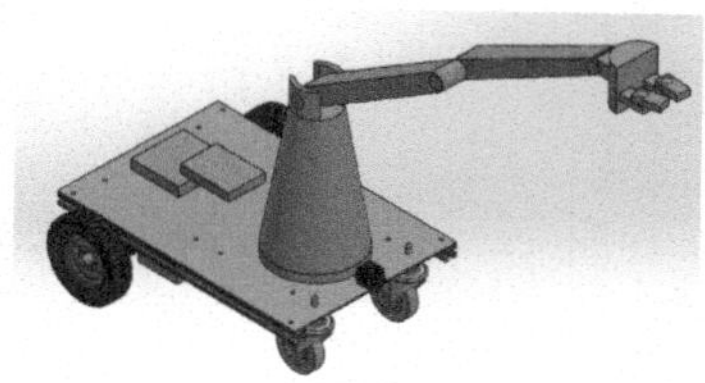

b. Mobile manipulator structure proposal.

Fig. 2. Design proposal.

The captured images will be sent to the data acquisition system, where it will be determined whether the robot should continue its trajectory or, if an object is detected, evaluate whether it is the hamburger bread to be collected; therefore the distance to the object is calculated to generate the control sequence that tells the actuators to move the manipulator and collect it. As well, it is necessary to select the type of power for supplying the necessary energy to the system, both for the mobile manipulator, as well as for the vision system (camera and the data processing system).

As a result of the applied methodology and the customer's requirements, a specific structure of the mobile manipulator was selected for developing the mathematical models of the control module, being selected an anthropomorphic arm of three degrees of freedom (DOF) and a mobile platform based on differential wheel. The proposed design that involves the placement of the camera on the mobile platform and not on the manipulator's wrist (see Fig. 2. b), was evaluated together with the vision algorithm, because the camera configuration does not move (see Fig. 2).

From the concept, the main modules for hardware and software architecture were defined. It is important to remember that the system is modular, in this way the programming of some modules could be modified to change the type of object to detected and/or the mobile manipulator.

2.3 Hardware Architecture Design

The hardware architecture is a USB500W02M-AF30 camera and a FIT-PC4 embedded system, with Linux operating system (64 bits) as work environment. The programming software as the Eclipse IDE and the OpenCV vision libraries for image processing were used, the main idea is to use free software tools.

2.4 Mathematical Model

The robotic arm for the object manipulation is the Mitsubishi 5 DOF anthropomorphic robotic arm RV-M1C, which has a clamp gripper that allows holding the object. The Denavit-Hartenberg (DH) methodology to obtain its forward kinematic model was used [10]. The reference systems of Mitsubishi manipulator could be observed in Fig. 3.

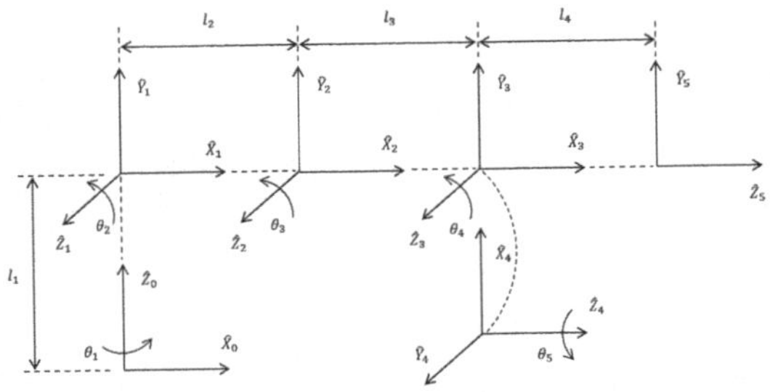

Fig. 3. Reference systems of Mitsubishi RV-M1C Robot.

The D-H parameters shown in Table 1, are substituted in Eq. (1) to create a homogeneous transformation matrix (HTM) for each DOF (Degree of Freedom) of the robot, these individual matrices are multiplied to obtain the final HTM.

Table 1. Forward kinematic parameters.

N	a_i	d_i	α_i	θ_i
1	0	d_1	90°	θ_1
2	a_2	0	0	θ_2
3	a_3	0	0	θ_3
4	0	0	90°	θ_4
5	0	d_5	0	θ_5

$$\begin{bmatrix} cos\theta_i & -sin\theta_i cos\alpha_i & sin\theta_i sin\alpha_i & a_i cos\theta_i \\ sin\theta_i & cos\theta_i cos\alpha_i & -cos\theta_i sin\alpha_i & a_i sin\theta_i \\ 0 & sin\alpha_i & cos\alpha_i & d_i \\ 0 & 0 & 0 & 1 \end{bmatrix} \tag{1}$$

Equation (2) corresponds to the $HTM_{4\times4}$.

$$HTM_{4\times4} = \begin{bmatrix} a_{11}\ a_{12}\ a_{13}\ a_{14} \\ a_{21}\ a_{22}\ a_{23}\ a_{24} \\ a_{31}\ a_{32}\ a_{33}\ a_{34} \\ a_{41}\ a_{42}\ a_{43}\ a_{44} \end{bmatrix} \tag{2}$$

where:

$$
\begin{array}{lll}
a_{11} : -C_1S_{234}S_5 - S_1C_5 & a_{22} : -S_1S_{234}C_5 - C_1S_5 & a_{14} : C_1(l_2C_2 + l_3C_{23} + l_5C_{234}) \\
a_{21} : -S_1S_{234}S_5 + C_1C_5 & a_{32} : C_{234}C_5 & a_{24} : S_1(l_2C_2 + l_3C_{23} + l_5C_{234}) \\
a_{31} : C_{234}S_5 & a_{13} : C_1C_{234} & a_{34} : l_1 + l_2S_2 + l_3S_{23} + l_5S_{234} \\
a_{41}, a_{42}, a_{43} : 0 & a_{23} : S_1C_{234} & a_{44} : 1 \\
a_{12} : -C_1S_{234}C_5 + S_1S_5 & a_{33} : S_{234} &
\end{array}
$$

To locate the position of the end effector on the Euclidean space, the position vector of $HTM_{4\times4}$ is required (elements a_{14}, a_{24}, a_{34}). This vector is sent to the control program to perform the inverse kinematics of the manipulator, positioning the end effector where the object is detected.

For modelling the movement of the mobile platform, the differential type wheeled locomotion is used, its kinematic model is represented by the Eq. (3), where V is the linear speed and W is the angular speed of the vehicle.

To obtain the angular velocity of each wheel (W_r, W_l), the HTM of Eq. (4) is used, by considering the values of wheel radius, ***r*** and the length between wheels spacing, ***l***, the wheel velocities are found.

$$
\begin{aligned}
\dot{x} &= v * \cos(\theta) \\
\dot{y} &= v * \sin(\theta) \\
\dot{\theta} &= w
\end{aligned}
\tag{3}
$$

$$
\begin{bmatrix} V \\ W \end{bmatrix} = \begin{bmatrix} r/2 & r/2 \\ r/2 * l & -r/2 * l \end{bmatrix} \begin{bmatrix} W_r \\ W_l \end{bmatrix}
\tag{4}
$$

3 Vision Algorithm

The fundamental part of this work focuses on the vision algorithm, because it is useful for the displacement of the mobile and object manipulation. The tests were carried out by using different environmental conditions in order to avoid future problems for the vision system.

The goal of the vision system is to locate an object; when it is located, the distance between it and the camera is calculated, after it is transforms into Euclidian coordinates. For adding robustness to the system, a brightness and contrast filter was implemented.

Figure 4 presents the general flow diagram of the system. The path tracking and vision algorithms are executed at the same time, since they complement each other. For example, if the vision algorithm is not working, the mobile manipulator will work without receive control actions. On the other hand, if only the vision system is working, the mobile manipulator will not move.

To achieve the bread detection, a method called segmentation was used, this is based on the HSV (Hue Saturation Value) color model and a threshold technique. This technique creates a binary image (black and white image), assigning a 1 in the image where the object is detected and a 0 in the area of least interest. Color-based recognition

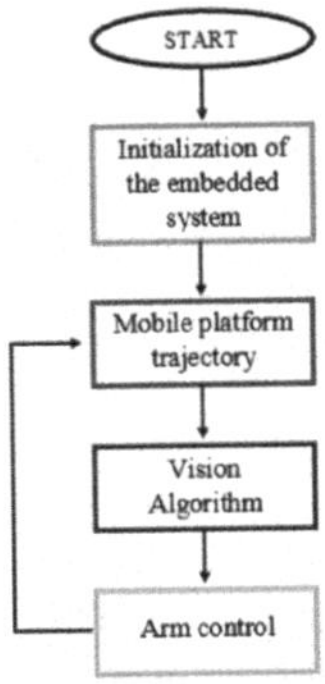

Fig. 4. General flow chart of system operation.

is done by converting the original RGB (Red Green Blue) image to HSV, the Eq. (5) performs a change of RGB values converting 8-bit values (0 to 255) to binary values (0 and 1). these values are transformed into HSV scale by the Eq. (6).

$$\begin{aligned} R' &= \frac{R}{255} G' = \frac{G}{255}, B' = \frac{B}{255} \\ C_{max} &= \max(R', G', B') \\ C_{min} &= max(R', G', B') \\ \Delta &= C_{max} - C_{min} \end{aligned} \tag{5}$$

$$H = \begin{cases} 60° \times \left(\frac{G'-B'}{\Delta} mod6\right), C_{max} = R' \\ 60° \times \left(\frac{B'-R'}{\Delta} + 2\right), C_{max} = G' \\ 60° \times \left(\frac{R'-G'}{\Delta} + 4\right), C_{max} = B' \\ 0° \\ \Delta = 0 \end{cases} \quad S = \begin{cases} 0, C_{max} = 0 \\ \frac{\Delta}{C_{max}}, C_{max} \neq 0 \end{cases} \tag{6}$$

$$V = C_{max}$$

After the color conversion, the parameters range to detect the brown color of the bread are: 21 for Hue, 222 for Saturation and finally 202 for Value. In Fig. 5 the values established for the brown color allow the proper identification of the object, highlighting its shape in a green circle and another of red color indicating the position of the centroid.

To achieve the video binarization, a grayscale color conversion is performed; later, the video is converted to black and white; finally, the binarized video is cleaned and any possible noise is eliminated.

In this way, when non-zero values are detected, the edges of the object will be displayed in white (see Fig. 6); this process is represented mathematically in Eq. (7).

$$dst(x, y) = \begin{cases} maxVal\, if\, src > T(x, y) \\ if\, not\, is\, equal\, to\, 0 \end{cases} \tag{7}$$

In computer vision and image processing, the centroid (center of mass) is the weighted average of the pixels that are the shape of an object. The location of the centroid in pixels is found by means of Eq. (8), using a relationship between the intensity of the pixels on both X and Y axes over the total intensity of the pixels of the whole image. As the position of the centroid is needed in cartesian coordinates, the Eq. (9) is used to perform that task.

$$C_x = \frac{M_{10}}{M_{00}} \quad C_y = \frac{M_{01}}{M_{00}} \tag{8}$$

$$Pos_x = \frac{M_{10}}{A_{obj}} \quad Pos_y = \frac{M_{01}}{A_{obj}} \tag{9}$$

Using the instruction *HoughCircles*, based on the Hough Transform, the red circle indicates the position of the centroid of the hamburger bread (see Fig. 7), and the green circle highlights the shape of the object. In this way if the manipulator moves from one place to another, both circles carry out the same action without losing the location of the object.

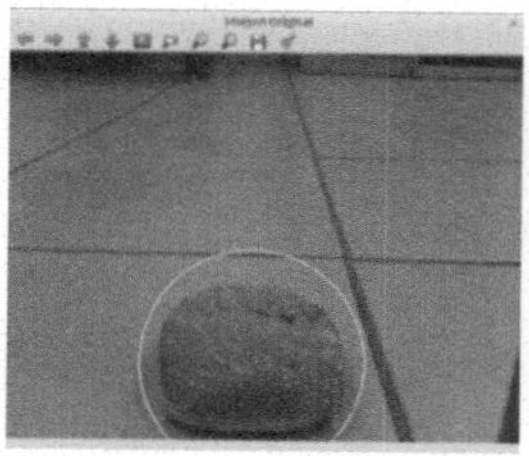

Fig. 5. Correctly identified.

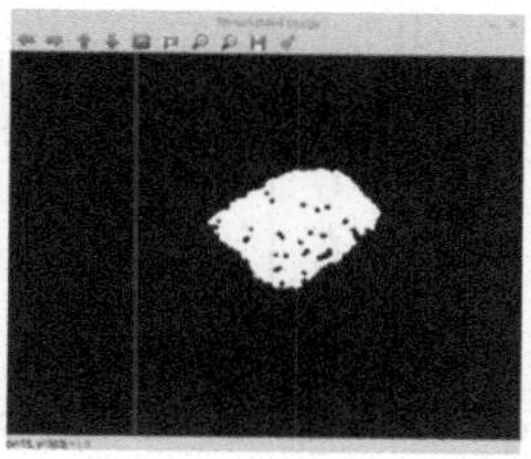

Fig. 6. Binary image.

Fig. 7. Hamburger bread centroid.

The calculation of the distance is essential, since it is required to carry out the navigation and manipulation actions. The distance between camera and object was performed by a technique called curve adjustment (Sum of Square Difference SSD), this method is based on the area of the object and the camera used. The distance is calculated by Eq. (10), where a multiplication is made between the size of the image in pixels, by the inverse of the square root of the area of the identified object (−0.468). By increasing or decreasing the area of the object the distance can change.

$$distance = \left(P_{img}\right)\left(A_{img}^{-0.468}\right) \tag{10}$$

Subsequently, the position of the object in Euclidean space must be calculated. First, the intrinsic parameters of the camera and the reference systems to the base of the manipulator and the camera are required. In order to obtain the intrinsic parameters of the camera. The intrinsic parameters were obtained by the camera calibration, using Matlab® and its Toolbox calibration. Table 2 shows the results of the camera calibration.

Table 2. Intrinsic parameters of the *USB500W02M-AF60* camera.

Focal distance	$fc = [345.688, 347.983]$
Main point	$cc = [147.581, 98.604]$

Before substituting values in Eq. (11), the rotation matrices and the translation of the camera to the base of the robot are obtained by Eq. (12) and Eq. (13) respectively, as mentioned in [11].

$$\begin{bmatrix} \lambda x_p \\ \lambda y_p \\ \lambda \end{bmatrix} = \begin{bmatrix} fk_x\vec{r_1} + C_x\vec{r_3} & fk_xt_x + C_xt_z \\ fk_x\vec{r_2} + C_y\vec{r_3} & fk_yt_y + C_yt_z \\ \vec{r_3} & t_z \end{bmatrix} \begin{bmatrix} X \\ Y \\ Z \\ 1 \end{bmatrix} \quad (11)$$

$$C_{R_M} = [R_{x(110°)}][R_{z(90°)}] = \begin{bmatrix} 0 & -1 & 0 \\ -0.3 & 0 & -0.9 \\ +0.9 & 0 & -0.3 \end{bmatrix} \quad (12)$$

$$c_{O_M} = \begin{bmatrix} t_x \\ t_y \\ t_z \end{bmatrix} = \begin{bmatrix} 0 \\ 0 \\ -t_z \end{bmatrix} \quad (13)$$

Using the complete model of Eq. (11), three equations are solved to obtain the points of the object in Euclidean space. This X, Y, Z point is sent to the program that controls the robotic arm, and the inverse kinematic is perform so that the bread be taken.

In Fig. 8 the initialization part of the vision algorithm is presented, where the characteristics of the object to be detected are programmed.

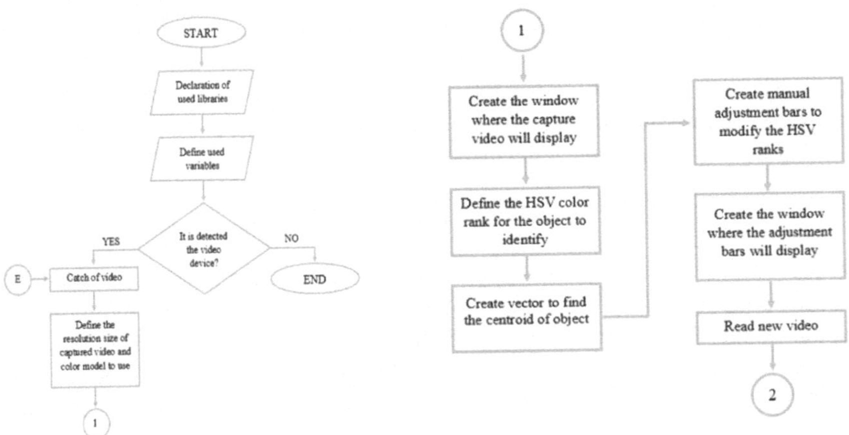

Fig. 8. Vision system flow diagram, part 1.

In Fig. 9 the algorithm continues with the application of the brightness and contrast filter, the color conversion from RGB to HSV, the binarization of the video and the calculation of the centroid of the object are performed. The final part of the vision algorithm begins with the visualization of the object in the binarized image.

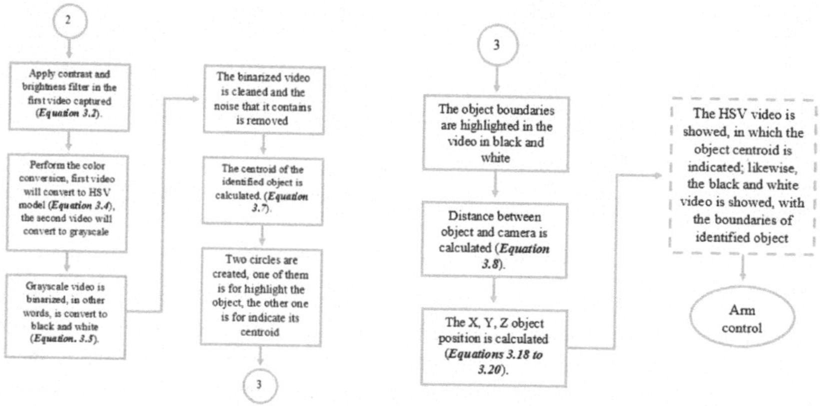

Fig. 9. Vision system flow diagram, part 2.

the distance between the camera and the identified object is calculated; finally, the position of the object in X, Y, Z coordinates is obtained, and the arm control algorithm is executed. The dotted line box is only executed when the user needs to make some adjustment to the algorithm and wants to view changes on a monitor.

4 Results of the Application of the Vision and Control Algorithm

In Fig. 10 the result of one of the experiments is presented. In Fig. 10a, the initial position of the mobile is displayed when the displacement trajectory starts to be executed, in this case a straight line. The flexometer indicates the distance between the camera and the bread. In the experiments corresponding to the first stage, the bread was placed at the minimum (30 cm) and the maximum (70 cm) separation from the chamber. It was observed that the technique used has an error of 1.6% (0.5 cm), with respect the real measurement. This difference is acceptable for the type of application since it is not necessary a high precision measurement. When the system is in spaces where the light directly incidents on the object or in places with low lighting, the error is in the order of 2.3% (error 0.7 cm), which is still acceptable. From the results obtained, it was observed that the segmentation technique has an accuracy of 98.3%, therefore it is a reliable method.

In the second stage, where the kinematics of the mobile are involved, the manipulator advances at 3 m/s once the bread is located (see Fig. 10b) and it stops at the minimum separation distance (30 cm), which allows the manipulator to grab the bread. Although there is a delay in milliseconds (10 ms) before the robot completely stops, this variation does not affect the manipulation of the object (see Fig. 10c).

a. Mobile manipulator initial path. **b.** Object detection and platform stop. **c.** Collection of the object. **d.** Placement of the object in the container.

Fig. 10. Result of one of the experiments

Finally, the execution time since the bread is taken until it's and placed in the container is 20 s, Fig. 10d. The workspace designated to execute the pick and place task was appropriate and an arm with fewer degrees of freedom could be used, since this task is easy to execute.

5 Conclusions

In this work, a vision system was developed for the recognition and location of an object in Euclidean space. This was implemented in a system embedded in the mobile robot. According with the defined problem to detect an object (hamburger bread) on the production plant floor and the purpose of the task for the mobile manipulator was to move to a point where it could be pick up and put it in a container. According to the results obtained from the experiments realized, the precision of the system segmentation technique is 98.3%.

The developed software architecture is modular, the user can modify the vision system to recognize objects of different shape and size, and only the HSV values should be adjusted depending on the color of the object. For this reason, the module offers versatility and can be used in different tasks. In a similar way, the type of mobile manipulator can be changed modifying the mathematical models of the robot (platform locomotion type and the forward and inverse manipulator kinematics), in the corresponding module of the software architecture.

Acknowledgements. We would like to thank to Instituto Politécnico Nacional for their support through the Secretaría de Investigación y Posgrado. The first author thanks the Consejo Nacional de Ciencia y Tecnología of México (CONACyT) for the scholarship awarded for Postgraduate studies at the IPN-ESIME Azcapotzalco.

References

1. Chua, P.Y., Ilschnner, T., Caldwell, D.G.: Robotic manipulation of food products – a review. Ind Robot Int J **30**(4), 345–354 (2003)

2. Arias M (2009) Navegación de un robot móvil en interiores utilizando flujo óptico a partir de imágenes panorámicas, Tesis de Maestría, Universidad Nacional de Colombia, Medellín
3. Se S, Lowe D, Little J (2001) Vision-based mobile robot localization and mapping using scale-invariant features. In: Proceedings of the 2001 IEEE international conference on robotics & automation, Seoul, Korea, May 2001, pp 1–8
4. Graefe V (1989) Dynamic vision systems for autonomous mobile robots. In: IEEE/RSJ international workshop on intelligent robots and systems 1989, Tsukuba, Japan, pp 1–12
5. Ishiguro H (1997) Distributed vision system: a perceptual information infrastructure for robot navigation, Sakyo-ku, Kyoto, Japan
6. Bulanon, DM, Kataoka, T, Okamoto, H, Hata, S (2004) Development of a real-time machine vision system for the apple harvesting robot. In: SICE annual conference in sapporo, Hokkaido Institute of Technology, Japan, 4–6 August 2004, pp 1-4
7. Wang, Y, Ying, Y, Chun, X, Xiaoyang Z, Weizhi, J (2012) A proposed approach to mechatronics design education: integrating design methodology, simulation with projects, Elsevier, pp 942–948
8. Zheng, JL, Duigou, Bricogne, M, Eynard, B (2013) Survey of design process models for mechatronic systems engineering. In: CIGI 2013, Compiègne Cedex, France, pp 1–6
9. Bergquist, K., Abeysekera, J.: Quality Function Deployment (QFD) - a means for developing usable products. Int J Ind Ergon **18**, 269–275 (1996)
10. Spong, MW., Vidyasagar, M (1989) Robot dynamics and control
11. Ibarra, JM, Iturbe, EJ (2017) Calibration and correction for radial distortion on vision systems for Robots, Robotic Laboratory, Control and Automation Department of CINVESTAV, Mexico, DF

Robust Control of a Master-Slave Manipulator Under Restricted Task-Space

D. Cruz-Ortiz[1(✉)], I. Chairez[2], and A. Poznyak[1]

[1] Automatic Control Department, CINVESTAV-IPN, Av. Instituto Politécnico Nacional 2508, Gustavo A. Madero, 07360 Mexico City, Mexico
{dcruz,apoznyak}@ctrl.cinvestav.mx

[2] Bioprocesses Department, UPIBI-IPN, Av. Acueducto 550, Gustavo A. Madero, 07340 Mexico City, Mexico
jchairezo@ipn.mx

Abstract. The aim of this manuscript is to develop a robust output-based controller for a master-slave robot manipulator (RM), under state constraints. The proposed controller satisfies a regular proportional-derivative (PD) structure with time-varying gains. This controller includes an integral compensation term which is guaranteeing the collision avoidance with possible obstacles placed within the boundaries of the task work-space. The attractive ellipsoid method justifies the design of the control strategy. The solution of a matrix inequality characterizes the zone of convergence and the sub-optimal control gains. A set of numerical simulations based on a two-link RM illustrates the advantages obtained with the proposed method.

Keywords: Robust control · Constraint system · Attractive ellipsoid

1 Introduction

Robotic manipulators (RMs) are electrical-mechanical devices with diverse applications in researching and industrial sectors [2,12]. The RMs are usually programmed to perform a repetitive and autonomous sequence of movements. These devices may exert tasks that can be dangerous by a human operator [1,5]. This fact favors the interest on developing new manipulator configurations, as well as, supplying them with new skills and degree of freedoms (dofs).

The most notable configuration of RM using the human experience and ability coupled with the potential of robotic systems [9] is a master-slave configuration (MSC). Nowadays, the MSC has been used in several technical applications due to its versatility for the movements of the final effector [6,13].

David Cruz-Ortiz is sponsored by Mexico scholarship from the CONACYT, scholar reference: 550817.

E. E. Hernandez et al. (Eds.): LASIRS 2019, MMS 86, pp. 240–248, 2020.
https://doi.org/10.1007/978-3-030-45402-9_23

The remote operation of RMs that realize their movements in a constrained space has encouraged the developing of automatic controllers which can ensure two conditions: (1) completing the tracking of a reference trajectory (provided by a Master Robot (MR)) and (2) guaranteeing the avoidance collisions with the objects located in the environment, as well as, satisfying the boundaries of the workspace.

The trajectory tracking problem between the states of the MR and the slave Robot (SR) can be solved by diverse controllers [7,10,11]. Linear feedback controllers or sliding modes have been widely used to solve the tracking problem for MSC [3,4]. However, these approaches may not satisfy the state constraints.

From the control point of view, notice that the RMs is a system affected by bounded perturbations (external). Additionally, it is unrealistic to expect that the mathematical models represent the nonlinear dynamics without uncertainties. Considering the previous argument, there are several results to get the minimization of the effects due to external perturbations and internal uncertainties. One of the most remarkable result is the attractive ellipsoid method (AEM) [8]. The AEM deals with unmatched perturbations. Also, the control gain is usually obtained by the solution of a linear minimization problem with a linear objective function and linear constraints.

This study develops a state feedback controller for MSC, aimed to satisfy restrictions in the states in presence of model uncertainties and external perturbations. The controller includes an integral compensation term to satisfy both the state constrains and collision avoidance for the SR. The application of the AEM estimates the gains of the controller.

1.1 Plant Description

Let define the dynamical model of the robotic manipulator, considering $x_{i1} \in \mathbb{R}^n$ and $x_{i2} \in \mathbb{R}^n$:

$$\dot{x}_{i1} = x_{i,2}$$

$$\dot{x}_{i2} = \Lambda_i(x_i) + \Theta_i(x_{i1})u_i + \eta_i(x_i) \tag{1}$$

$$y_i = \Upsilon_i x_i$$

where $x_i = \left[x_{i1}^\top \, x_{i2}^\top\right]^\top, x_i \in \mathbb{R}^{2n}$. The subscript $i = \{m, s\}$ describe the MR and SR. The drift function $\Lambda_i : \mathbb{R}^{2n} \times \mathbb{R}^+ \to \mathbb{R}^n$ gathers all the terms associated with gravitational effects and components of Coriolis matrix. This function admits the following restriction

$$\|\Lambda_i(z_1) - \Lambda_i(z_2)\|^2 \leq L_{i1} + L_{i2}\|z_1 - z_2\|^2 \tag{2}$$

with $z_1 \in \mathbb{R}^{2n}$ and $z_2 \in \mathbb{R}^{2n}$ and L_{i1} and L_{i2} are positive constants.

The state dependent matrix $\Theta_i : \mathbb{R}^n \times \mathbb{R}^+ \to \mathbb{R}^{n \times n}$ is a symmetric positive matrix, that represents the robot inertial term [1]. The matrix Θ_i has an inverse matrix Θ_i^{-1} $\forall t \geq 0$, also has bounded norm $\|\Theta_i(z_3)\|^2 \leq g_i^+$.

with g_i^+ a positive scalar and $z_3 \in \mathbb{R}^n$. Because of the positiveness and symmetry of matrix Θ_i, the fully actuated system concept applies for the master device. Based on the previous fact, the following assumption is valid.

Assumption 1 *The non-linear system* (1) *is controllable.*

The term $\eta_i : \mathbb{R}^{2n} \times \mathbb{R}^+ \rightarrow \mathbb{R}^n$ represents the effect of external perturbations and internal uncertainties and admits the upper bound

$$\|\eta_i(x_i)\|^2 \leq L_{i\eta_1} + L_{i\eta_2}\|x_i\|^2 \tag{3}$$

where $L_{i\eta_1}$ and $L_{i\eta_2}$ which are positive constants.

The vector $y_i \in \mathbb{R}^n$ is the measurable output with $\Upsilon_i = \begin{bmatrix} I_n \ 0_n \end{bmatrix} \in \mathbb{R}^{n \times 2n}$ and the term $u_i \in \mathbb{R}^n$ represents the control input. Here it is denoted the set Ω_M that represents the space where the trajectories of the MR are restricted. The set Ω_1 represents the state constraints for the SR. The obstacles may exist within the workspace Ω_1. These obstacles correspond to Ω_2. Therefore, the real workspace of SR is $\Omega_W = \Omega_1 - \Omega_2$.

2 Problem Statement

Usually, the reference trajectories for the MR came from a human operator (open-loop). However, in some cases, these reference trajectories can be obtained as a result of a feasible trajectory tracking problem between the states of MR and some desired reference trajectories (close-loop). This problem can be reformulated as the design of the control input (u_m) for the MR such that

$$u_m \triangleq \{u_m : \overline{\lim}\|x_{m1} - x_{m1}^*\| = 0 \quad \forall t \geq T\} \tag{4}$$

where $x_{m1}^* \in \mathbb{R}^n$ describes the reference trajectory of the MR. Here, it is considered that each component of x_{m1}^* is a C^1 function. The finite-time convergence of this system is described by T.

From the physical point of view, the state x_{m1} of MR is bounded as a consequence of mechanical restrictions in the RM. Therefore, it is possible to establish the upper bounded from the state x_m as a constant x_m^+, such that:

$$x_m \in dom\{x_m \mid \|x_m\|^2 \leq x_m^+, \quad x_m^+ \in \mathbb{R}^+\} \tag{5}$$

Considering that the reference trajectories of SR are given by the movements executed by the MR, the set of admissible trajectories for the SR must be upper bounded by x_s^+, that is:

$$x_s \in dom\{x_s \mid \|x_s\|^2 \leq x_s^+, \quad x_s^+ \in \mathbb{R}^+\} \tag{6}$$

The SR trajectories must be restricted by the boundaries of Ω_W, avoiding the collision with the obstacles. The previous fact can be reformulated as a trajectory tracking problem with state constraints defined as follows:

$$\|x_{m1} - x_{s1}\| \leq \beta, \quad \beta \in \mathbb{R}^+$$
$$\text{subjected to} \tag{7}$$
$$\|x_{s1} - x_f\| > \epsilon, \quad \epsilon \in \mathbb{R}^+$$

Here $x_f \in \mathbb{R}^n$ defines the vector corresponding the distance between each $\sigma - ith$ component of state x_{s1} and the boundary $\partial\Omega_W$ of the set Ω_W. Notice that $\partial\Omega_W = \partial\Omega_1 \cup \partial\Omega_2$ where $\partial\Omega_1$ is the boundary of the set Ω_1 and $\partial\Omega_2$ is the boundary of the set Ω_2. Also, $0 \notin \partial\Omega_W$. Then, the norm used in the inequality (7) corresponds to the Hausdorff distance, that is $d(x_{s1}, \partial\Omega_W) = \inf_{\zeta \in \partial\Omega_W}\{d(x_{s1}, \zeta)\}$

The objective of the controller is designing the admissible control $u_s \in U^{adm}$ aimed to solve the tracking trajectory between x_{m1} and x_{s1} fulfilling that $x_{s1} \subset \Omega_W$, subject to the state constraint defined in (7). The control input belongs to the admissible set U^{adm} characterized as:

$$U^{adm} = \{u_s : \|u_s\|^2 \leq u_{s0} + u_{s1}\|x\|^2\}, \quad u_{s0}, u_{s1} \in \mathbb{R}^n \tag{8}$$

The condition (8) allows to design an output feedback controller based on the AEM [2].

3 Controller Design

3.1 Master Device

In this case of study the sliding mode controller is proposed to regulate the trajectory tracking for the MR. However, this controller does not contemplate the state constraints. The previous fact justifies the controller design with the state constraints for the SR. The sliding mode controller for the MR is described by the following expression

$$u_m = -K_m Sign(s) \tag{9}$$

The variable $s \in \mathbb{R}^n$ is the sliding surface defined by $s = \Delta_m + \tilde{C}\dot{\Delta}_m$. The variable $\Delta_m \in \mathbb{R}^n$, defined as $\Delta_m = x_{m1} - x^*_{m1}$, describes the tracking error and $\dot{\Delta}_m \in \mathbb{R}^n$ is the derivative of the tracking error, with the matrices $K_m \in \mathbb{R}^{n\times n}$ and $\tilde{C} \in \mathbb{R}^{n\times n}$. The function $Sign(s)$ satisfies the following definition:

$$Sign(s) \triangleq \left[sign(s_1)\ sign(s_2)\ ...\ sign(s_n)\right]^\top \tag{10}$$

The problem of the tracking trajectory between the master and the slave devices can be solved by some well-developed control schemes. The most common manner for solving the controller design is the so-called PD controller. However, this kind of control, cannot solve the state constraints considered in the problem formulation. In this section, a stable PD controller with state constrains can be developed. This controller applied on the SR and allowed ensuring an adequate trajectory tracking.

3.2 Slave Device

Let us consider the trajectory tracking error $\Delta \in \mathbb{R}^{2n}$ defined as $\Delta = x_s - x_m$. To satisfy simultaneously the control objective and the state constraint defined in (7), suggest the control input $u_s \triangleq u_{s0} + u_{s1}$ where $u_{s0} \in \mathbb{R}^n$ is the nonlinear compensation and $u_{s1} \in \mathbb{R}^n$ is the term that ensures the state constraints. This control is defined as $u_{s1} = \Theta_s^{-1}\tilde{u}_{s1}$ where $\tilde{u}_{s1} \in \mathbb{R}^n$ satisfies:

$$\tilde{u}_{s1} = -\Phi_1 \Delta - \Phi_2 \kappa \tag{11}$$

Here, the control gains $\Phi_1 \in \mathbb{R}^{n\times 2n}$, $\Phi_2 \in \mathbb{R}^{n\times n}$ should be designed. The term $\kappa \in \mathbb{R}^n$ is proposed as $\kappa = \Lambda_s + \Theta_s u_{s0}$.

3.3 Stable Equilibrium Point for Tracking Trajectories

Consider the tracking error Δ, the full time derivative is $\dot{\Delta} = \dot{x}_s - \dot{x}_m$, that can be rewritten as $\dot{\Delta} = \overline{\Lambda}(\Delta) + \overline{\Theta}_s(u_{s0} + u_{s1}) + \overline{\eta}$. The function $\overline{\Lambda}$ satisfy:

$$\overline{\Lambda}(\Delta) := \chi^\top \Lambda + A\Delta + \begin{bmatrix} \chi \\ 0_{n\times 2n} \end{bmatrix} \Delta. \tag{12}$$

The matrix $A \in \mathbb{R}^{2n\times 2n}$ is *Hurwitz*, the matrix $\chi \in \mathbb{R}^{n\times 2n}$ has the following structure $\chi = \begin{bmatrix} 0_n \ I_n \end{bmatrix}$, the function $\overline{\Theta}_s$ is $\overline{\Theta}_s := \chi^\top \Theta_s$ and the function $\overline{\eta}$ is governed by: $\overline{\eta} := \chi^\top \eta - A\Delta$, with $\Lambda := \Lambda_s - \Lambda_m - \Theta_m u_m$ and $\eta = \eta_s - \eta_m$. The control inputs with the control gain $\Phi_0 \in \mathbb{R}^{n\times 2n}$ are proposed, to design a stabilizing control:

$$\begin{aligned} u_{s1} &= -\Theta_s^{-1}(\Phi_1 \Delta + \Phi_2 \kappa) \\ u_{s0} &= -\Theta_s^{-1}(\Lambda + \Phi_0 \Delta) \end{aligned} \tag{13}$$

The next equation define the time varying gain matrices

$$\begin{aligned} \Phi_1 &:= \frac{\rho}{2}[1 + sign(\Delta^\top P \overline{\Theta}_s \Phi_2 \kappa)] \\ \Phi_2 &:= \|\Delta\|^2 \rho^\top \end{aligned} \tag{14}$$

where the gain $\rho \in \mathbb{R}^{n\times 2n}$ and $P \in \mathbb{R}^{n\times 2n}$ a positive definite matrix.

Considering the control gains and substituting the control input u_s, the ful-time derivative of Δ satisfy

$$\dot{\Delta} = \tilde{A}\Delta - \chi^\top(\Phi_1 \Delta + \Phi_2 \kappa) + \overline{\eta} \tag{15}$$

where the matrix $\tilde{A} \in \mathbb{R}^{2n\times 2n}$ is also *Hurwitz*. This matrix is $\tilde{A} = A + \begin{bmatrix} \chi^\top & -\Phi_0^\top \end{bmatrix}^\top$.

The next theorem, provides the sub-optimal values of the control gain guaranteeing a minimal invariant attractive ellipsoid.

Theorem 1. *If there exist positive definite matrix* P *and* ρ*, as well as the positive constants* α *and* ϵ *such that* $\Gamma(P, \rho \mid \alpha, \epsilon) \leq 0$*, then the set* $E_s(P\frac{\alpha}{\beta})$ *is an invariant attractive ellipsoid for the state trajectories, that is:*

$$\lim_{t\to\infty} sup \quad \Delta^\top \left(P\frac{\alpha}{\beta}\right)\Delta \leq 1 \tag{16}$$

where $\beta = \epsilon(1 + 2v^+ + 2\varrho_4^+)$ *and the matrix* $\Gamma(P, \rho \mid \alpha, \epsilon)$ *is:*

$$\begin{bmatrix} \Gamma_{11} & -0.5P\chi^\top\rho & -P\chi^\top\rho & P \\ -0.5\rho^\top\chi P & -\epsilon I_n & 0 & 0 \\ -\rho^\top\chi P & 0 & -\epsilon I_n & 0 \\ P & 0 & 0 & -\epsilon I_n \end{bmatrix} \tag{17}$$

$$\begin{aligned} \Gamma_{11} &= P(\tilde{A} + 0.5\alpha I_{2n}) + (\tilde{A}^\top + 0.5\alpha I_{2n})P \\ &-0.5(P\chi^\top\rho + \rho^\top\chi P) + \epsilon I_{2n\times 2n}(1 + 2v^+ + 2\varrho_4^+) \end{aligned} \tag{18}$$

4 Numerical Results

The numerical evaluation of the suggested controller considered the two-link manipulator as benchmark system (Fig. 1). The dynamic model of this device satisfies:

$$M(q)\ddot{q} + C(q, \dot{q})\dot{q} + P_q(q) = u \tag{19}$$

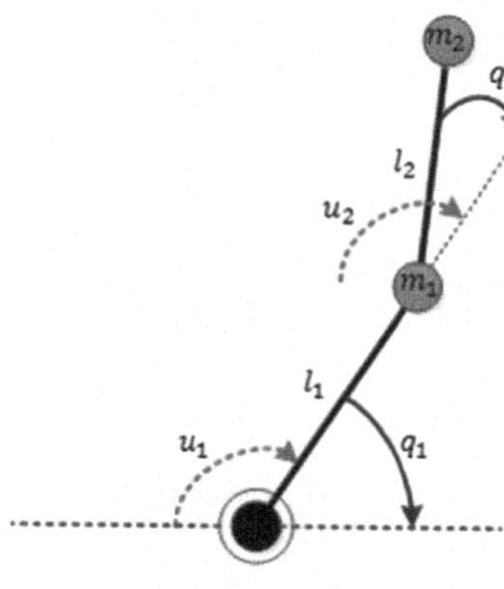

Fig. 1. Two-link manipulator.

The vector $q \in \mathbb{R}^n$, defined as $q = \begin{bmatrix} q_1 \; q_2 \end{bmatrix}^\top$, characterizes the generalized position of the mechanic system, the variables l_i and m_i ($i = \{1, 2\}$) represent the length and the mass of each link respectively. The term u represents the control input with $u = \begin{bmatrix} u_1 \; u_2 \end{bmatrix}^\top$. The term g describes the gravitational acceleration.

The inertia matrix $M(q)$ is:

$$M(q) = \begin{bmatrix} \alpha_1 + \alpha_2 + 2\alpha_3 \cos(q_2) & \alpha_2 + 2\alpha_3 \cos(q_2) \\ \alpha_2 + 2\alpha_3 \cos(q_2) & \alpha_2 \end{bmatrix} \tag{20}$$

whit α_1, α_2 and α_3 are defined as follows

$$\begin{aligned} \alpha_1 &= (m_1 + m_2)l_1^2 \\ \alpha_2 &= m_2 l_2^2 \\ \alpha_3 &= m_2 l_1 l_2 \end{aligned} \tag{21}$$

The matrix $C(q, \dot{q})$ satisfies:

$$C(q, \dot{q}) = \alpha_3 \cos(q_2) \begin{bmatrix} -\dot{q}_2 & -\dot{q}_1 - \dot{q}_2 \\ \dot{q}_1 & 0 \end{bmatrix} \tag{22}$$

And the matrix $P_q(q)$ corresponds to

$$P_q(q) = \begin{bmatrix} \alpha_4 \cos(q_1) + \alpha_5 \cos(q_1 + q_2) \\ \alpha_5 \cos(q_1 + q_2) \end{bmatrix} \tag{23}$$

where the variables α_4 and α_5 are:

$$\alpha_4 = (m_1 + m_2)g l_1, \quad \alpha_5 = m_2 g l_1 \tag{24}$$

The complete master-slave robotic system uses two manipulators like the one depicted in Fig. 1. Both manipulator systems are simulated considering the following parameters $m_1 = 0.2\,\mathrm{kg}$, $m_2 = 0.6\,\mathrm{kg}$, $l_1 = 0.6\,\mathrm{m}$, $l_2 = 0.3\,\mathrm{m}$, $g = 9.8\,\frac{\mathrm{m}}{\mathrm{s}^2}$.

The application of the proposed controller for the SR case, considered the matrix $\tilde{A}$ as follows.

$$\tilde{A} = \begin{bmatrix} 0 & 0 & 0.95 & 0 \\ 0 & 0 & 0 & 0.95 \\ -475 & -38 & -142.50 & -38 \\ -38 & -161.5 & -36.10 & -152 \end{bmatrix} \tag{25}$$

The output perturbation was suggested as follows:

$$\eta_s(t) = (0.005 + 0.15 \sin(t)) \begin{bmatrix} 1 & 1 \end{bmatrix}^\top \tag{26}$$

The initial conditions for the system are selected as

$$x_s(0) = \begin{bmatrix} 0.1 & 0.5 \end{bmatrix}^\top \tag{27}$$

Selecting the parameters $\alpha = 0.4$, and $\epsilon = 0.4$, the matrix obtained from the solution of (17) is

$$P^{-1} = \begin{bmatrix} 136.4498 & -3.7838 & 36.1858 & -2.1982 \\ -3.7838 & 172.1710 & 0.0170 & 11.5184 \\ 36.1858 & 0.0170 & 246.1828 & 9.3179 \\ -2.1982 & 11.5184 & 9.3179 & 228.2731 \end{bmatrix} \tag{28}$$

This value of the matrix P leads to the control gain:

$$\rho = \begin{bmatrix} -0.0026 & -0.0002 & -0.0001 & -0.0002 \\ -0.0000 & -0.0013 & -0.0001 & -0.0010 \end{bmatrix} \tag{29}$$

Figure 2 depict the results obtained if the control proposed in this manuscript is applied. The tracking performances comparison between the suggested controller with state constraints (PD_C) and the classical PD. The performance and the corresponding zoom for the tracking trajectory obtained for the first link of SMR is depicted in Fig. 3, where $q_{m,1}$ is the reference provided by the MR and the constraints are defined by C_1 and C_2.

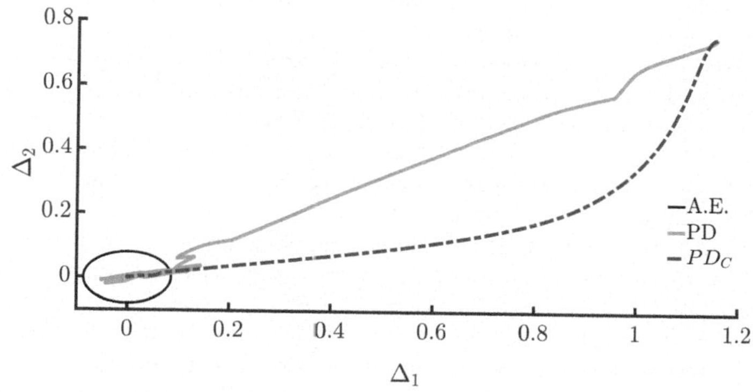

Fig. 2. Tracking errors and attractive ellipsoid.

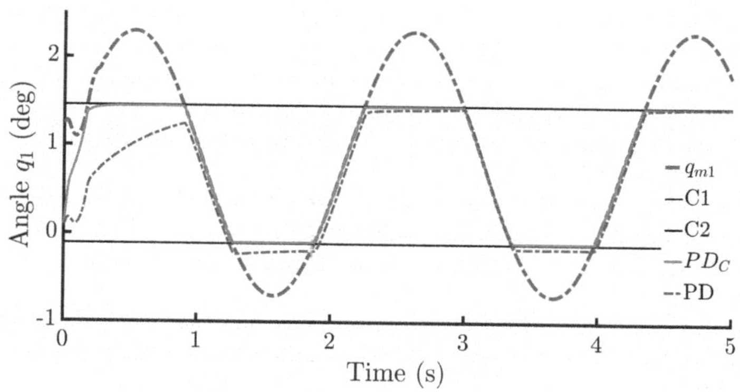

Fig. 3. Tracking trajectory of the first link of SR.

5 Conclusions

The controller design presented in this paper guarantees the trajectory tracking between two manipulator systems configured in a master-slave form. The principal contribution of this paper is the inclusion of state constrains and collision avoidance for the SR in the designed control based on the novel integral compensation form.

References

1. Batlle J, Ridao P, Salvi J (1997) Integration of a teleoperated robotic arm with vision systems using corba compatible software. In: 30th International Symposium on Automotive Technology and Automation, Robotics, Motion and Machine Vision in the Automotive Industries, pp 371–378
2. Capolei MC, Wu H, Andersen NA, Ravn O (2017) Positioning the laparoscopic camera with industrial robot arm. In: 3rd International Conference on Control, Automation and Robotics, pp 138–143. IEEE (2017)
3. Cruz-Ortiz D, Luviano-Juarez A, Chairez I (2014) Output sliding mode controller to regulate the gait of gecko-inspired robot. In: XVI Congreso Latinoamericano de Control Automático
4. Huang C, Xie L, Liu Y (2012) PD plus error-dependent integral nonlinear controllers for robot manipulators with an uncertain jacobian matrix. ISA Trans 51(6):792–800
5. La HM, Lim RS, Basily BB, Gucunski N, Yi J, Maher A, Romero FA, Parvardeh H (2013) Mechatronic systems design for an autonomous robotic system for high-efficiency bridge deck inspection and evaluation. IEEE/ASME Trans Mechatron 18(6):1655–1664
6. Nguyen KD, Dankowicz H (2016) Cooperative control of networked robots on a dynamic platform in the presence of communication delays. Int J Robust Nonlinear Control 27:1433–1461
7. Pchelkin SS, Shiriaev AS, Robertsson A, Freidovich LB, Kolyubin SA, Paramonov LV, Gusev SV (2017) On orbital stabilization for industrial manipulators: case study in evaluating performances of modified PD+ and inverse dynamics controllers. IEEE Trans Control Syst Technol 25(1):101–117
8. Poznyak A, Polyakov A, Azhmyakov V (2014) Attractive Ellipsoids in Robust Control. Springer, Cham
9. Rebelo J, Schiele A (2012) Master-slave mapping and slave base placement optimization for intuitive and kinematically robust direct teleoperation. In: 2012 12th International Conference on Control, Automation and Systems (ICCAS), pp 2017–2022. IEEE
10. Sharma R, Gaur P, Mittal A (2015) Performance analysis of two-degree of freedom fractional order pid controllers for robotic manipulator with payload. ISA Trans 58:279–291
11. Su Y (2009) Global continuous finite-time tracking of robot manipulators. Int J Robust Nonlinear Control 19(17):1871–1885
12. Tortora G, Dario P, Menciassi A (2014) Array of robots augmenting the kinematics of endocavitary surgery. IEEE/ASME Trans Mechatron 19(6):1821–1829
13. Truong DQ, Truong BNM, Trung NT, Nahian SA, Ahn KK (2017) Force reflecting joystick control for applications to bilateral teleoperation in construction machinery. Int J Precis Eng Manuf 18(3):301–315

On the Dynamics and Control of a Single-Wheel Robot with Inertial Locomotion

Mario E. Herrera-Cordero[1], Manuel Arias-Montiel[1(✉)], Marco Ceccarelli[2], and Esther Lugo-González[1]

[1] Institute of Electronics and Mechatronics, Universidad Tecnológica de la Mixteca, Huajuapan de León, Oaxaca 69000, México
herrera.enrique09@gmail.com, {mam,elugog}@mixteco.utm.mx
[2] LARM2: Laboratory of Robot Mechatronics, University of Rome, Tor Vergata 00133, Italy
marco.ceccarelli@uniroma2.it

Abstract. This work presents the mathematical modeling and the control design of a single-wheel robot with inertial locomotion. The forward-backward robot motion is obtained by the controlled angular displacement of a single pendulum, while the robot's mathematical model is obtained by using the Euler-Lagrange formalism. The resulting mathematical model is represented by highly coupled non-linear differential equations which are numerically solved and the obtained results are validated by comparison with data from SolidWorks simulations. The mathematical model is linearized around the operation point in order to synthesize a control law to regulate the linear velocity of the robot. The control scheme is based on the root locus method for the linearized model. The control system performance is evaluated by numerical simulations of the non-linear model in closed-loop. The obtained results show the viability for the experimental implementation of the proposed control system which is proposed as a future work for this research.

Keywords: Single-wheel robot · Dynamic model · Velocity control

1 Introduction

Single-wheel and unicycle robots have emerged as an alternative to multi-wheel robots in order to reduce the size and the required space for operating. Single-wheel robots are compact and they are able to operate in a flexible manner as there is only one point of contact with the surface. This kind of robot can achieve zero rotation radius and has potential applications, such as free movement and load transportation in narrow spaces [1]. One of the most important issues in single-wheel robots is the problem of lateral equilibrium. Different approaches have been proposed to solve this problem: horizontal inertial wheels [2], lateral

E. E. Hernandez et al. (Eds.): LASIRS 2019, MMS 86, pp. 249–260, 2020.
https://doi.org/10.1007/978-3-030-45402-9_24

inertial wheels [3] and high-speed gyroscopes [4]. Control strategies for single-wheel and unicycle robots are first developed for straight trajectories as established by Lee et al. [5]. In the context of single-wheel robots, a representative investigation about the design and control of single-wheel robots is the work of Jung et al. In [6] authors proposed a practical mechatronics approach for the development of a the GYROBO. This mechatronic approach solves some problems regarding the system integration and allows the implementation of linear controllers for both roll and yaw angles. In [7], Lee and Jung analysed the stability of the vertical mode of GYROBO and test the balancing control performance of GYROBO empirically to verify the comparison between the tested stability bandwidth and the determined offset range. They only apply control for the balancing or the GYROBO.

This paper is an extension of the work presented in [8] and presents the mathematical modeling and the control design of a single-wheel robot with inertial locomotion. The system dynamics is analyzed by an energetic approach and the robot's mathematical model is obtained by using the Euler-Lagrange formalism. The resulting mathematical model is represented by highly coupled non-linear differential equations which are numerically solved and the obtained results are validated by comparison with data from SolidWorks simulations. The mathematical model is linearized around the operation point in order to synthesize a control law to regulate the linear velocity of the robot. The control scheme is based on the root locus method for the linearized model. The control system performance is evaluated by numerical simulations of the non-linear model in closed-loop. The obtained results show the viability for the experimental implementation of the proposed control system which is proposed as a future work for this research.

2 Robot Description

The proposed design is based on the barycenter offset (BCO) driving principle. In this kind of system, the mass distribution of the robots is shifted by an internal mechanism in order to move the robot from one equilibrium position to another. This principle has been implemented in diverse locomotion systems for spherical robots [9].

In this project phase, the robot is only designed for forward and backward movement on a straight trajectory and the mass distribution is shifted by the movement of an inertial pendulum driven by a CD micromotor and internal spur gears mechanisms. The mechanical elements of the single-wheel robot are shown in Fig. 1. A more detailed description of the robot design and operation principle is given in [8].

3 System Dynamics

The dynamic model of the single-wheel robot is obtained based on mechanical elements shown in Fig. 1 and the following assumptions:

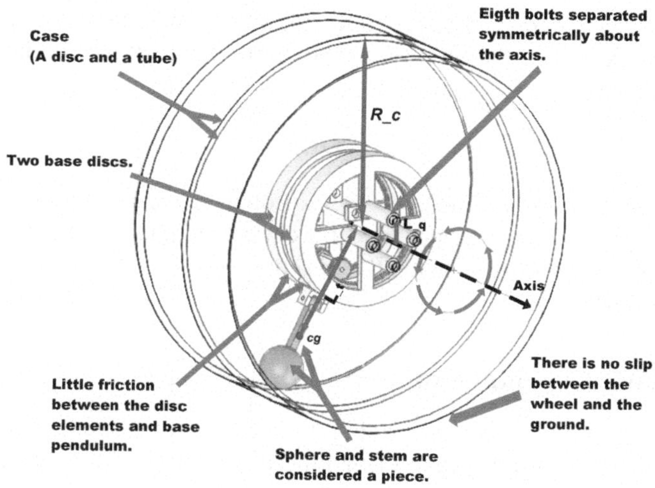

Fig. 1. Single-wheel pendulum robot configuration.

1. No slipping between the wheel and ground.
2. The pendulum is formed by a sphere and a solid rod, as they are considered as one solid body.
3. Eight bolts distributed symmetrically with respect to the rotation axis are considered.
4. Friction between the two base discs and pendulum is neglected.
5. The mass and moment of inertia of the pinion is neglected as these values are considerably smaller than the values for the other mechanical elements.

In Fig. 2 a free-body diagram of a simplified view of the system is shown, where θ is the robot angular displacement, τ is the input torque provided by the

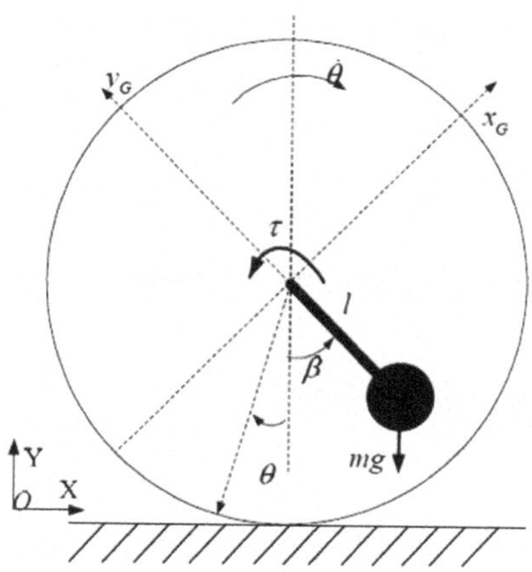

Fig. 2. Free-body diagram of the single-wheel pendulum robot.

micromotor, $\theta_2 = \theta - R\alpha$ is the pendulum angle, α is the pinion angular displacement and R is the transmission ratio of the internal spur gears mechanism.

The linear velocity of the robot is

$$\dot{X} = R_c\dot{\theta} \tag{1}$$

where R_c is the housing radius and $\dot{\theta}$ is the robot angular velocity.

To obtain the system's kinetic and potential energies, we must analyse the elements depending on the variable θ and separately, the elements depending on θ and α as follows.

3.1 Elements Depending on θ

The elements depending on the variable θ are the robot housing, the base discs, the bolts and the internal gears. Their kinetic and potential energies are given by:

$$T_1 = \frac{1}{2}[m_c R_c^2 + m_d R_c^2 + m_{cor} R_c^2 + I_c + I_{cor} + 8(m_{bolt}(L_q^2 + R_c^2) + I_{bolt})]\dot{\theta}^2 \tag{2}$$

$$V_1 = m_c g R_c + m_d g(R_c - R_d) + m_{cor} g(R_c - R_{cor}) + 8 m_{bolt} g L_q \dot{\theta}^2 \tag{3}$$

where L_q is the distance from the bolt axis to the rotation axis of the wheel, m_c is the robot housing mass, m_d is the base discs mass, m_{cor} is the internal gears mass, m_{bolt} is a bolt mass and I_c, I_d, I_{cor} and I_{bolt} denote the respective moment of inertia.

3.2 Elements Depending on α and θ

The elements depending on the variables α and θ are the pendulum and its base (the composed pendulum) with mass center in C_{tm} given by the coordinates X_{ctm} and Y_{ctm} (see Fig. 3)

The kinetic and potential energies of the pendulum are [8]

$$T_2 = \frac{1}{2} m_{comp} R_c^2 \dot{\theta}^2 + m_{comp} \dot{X} D_{ctm} cos(\theta_2)\dot{\theta_2} + \frac{1}{2} m_{comp} D_{ctm}^2 \dot{\theta_2}^2 + \frac{1}{2} I_{ctm} \dot{\theta_2}^2 \tag{4}$$

$$V_2 = m_{comp} g D_{ctm}(1 - cos(\theta_2)) \tag{5}$$

where m_{comp} is the mass of the composed pendulum and I_{ctm} is the moment of inertia of the pendulum.

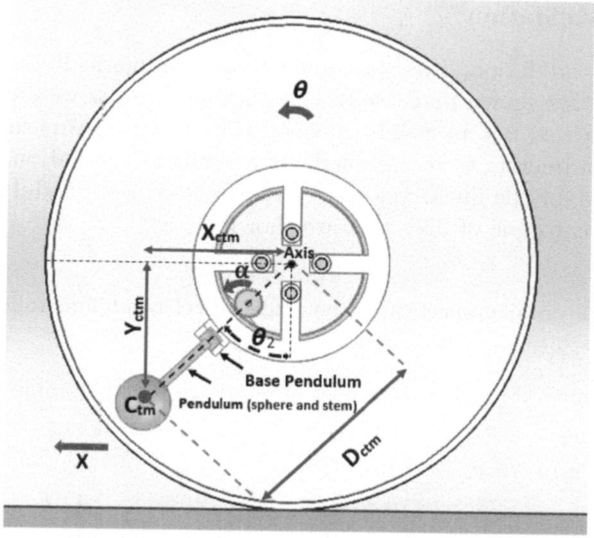

Fig. 3. Variables for the calculation of kinetic and potential energies of the pendulum.

3.3 Dynamic Model

The total kinetic and potential energies of the single-wheel robot are obtained by Eqs. (2), (4), (3) and (5), respectively.

$$T_{total} = \frac{1}{2}[m_c R_c^2 + m_d R_c^2 + m_{cor} R_c^2 + I_c + I_{cor} + 8(m_{bolt}(L_q^2 + R_c^2) + I_{bolt}) + m_{comp} R_c^2]\dot{\theta}^2 - m_{comp}\dot{X} D_{ctm} cos(\theta_2)\dot{\theta}_2 + \frac{1}{2} m_{comp} D_{ctm}^2 \dot{\theta_2}^2 + \frac{1}{2} I_{ctm} \dot{\theta_2}^2 \tag{6}$$

$$V_1 = m_c g R_c + m_d g(R_c - R_d) + m_{cor} g(R_c - R_{cor}) + 8 m_{bolt} g L_q \dot{\theta}^2 + m_{comp} g D_{ctm}(1 - cos(\theta_2)) \tag{7}$$

The Euler-Lagrange procedure can be applied to obtain the dynamic model of the single-wheel pendulum robot described by the following expressions:

$$\ddot{\theta} = [m_{comp} D_{ctm} R_c sin(b\alpha - \theta)(b\dot{\alpha} + \dot{\theta})^2 + m_{comp} g D_{comp} sin(b\alpha - \theta) - [m_{comp} D_{ctm} R_c cos(b\alpha - \theta) - I_{comp} - m_{comp} D_{comp}^2] b\ddot{\alpha} - c\dot{\theta}]\frac{1}{W} \tag{8}$$

$$\ddot{\alpha} = [\frac{T_m}{b} - [m_{comp} D_{ctm} R_c cos(b\alpha - \theta) - I_p - m_{comp} D_{ctm}^2]\ddot{\theta} - m_{comp} g D_{ctm} sin(b\alpha - \theta)] \quad \frac{1}{[I_{comp} + m_{comp} D_{ctm}^2] b} \tag{9}$$

where T_m is the input torque provided by the motor, c is the friction coefficient between the wheel and the ground, $b = \frac{R_{pinion}}{R_{cor}}$, $W = e - 2m_{comp} D_{ctm} R_c cos(b\alpha - \theta) + I_{comp} + m_{comp} D_{ctm}^2$ and $e = m_c R_c^2 + m_d R_c^2 + m_{cor} R_c^2 + 4(m_{bolt}(L_q^2 + R_c^2)) + I_c + I_d + I_{cor} + 8I_{bolt}$.

3.4 Model Validation

For the model validation, Eqs. (8) and (9) were numerically solved with the physical properties shown in Table 1. The obtained results were compared with a simulation carried out in Solidworks software where the material properties and friction parameters were assigned for a realistic simulation. In Fig. 4 the obtained results for the linear velocity of the single-wheel pendulum robot with a constant input torque of $2N \cdot mm$ are shown.

Table 1. The physical properties of the single-wheel pendulum robot's mechanical elements.

Element	Mass (g)	Moment of inertia $(g \cdot cm^2)$	Radius (cm)
Housing	583.44	73884.32	13.5
Base discs	70.42	1548.73	–
Bolts	1.87	0.35	1.64
Pendulum	122.36	586.611	10.3441
Pinion	Negligible	Negligible	0.75
Gear	4.199	7112.644	3.975

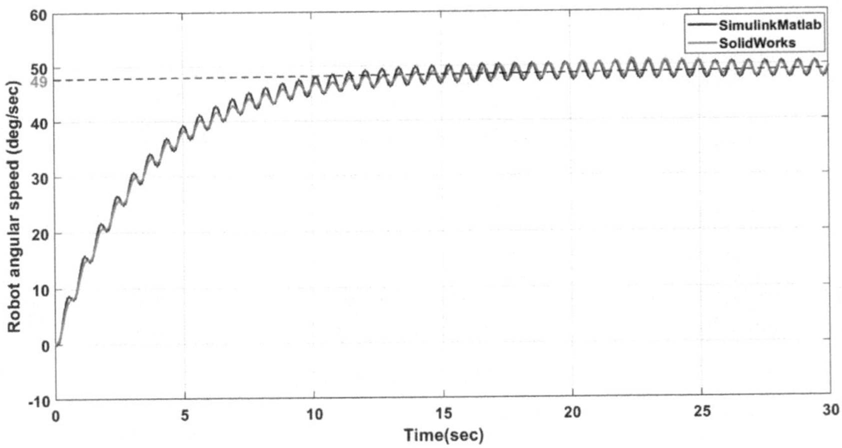

Fig. 4. Robot linear velocity with a constant input torque, numerical solution of the non-linear equations and SolidWorks simulation.

As can be observed in Fig. 4, the results obtained by evaluating Eqs. (8) and (9) are consistent with the numerical simulation carried out in the Solidworks environment which is interpreted as the dynamic model validation.

4 Linearization, Transfer Function and Velocity Control

From Eqs. (8) and (9), it is possible to linearize the system model around an operating point. This point is considered with the pendulum in a vertical position down (which is also an equilibrium position), this implies $\theta_2 = 0$ and $\theta = b\alpha$. Moreover, the equilibrium input is 0 and as a consequence the system has a constant velocity, $\ddot{\theta} = 0$ and $\ddot{\alpha} = 0$. The state variables are defined as $x_1 = \theta_{eq}$, $x_2 = \dot{\theta}_{eq}$, $x_3 = \alpha_{eq}$, $x_2 = \dot{\alpha}_{eq}$ and $U_{eq} = 0$.

With the equilibrium conditions, it is possible to linearize the model by using the Taylor series expansion [10] and considering the numerical values of Table 1, the linearized model in space state is given by

$$\begin{bmatrix} \dot{x}_{1\delta} \\ \dot{x}_{2\delta} \\ \dot{x}_{3\delta} \\ \dot{x}_{4\delta} \end{bmatrix} = \underbrace{\begin{bmatrix} 0 & 1 & 0 & 0 \\ -3.7951 & -0.3034 & 0.7161 & 0 \\ 0 & 0 & 0 & 1 \\ 486.0901 & 0.4006 & -91.7151 & 0 \end{bmatrix}}_{Matrix_A} \begin{bmatrix} x_{1\delta} \\ x_{2\delta} \\ x_{3\delta} \\ x_{4\delta} \end{bmatrix} + \underbrace{\begin{bmatrix} 0 \\ -32.3079 \\ 0 \\ 2.0577e04 \end{bmatrix}}_{Matrix_B} [T_m] \tag{10}$$

$$\begin{bmatrix} y_1 \\ y_2 \\ y_3 \\ y_4 \end{bmatrix} = \underbrace{\begin{bmatrix} 1\,0\,0\,0 \\ 0\,1\,0\,0 \\ 0\,0\,1\,0 \\ 0\,0\,0\,1 \end{bmatrix}}_{Matrix_C} \begin{bmatrix} x_{1\delta} \\ x_{2\delta} \\ x_{3\delta} \\ x_{4\delta} \end{bmatrix} + \underbrace{\begin{bmatrix} 0 \\ 0 \\ 0 \\ 0 \end{bmatrix}}_{Matrix_D} [T_m] \tag{11}$$

From the linearized system of Eqs. (10) and (11) the transfer function can be obtained as

$$G(s) = \mathbf{C}(s\mathbf{I} - \mathbf{A})^{-1}\mathbf{B} + \mathbf{D} \tag{12}$$

$$G(s) = \begin{bmatrix} \frac{-32.3079s^2+1.1772e+04}{s^4+0.3034s^3+95.5102s^2+27.5410s} \\ \\ \frac{-32.3079s^2+1.1772e+04}{s^3+0.3034s^2+95.5102s+27.5410} \\ \\ \frac{2.0577e+04s^2+6.2302e+03s62387}{s^4+0.3034s^3+95.5102s^2+27.5410s} \\ \\ \frac{2.0577e+04s^2+6.2302e+03s62387}{s^3+0.3034s^2+95.5102s+27.5410} \end{bmatrix} \tag{13}$$

It can noted that the second row from the right side of Eq. (13) corresponds to the transfer function of single-wheel robot velocity.

$$G(s_{\dot{\theta}}) = \frac{-32.3079s^2 + 1.1772e + 04}{s^3 + 0.3034s^2 + 95.5102s + 27.5410} \tag{14}$$

The transfer function given by (14) will be used to develop the control scheme for the single-wheel robot velocity. In order to validate this function, a numerical simulation with a constant input torque of $2N \cdot mm$ is carried out. The obtained

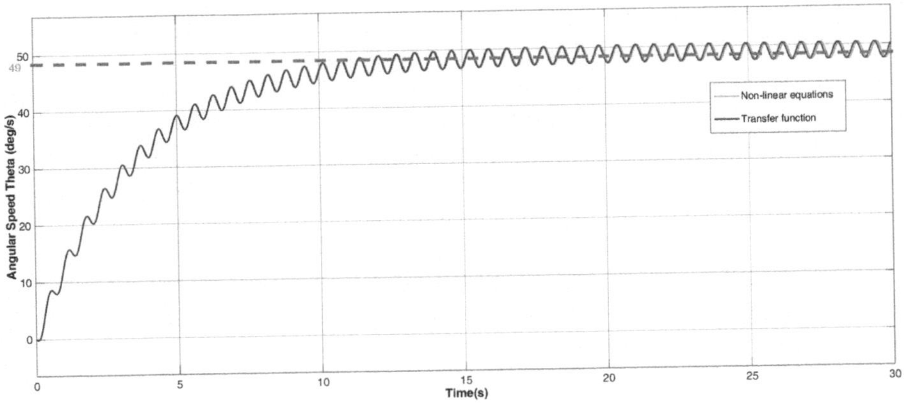

Fig. 5. Comparison of linear velocity of the single-wheel pendulum robot with a constant input torque, numerical solution of the non-linear equations and transfer function response.

results are compared with the non-linear system response and this comparison is presented in Fig. 5.

As can be appreciated, the obtained response from the transfer function is practically the same as that the obtained from non-linear equations. These results validate the linearization process and justify the use of the transfer function to synthesize the velocity control by linear techniques.

Before designing the controller, it is necessary to analyse the pole and zeros placement in open and closed loop. In Fig. 6 the distribution of pole and zeros in open loop is depicted. The open loop system is stable due to both poles having negative real parts.

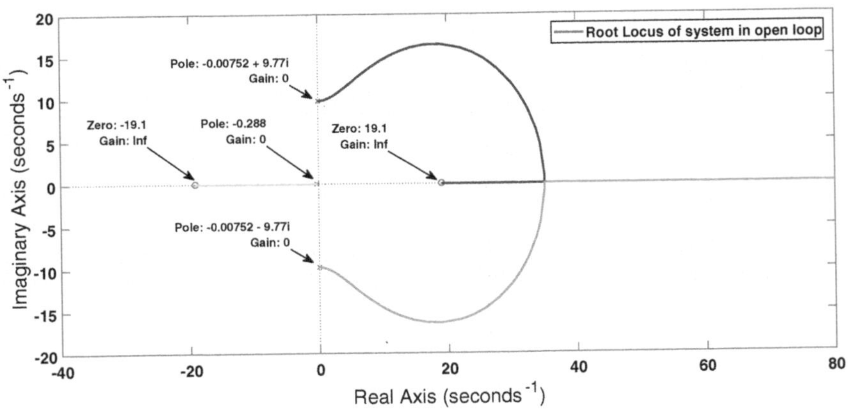

Fig. 6. Pole and zeros placement in open loop.

The closed loop transfer function is obtained as

$$\frac{C(s)}{R(s)} = \frac{G(s)}{1 + G(s)H(s)} = \frac{-32.31s^2 + 11772}{s^3 - 32s^2 + 95.51s + 11800} \tag{15}$$

From the transfer function (15) it is possible to obtain the placement of pole and zeros in closed loop which is shown in Fig. 7. There is a pair of conjugated poles in the right semi-plane, which means the closed loop system is unstable even if there are variations in the gain.

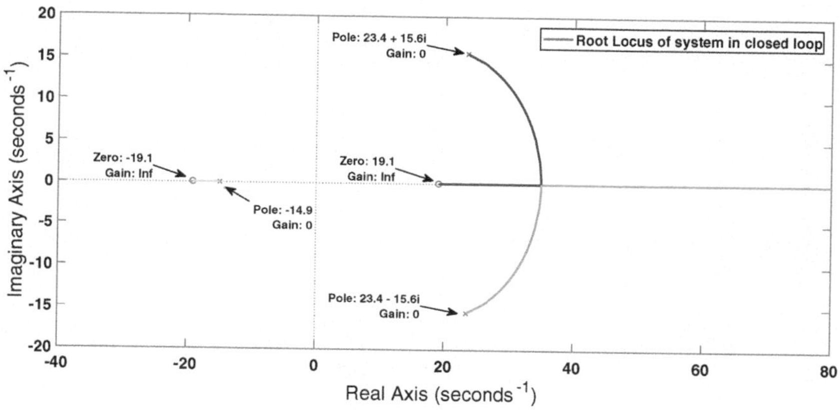

Fig. 7. Pole and zeros placement in closed loop.

It is possible to apply the root-locus control technique in order to impose a desired behavior in the closed loop system. In this case, the proposed values for the damping coefficient and natural frequency are $\zeta = 0.5$ and $\omega_n = 1\frac{rad}{s}$, respectively. With these values, the dominant poles are defined by $s_{1,2} = \zeta\omega_n \pm \sqrt{1-\zeta^2}i \Rightarrow s_{1,2} = -0.5 \pm 0.866i$.

The closed loop instability is caused by the dominant pole in the right semi-plane and the zero in the left semi-plane (see Fig. 7). A direct compensator is proposed to cancel these pair of conjugated poles and the zero in the left side of the complex plane.

$$\begin{aligned} G(s) * G(s)_{supr} &= \frac{-32.3079s^2 + 11772}{s^3 + 0.3034s^2 + 95.5102s + 27.5410} \\ &* \frac{s^2 + 0.01503027s + 95.49204}{(s + 19.088454)(s)} \\ &= \frac{-32.3079(s - 19.088454)}{(s + 0.288)(s)} \end{aligned} \tag{16}$$

The proposed controller must compensate for the deficiency angle introduced by the replacement of the dominant closed loop pole. By following the graphic method, the transfer function of the compensated closed loop system is given by

$$G_T = G(s)_s * G(s)_{comp} = \frac{-32.3079(s - 19.088454)}{(s + 0.288)(s)} * \frac{K(s + 0.604129)}{s + 1.65520103} \tag{17}$$

where the calculated gain is $K = 2.32967 \times 10^{-3}$.

The placement of pole and zeros of the compensated closed loop system is presented in Fig. 8. The placement of the dominant pole in the desired location can be observed.

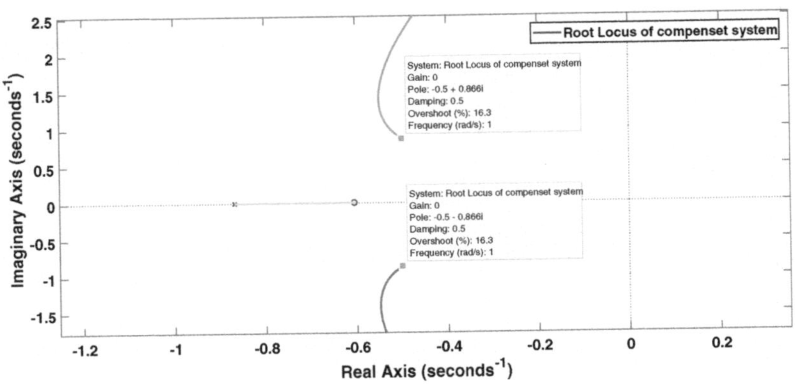

Fig. 8. Pole and zeros placement in closed loop.

The compensated closed loop system given by Eq. (17) was numerically simulated in order to appreciate the robot velocity behavior. The obtained results from this simulation are presented in Fig. 9. In this figure, a comparison between

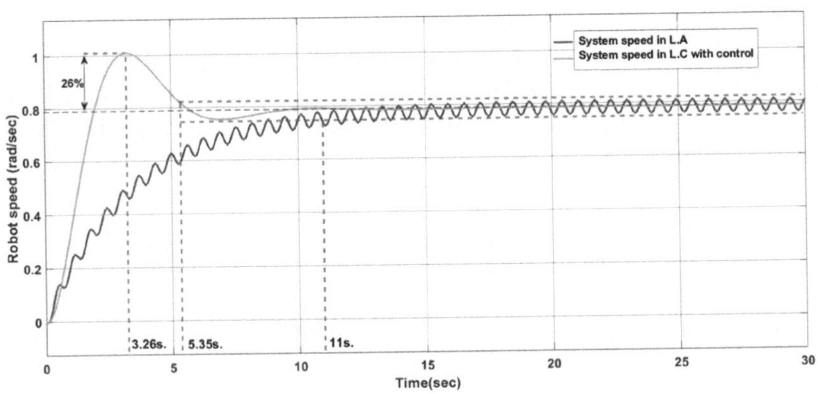

Fig. 9. Comparison of the linear velocity of the single-wheel pendulum robot with a constant input torque and with the proposed controller.

the open loop system with a constant input torque and the controlled system is shown. As can be observed, the controller performance presents an overshoot of about 26% and the settling time is around 205% faster than the open loop system. Moreover, oscillations are suppressed.

The compensated closed loop system response could be improved by modifying the requirements for the controller. However, it is important to take into account the physical limitations of the micromotor proposed to drive the single-wheel robot. In Fig. 10 the required torque for the robot control is shown. As can be observed, the maximum required torque is $5.325N \cdot mm$. This value can be used to select the micromotor with the adequate characteristics for this application.

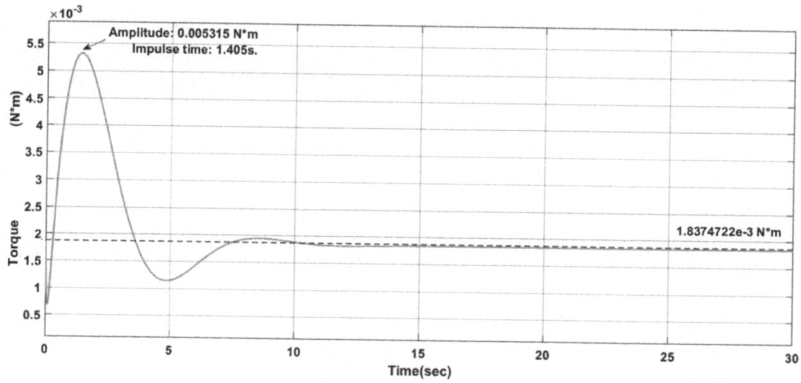

Fig. 10. Required torque for the single-wheel robot control.

5 Conclusions

In this work, the development of the dynamic model of a single-wheel robot with inertial pendulum locomotion was presented. The obtained mathematical model was validated by comparing the equations solution with numerical simulations carried out with the virtual prototype of the robot. The dynamic model was linearized around an operating point and the linearized model was used to synthesize a closed-loop controller for the forward robot velocity based on the root locus technique. The designed controller was proved by numerical simulations on the original non-linear system equations. The closed-loop system response shows the controller performance achieves the specifications for the robot operation. At the present time, a first experimental prototype of the single-wheel robot is being completed in order to experimentally validate the system dynamic model and the proposed linear controller. The proposed configuration limits the robot's movement along a single axis, and as a future work, an extended configuration to give the robot the capacity of movement in a plane could be proposed.

References

1. Jin H, Wang T, Yu F, Zhu Y, Zhao J, Lee J (2016) Unicycle robot stabilized by the effect of gyroscopic precession and its control realization based on centrifugal force compensation. IEEE/ASME Trans. Mechatron. 21(6):2737–2745. https://doi.org/10.1109/TMECH.2016.2590020
2. Vos WD, Von Flotow AH, Zhuang H (1990) Dynamics and nonlinear adaptive control of an autonomous unicycle: theory and experiment. In: Proceedings of the 29th IEEE Conference on Decision and Control, pp. 182–187. IEEE Press, New York. https://doi.org/10.1109/CDC.1990.203573
3. Jin H, Hwang J, Lee J (2011) A balancing control strategy for a one-wheel pendulum robot based on dynamic model decomposition: simulations and experiments. IEEE/ASME Trans. Mechatron. 16(4):763–768. https://doi.org/10.1109/TMECH.2010.2054102
4. Zhu Y, Gao Y, Xu C, Zhao J, Jin H, Lee J (2015) Adaptive control of a gyroscopically stabilized pendulum and its application to a single-wheel pendulum robot. IEEE/ASME Trans. Mechatron. 20(5):2095–2106. https://doi.org/10.1109/TMECH.2014.2363090
5. Lee J, Han S, Lee J (2013) Decoupled dynamic control for pitch and roll axes of the unicycle robot. IEEE Trans. Ind. Electron. 60(9):3814–3822. https://doi.org/10.1109/TIE.2012.2208431
6. Park JH, Jung S (2013) Development and control of a single-wheel robot: practical mechatronics approach. Mechatronics 23(6):594–606. https://doi.org/10.1016/j.mechatronics.2013.05.010
7. Lee SD, Jung S (2015) Experimental verification of stability region of balancing a single-wheel robot: an inverted stick model approach. In: Proceedings of Industrial Electronics Society, IECON 2015 - 41st Annual Conference of the IEEE, pp. 4556–4561. IEEE Press, New York. https://doi.org/10.1109/IECON.2015.7392810
8. Herrera-Cordero ME, Arias-Montiel M, Lugo-Gonzalez E (2019) Design and dynamic modeling of a novel single-wheel pendulum robot. In: Ceccarelli M, Gasparetto A (eds) Mechanism Design for Robotics. Springer, Cham, pp 353–360. https://doi.org/10.1007/978-3-030-00365-4_42
9. Chase R, Pandya A (2012) A review of active mechanical driving principles of spherical robots. Robotics 1(1):3–23. https://doi.org/10.3390/robotics1010003
10. Roubal J, Husek P, Stecha J (2010) Linearization: students forget the operating point. IEEE Trans. Educ. 53(3):413–418. https://doi.org/10.1109/TE.2009.2026427

Analysis and Simulation of the Oscillations in a Pulley-Ribbed Belt System

Ernesto Alonso Villegas-Jiménez[1(✉)], Jorge Díaz-Salgado[1], Oliver Marcel Huerta-Chávez[1], Fernando Macedo-Chagolla[2], and Edgar Arturo Dávila-Martínez[1]

[1] Tecnológico de Estudios Superiores de Ecatepec, Av. Tecnológico S/N, Col. Valle de Anáhuac, Sección Fuentes, 55210 Ecatepec de Morelos, Estado de México, Mexico
llanus_2@yahoo.com.mx

[2] Facultad de Estudios Superiores Aragón, Av. Hacienda de Rancho Seco S/N, Impulsora Popular Avícola, 57130 Nezahualcóyotl, Estado de México, Mexico

Abstract. This paper presents the analysis and computer simulation of the rotational and transverse oscillations present in a circular motion transmission system through the relationship of pulley-ribbed belt, widely used in the industrial robots. The analysed system is a pulley-ribbed belt coupled with rigid pulleys. The analysis is based in the mathematical model developed for serpentine belt drive systems. The analysis includes: (i) the geometric characteristics, (ii) the behavior of the transmission in steady state, and (iii) the dynamical response of the system. Dynamic analysis of these type of systems is complex because: (i) it has coupled rotational and transverse vibration modes in a non-linear form, and (ii) includes a circular motion transmission system acceleration-deceleration profile. The proposed analysis allows to determine a priori the frequency, the phase shift and the oscillation amplitude of a section of the ribbed belt excited by any forced velocity profile.

Keywords: Dynamic analysis of a pulley-ribbed belt system · Pulley-ribbed belt system · Rotational modes · Transverse modes

1 Introduction

The rotational motion transmission system based on pulley-ribbed belt coupling, is a flexible coupled mechanical system with sliding elements, used widely in industrial robotics, because of the advantages with respect to other types of mechanical couplings. Although it has been used for more than forty years to transmit mechanical energy through a circular movement, this system still presents important challenges as: (i) a detailed mathematical analysis, (ii) the implementation of computer aided design, and (iii) the study of new applications.

Circular motion transmission systems must operate reliably with minimal noise and vibration. The problems that normally arise are due to resonance modes that generates

E. E. Hernandez et al. (Eds.): LASIRS 2019, MMS 86, pp. 261–268, 2020.
https://doi.org/10.1007/978-3-030-45402-9_25

noise and fatigue mechanics effects. Since 1979 the belts in "v" have been largely replaced by serpentine belt drives with dynamic tensioners in transmissions systems [1].

Nowadays, these systems use multichannel ribbed belts, to reduce flexion and upgrade damper stiffness. Ribbed type transmissions are reliable, reduce the tension of the belt and the generation of friction heat when operating near the constant tension of the belt [2].

The use of pulley-ribbed belt system reduces noise and vibration levels. There is a great interest in the full understanding of the dynamic and static behavior of this nonlinear mechanical system. Effective tools for the model have been developed over several years and predict its response. Rotational and vibration transverse modes occur in all transmission systems that involve belts with fixed pulleys [3].

In rotational modes driven pulleys rotate on their axes and belt sections act as axial springs. In the transverse modes sections of belt vibrate transversally, similar to a stretched rope. Both degrade the performance of the system. Rotational modes of sheaves induce dynamic tension in the belt, dynamic reactions, fatigue effect on the elements and noise in the structure. The transverse vibrations also induce dynamic tensions and can radiate directly noise.

Rotational modes in a ribbed belt have been studied during the last twenty five years. In [4] rotational modes are studied as a stationary system including the support and the belt damping. In [5] is presented the study of a ribbed belt with a dynamic tensor under the rapid acceleration of the machine. Finally, in [6] the transient response of a similar system is examined considering rapid acceleration of the engine, to predict the sliding of the belt. However, all these studies ignore the transverse dynamics of the belt. In [7] a dynamic tensioner coupled with a transmission system is studied.

In this work an appropriated mathematical model is proposed. This model must be at the same time: (i) as simple as possible and (ii) reflect accurately the static and dynamic behavior of the system. The document is organized as follows: in section number one the mathematical model is proposed. The computational method used in the static analysis is described in section two. Results and the respective analysis are presented in section three. Finally some conclusions are presented.

2 Mathematical Model Deduction

The ribbed pulley-belt system (Fig. 1) is an example of a flexible coupling with sliding elements, capable of reducing rotational and transverse oscillations. This system was modeled by [8] under fairly realistic modeling considerations and uses the minimum of elements to model its behavior which are: (i) a driving pulley, (ii) a driven pulley, (iii) a dynamic tensor, and (iv) a belt.

Practical considerations for analysis are: (i) the damping factor is considered negligible, (ii) the belt is modeled with negligible flection stiffness, (iii) the belt speed is constant and uniform for axial translation, (iv) the belt has an insignificant slip factor.

The transverse vibrations present in the two sections of the belt when attaching a dynamic tensor are first studied in [8].

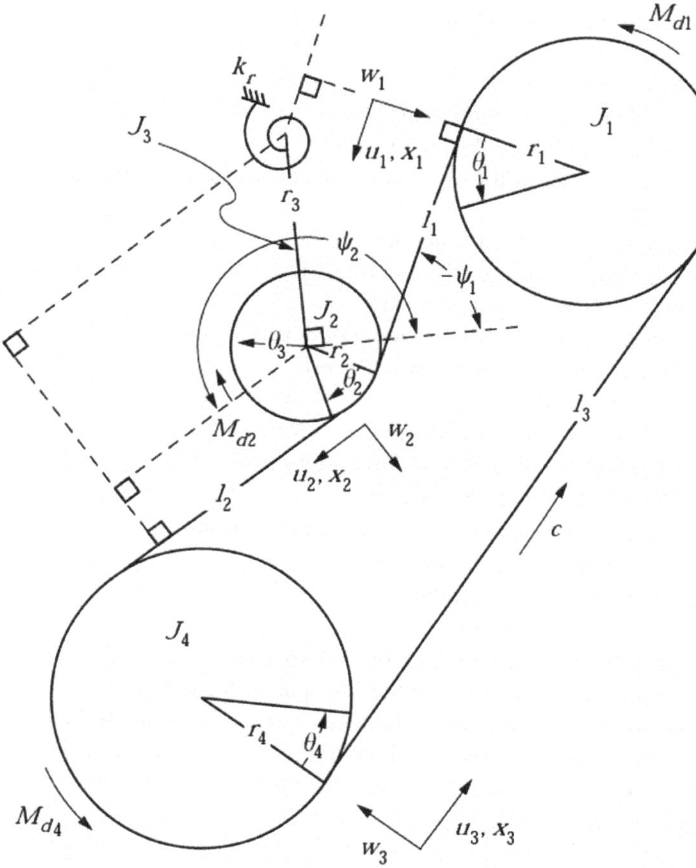

Fig. 1. Diagram of the drive system by means of a belt and three pulleys.

Previous research assumes that in the linear response: rotational and transverse movements are decoupled. This approximation is good for systems with fixed elements, but because the rotation of the tensor arm linearly couples the transverse and rotational movements, it cannot be used for systems where there is a dynamic tensor.

Hamilton principle is used to derive the non-linear in steady-state equations to describe the system behavior at constant speed. The nonlinearities considered are geometrical describing a finite stretch of the sections of the belt. The steady-state assessment is presented in [8] and [9]. The non-linear equation of transverse movement for the three sections of belt is [10]:

$$m\left(w_{i,tt} + 2cw_{i,xt}\right) - P_{ti}w_{i,xx} = P_{di}w_{i,xx} \qquad (i = 1, 2, 3) \tag{1}$$

Where: c is the belt travel speed constant, i is the belt section number and the subscripts, x, and t, are used to denote the partial derivative with respect to the coordinate section, x, and time t, respectively. The $2cw_{i,xt}$ term is the Coriolis acceleration component.

The nomenclature corresponds to:

c = steady state belt speed $[m/s]$
EA = longitudinal belt stiffness modulus $[N/m]$
F_{di} = dynamic force applied to the pulley i $[N]$
J_i = mass moment of inertia for *i-th* discrete element, for tensioner arm it is $J_{arm} + m_{pulley} r_{arm}^2 [kgm^2]$
k_4 = translational stiffness of tensioner arm $[N/m]$
k_{gr} = tensioner geometric spring stiffness $[N/m]$
k_i = longitudinal stiffness of span i $[N/m]$
k_r = rotational spring constant adimensional of tensioner arm
K_s = stiffness rotational of tensioner arm $[N/m]$
l_i = length of belt span i $[m]$
M_{di} = dynamic moment applied in the pulley i $[Nm]$
m = belt mass per unit length, constant throughout system $[kg]$
P_c = balancing centrifugal pressure $[kg/s^2]$
P_{ti} = traction tension in the spans in equilibrium operation $[kg/s^2]$
P_{di} = dynamic tension in span *i-th* $[kg/s^2]$
r_i = *i-th* system discrete element radius $[m]$
t = time $[s]$
$u_i(x_i, t)$ = longitudinal displacement in the *i-th* span of the belt $[m]$
$\omega_i(x_i, t)$ = transversal displacement in the *i-th* span of the belt $[m]$
x_i = local coordinate in the longitudinal direction of the *i-th* span of the belt $[°]$
z_i = local coordinate in the transversal direction of the *i-th* span of the belt $[°]$
θ_i = rotation of *i-th* discrete element in system from equilibrium $[°]$
ψ_1, ψ_2 = alignment angles $[°]$
ω = frequency of oscillation $[s]$

P_{ti} is the traction stress in the *i-th* band element in steady state. The component of centripetal acceleration $c^2 w_{i,xt}$ is included in the term $P_{ti} w_{i,xx}$. Pulleys motion is described by the next linear equations [11]:

$$P_{d1} - P_{d3} = m_1 \ddot{x}_1 \tag{2}$$

$$P_{d2} - P_{d2} = m_2 \ddot{x}_2 \tag{3}$$

$$P_{d3} - P_{d3} = m_3 \ddot{x}_3 \tag{4}$$

Where $x_i = r_i \theta_i$, $m_i = J_i / r_i^2$ and P_{di} are the dynamic tensions in each section, induced by an infinitesimal pulley and a rotational arm tensioner.

Well then:

$$P_{d1} = k_1 (x_3 \cos \psi_1 + x_2 - x_1) \tag{5}$$

$$P_{d2} = k_2 (x_3 \cos \psi_2 + x_4 - x_2) \tag{6}$$

$$P_{d3} = Pk_3(x_1 - x_4) \tag{7}$$

Here $k_i = EA/(l_i)$, (l_i) it is the length of the section i of the band, and 1, 2, are the angles of alignment between the movement of the tension arm and the adjacent sections of the band in balance.

The linearized equation of motion of the tension arm is:

$$\begin{aligned} &\left(-P_{t1}\omega_{1,x}(l_1,t) + mc\omega_{1,t}(l_1,t)\right) \sin \psi_1 \ldots \\ &+\left(P_{t2}\omega_{2,x}(0,t) - mc\omega_{2,t}(0,t)\right) \sin \psi_2 \ldots \\ &-k_1(x_3 \cos \psi_1 + x_2 - x_1) \cos \psi_1 \ldots \\ &-k_2(x_3 \cos \psi_2 + x_4 - x_2) \cos \psi_2 \ldots \\ &-\left(k_s + k_{gr}\right)x_3 = m_3\ddot{x}_3 \end{aligned} \tag{8}$$

Where, $k_s = k_r/r_3^2$, k_r, are the rotational stiffness of the tensioner spring and

$$k_{gr} = (P_{t1} \sin \psi_1 - P_{t2} \sin \psi_2)/r_3 \tag{9}$$

is tensioner arm stiffness when changing your position.

3 Computational Method

A computational method is established for the stationary analysis. The method consists in determining the roots of the vibration modes function obtained by the modal analysis. Figure 2 shows the frequency function in steady state regime, this is, with constant rotational velocity.

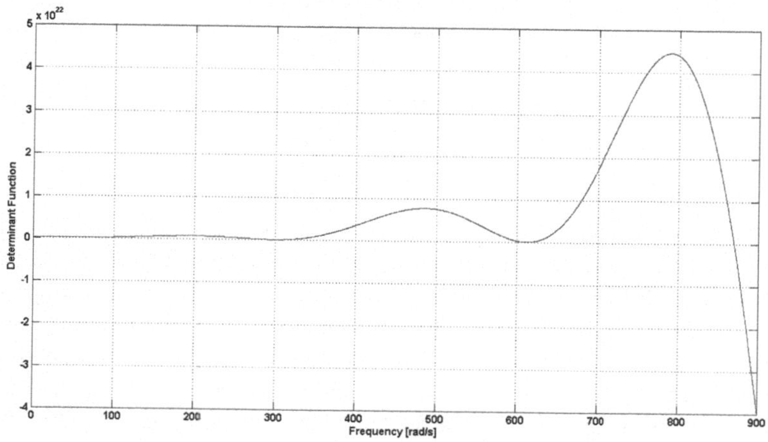

Fig. 2. Determinant function vs. Frequency

The function roots can be obtained: (i) in a graphical form or (ii) using an appropriate numerical method. The proposed algorithm is described next:

1. Determine the explicit function for a valid set of parameters.
2. Run an outer loop varying different speed of rotation values within a range of interest.
3. Run an internal loop varying an initial search range determined by user.
4. Using an adequate root search function a vector of founded roots is formed within the range of interest.
5. Run a last loop to determine the changes between the founded roots, the number of founded roots and the root chart.

4 Analysis of Results

Figure 3 shows the effect of the variation in the transport speed in the first six oscillation modes of the system, according to the following parameters:

$$
\begin{aligned}
r_1 &= 0.0889[m] \\
r_2 &= 0.4520[m] \\
r_3 &= 0.0970[m] \\
r_4 &= 0.0269[m] \\
l_1 &= 0.1548[m] \\
l_2 &= 0.3449[m] \\
l_3 &= 0.5518[m] \\
\psi_1 &= 44.21[^\circ] \\
\psi_2 &= 178.74[^\circ] \\
J_1 &= 0.072480\left[kg \cdot m^2\right] \\
J_2 &= 0.000293\left[kg \cdot m^2\right] \\
J_3 &= 0.001165\left[kg \cdot m^2\right] \\
J_4 &= 0.000293\left[kg \cdot m^2\right] \\
k_r &= 54.37 \\
m &= 0.1029[kg] \\
EA &= 170\left[\tfrac{N}{m}\right] \\
P_c &= 300\left[kg/s^2\right]
\end{aligned}
$$

The results show that: (i) the first and fourth mode remains quasi-invariant along the transport speed variation, (ii) the second mode decrease smoothly as transport speed increases, (iii) the third, fifth and sixth modes drastically decrease as the transport speed increases, (iv) the behavior of the first six dominant modes of oscillation can be observed with respect to the transport speed of the driving pulley using three different sets of angles for the position of the tension arm ($\psi 1$ and $\psi 2$) and (v) the curves show that the dominant modes of oscillation are practically independent of the tensioning angles of the tensioner and dependent on the speed of rotation of the driving pulley. The higher modes are not included since they are harmonics of the fundamental ones.

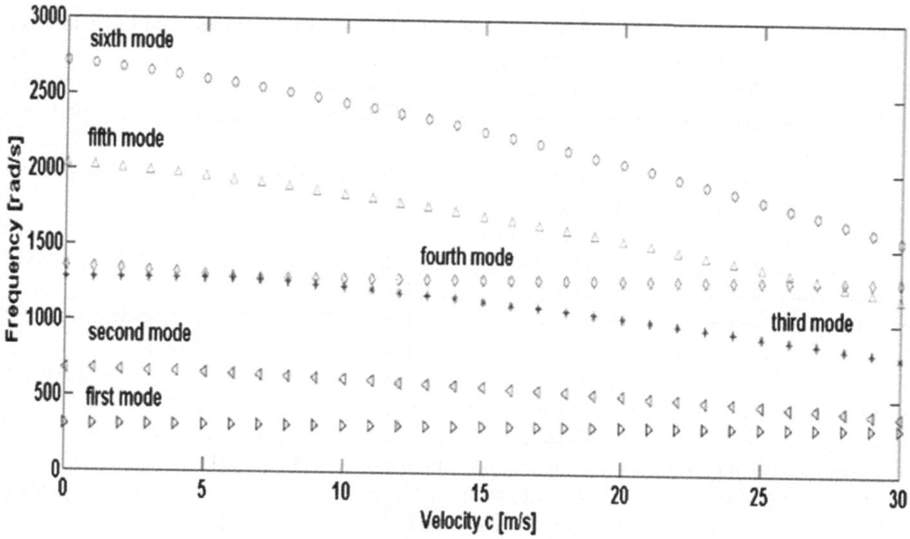

Fig. 3. Effect of transport speed in natural frequencies

5 Conclusions

The results obtained in this research show the dependence of the modes of oscillation as function of the rotational speed of the conductive pulley.

Three algorithms are proposed for the numerical solution of steady state which allow an analysis of behavior of different system parameters, so that the influence of each parameter can be seen graphically and quantitative in the oscillation modes of the system. In particular two types of parametric analysis were proposed: (i) the behavior of the oscillation modes in the orientation of the tensioner arm and (ii) the behavior of the first mode of oscillation in the initial tension of the belt sections.

The variation of any of the system parameters affects the modes of vibration, since they are function of the system eigenvalues, and these, in turn, are a function of the system parameters.

Although, in principle, the designer can vary freely the values of the parameters, by the geometric arrangement of the system elements, not any set of parameters is physically achievable.

Therefore, to determine whether a set of parameters is physically achievable, you must: (i) construct the system or, (ii) draw the system at scale and determine it graphically. This process can be angry and time consuming, so it is desirable to have a method that allows you to carry out the system stroke in a simple way.

References

1. Beikmann RS, Perkins N, Ulsoy A (1996) Free vibration of serpentine belt drive systems. J Vib Acoust 118(3):406–413

2. Casidy RL, Fan SK, Macdonald RS, Samson WF (1979) Serpentine-extended life accessory drive. In: SAE international, United States
3. Houser DR, Oliver L (1975) Vibration of V-belt drives excited by lateral and torsional inputs. In: Proceedings of the fourth world congress on the theory of mechanisms, Newcastle Upon Tyne, England, vol 4, pp 899–905
4. Gaspar RGS, Hawker LE (1989) Resonance frequency prediction of automotive serpentine belt drive systems by computer modeling. In: Proceedings of ASME conference on mechanical vibration and noise, vol 18, no 12, pp 13–16
5. Barker CR, Oliver LR, Breig WF (1991) Dynamic analysis of belt drive tension forces during rapid engine acceleration. In: SAE international, United States
6. Hwang SJ, Perkins NC, Ulsoy AG, Meckstroth R (1994) Rotational response and slip prediction of serpentine belt drive systems. ASME J Vib Acoust 116(1):71–78
7. Hawker LE (1991) A vibration analysis of automotive serpentine accessory drive systems. PhD dissertation, University of Windsor, Ontario, Canada
8. Beikmann RS (1992) Static & dynamic behavior of serpentine belt drive systems: theory and experiment. PhD dissertation, University of Michigan, Ann Arbor, MI
9. Beikmann RS, Perkins NC, Ulsoy AG (1997) Design and analysis of automotive serpentine belt drive systems for steady state performance. J Mech Des 119(2):162–168
10. Wickert JA, Mote CD Jr (1998) Current research on the vibration and stability of axially-moving materials. Shock Vib Dig 20(5):3–13
11. Li X, Chen L (2008) Modal analysis of coupled vibration of belt drive systems. Appl Math Mech 29(1):9–13

Attitude Control of Torpedo Anchor with Vectorial Thrust for Use in a Pseudo-Submersible Platform

G. J. Garnica-Castro[1](✉), O. M. Huerta-Chávez[2], J. Cruz-Cruz[2], and M. Toledo-Velázquez[2]

[1] Tecnológico de Estudios Superiores de Ecatepec Div. Ing. Aeronáutica, Ecatepec de Morelos, Estado de México, Mexico
jgibran@tese.edu.mx

[2] Instituto Politécnico Nacional, Sección de Estudios de Posgrado ESIME Zacatenco, Ciudad de México, Mexico

Abstract. In the present days, the hydrocarbons in the ground are limited by the high human consumptions. Therefore, a necessity of exploring in deep ocean becomes a reality with huge challenges, one of these challenges is the anchorage requirement, of course a problem with its installation. The present work deals on one way to anchorage, the torpedo anchor. This anchorage system works with a body forces effects added to hydrodynamics effects, due by the interaction between torpedo's surface and water. On the body force we have gravity and flotation (constants), on the other hand, we have a nonconstants hydrodynamics forces that can change the trajectory during the fall, and producing a non-successful installation of the torpedo on the seabed. So the dynamics of the torpedo anchor are derived by doing a rigid body analysis, computing the general equations that describe the torpedo trajectory, second, performs a CFD analysis to obtain the hydrodynamics forces, and by the last, add on an extra vectoring control force (to correct the orientation).

Keywords: Computational fluid dynamics · Deep penetrating anchor · RANS · Deep ocean platform

1 Introduction

For the anchorage presses, the current system of the torpedo is under the action of the gravity force and the active changes in the hydrodynamics proprieties, this must be analyzed to guarantee the penetration on the seabed.

This work includes a dynamic analysis of Deep Penetration Anchor (DPA), added a vectorial "control" force to correct the trajectory. So, it's necessary perform a mathematical analysis to obtain a mathematical model for describe the dynamical trajectory in function of the gravity and hydrodynamic proprieties.

E. E. Hernandez et al. (Eds.): LASIRS 2019, MMS 86, pp. 269–278, 2020.
https://doi.org/10.1007/978-3-030-45402-9_26

1.1 Mathematical Model

As we know, the full description of any 3D rigid body has 6 Degrees of Freedom, so, it's necessary obtain the 6 equations that describe the dynamic motion of the torpedo. According to [1], the vectorial form of this equations is:

$$M\dot{v} + C(v)v + D(v)v + g(\eta_1) = \tau + w \tag{1}$$

$$\dot{\eta} = J(\eta_2)v \tag{2}$$

$$y = \eta + u \tag{3}$$

Where η *is* the earth-fixed position vector (that describe, of course, the attitude), v the body-fixed vector of linear and angular velocities, τ are the vector control forces, and w the disturbance vector. Therefore, the DPA motion is defined in the body-fixed frame on relation to the earth-fixed frame (considered as inertial frame). Thus, its necessary do a relation between both frames. The relation of these frames is show in the next Fig. 1.

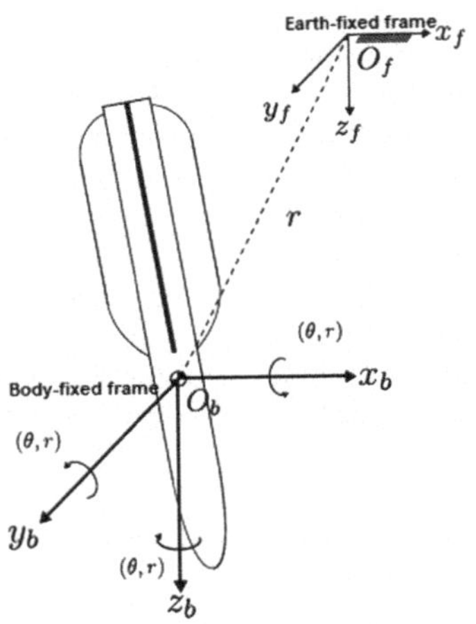

Fig. 1. Torpedo's frame of reference related to the earth fixed frame.

Do [2] proposes to divide the analysis in its kinematic and dynamic parts. For the kinematic analysis, we use the Eq. (3), where η_2 is the angle vector defined as $\eta_2 = [\phi, \theta, \psi]^T$, so

$$J(\eta_2) = \begin{bmatrix} R_b^i(\eta_2) & 0_{3x3} \\ 0_{3x3} & T^{-1}(\eta_2) \end{bmatrix} \tag{4}$$

$$T^{-1}(\eta_2) = \begin{bmatrix} 1 & \sin\varphi \tan\theta & \cos\varphi \tan\theta \\ 0 & \cos\varphi & -\sin\varphi \\ 0 & \sin\varphi \sec\theta & \cos\varphi \sec\theta \end{bmatrix} \tag{5}$$

$$R_b^i(\eta_2) = \begin{bmatrix} \cos\varphi\cos\theta & \cos\theta\sin\psi & -\sin\theta \\ -\cos\varphi\sin\psi + \cos\varphi\sin\theta\sin\psi & \cos\varphi\cos\psi + \sin\varphi\sin\theta\sin\psi & \cos\varphi\sin\theta \\ \cos\varphi\sin\theta\sin\psi + \sin\varphi\sin\psi & \cos\varphi\sin\theta\sin\psi - \cos\varphi\sin\psi & \cos\varphi\sin\theta \end{bmatrix} \quad (6)$$

As we know, the dynamical analysis follows the Newton's second law

$$F_b^i = m(\dot{v}_1 + v_2 v_b^i) \quad (7)$$

$$M_b^i = m(h_b^i + v_2 h_b^i) \quad (8)$$

Where $v_1 = \left[u\ v\ w\right]^T$, $v_2 = \left[p\ q\ r\right]^T$ y $h = I v_2$, and I is the identity matrix.

1.2 Thrust Force

The following diagram shows the thrust force generated by the nozzle in the body-fixed frame, noted by F_p (Fig. 2).

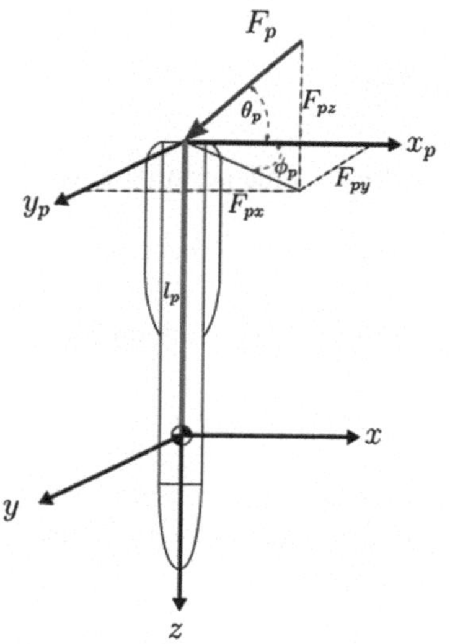

Fig. 2. Free body diagram with the thrust force acting on the torpedo.

The force and moment by the thrust are

$$F_p = F_{px}\hat{i} + F_{py}\hat{j} + F_{pz}\hat{k} \quad (9)$$

$$M_p = M_{px}\hat{i} + M_{py}\hat{j} + 0\hat{k} \quad (10)$$

Of course, the components are the projections in each axis, i.e. $F_{px} = F_p \cos\theta_p \cos\varphi_p$, $F_{py} = F_p \cos\theta_p \sin\varphi_p$ $F_{pz} = F_p \sin\theta_p$, $M_{px} = (F_p \cos\theta_p \cos\varphi_p)(l_p)$ and $M_{py} = (F_p \cos\theta_p \sin\varphi_p)(l_p)$, where l_p is the distance.

1.3 Hydrodynamics Perturbations

The forces and moments by the hydrodynamics perturbations are described in the Fig. 3.

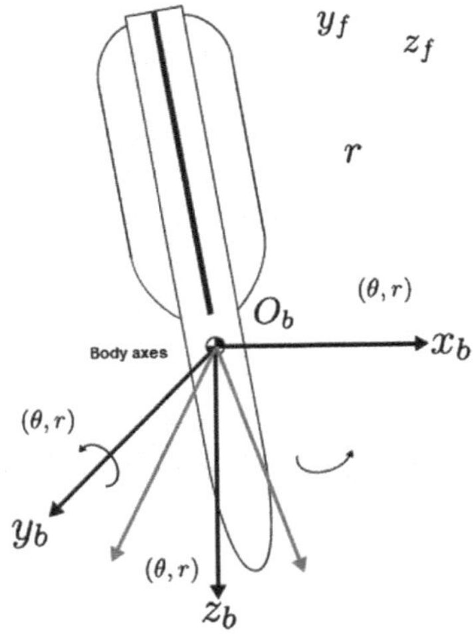

Fig. 3. Hydrodynamics forces

Therefore, the forces of hydrodynamic perturbations are

$$F_{Lx} = -q_{\infty} C_L A \cos \psi_h \tag{11}$$

$$F_{Ly} = -q_{\infty} C_L A \sin \psi_h \tag{12}$$

$$F_D = -q_{\infty} C_D A \tag{13}$$

Where C_L and C_D are the hydrodynamic coefficients (that are obtained by a CFD analysis), and q_{∞} is the dynamic pressure, which is defined by $q_{\infty} = \frac{1}{2}\rho(v_h^b)^2$.

2 CAD-CFD

The analysis is performed in a conventional torpedo anchor used by Petrobras in Brazil [3], shown in the Fig. 4, and its characteristics are summarized in Table 1.

We perform a classical CFD approach, hence the control volumen and mesh must be in accordance to the problem.

The control volumen is in a pastel shape, this based on that the above equations are derived in the body-fixed axis, so the calculated hidrodynamics forces does not need to be transformed. This control volume is indicated in the Fig. 5.

In this figure, we can observe the inlet velocity at different angles, this can be performed easily in a pastel control volumen.

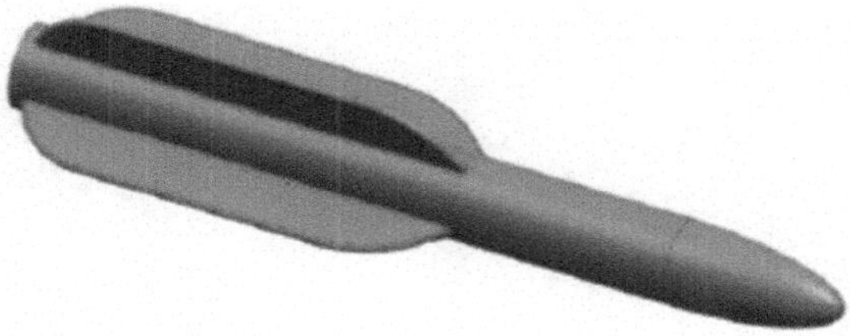

Fig. 4. Isometric view of the torpedo model.

Table 1. Physical Propeties of the torpedo

Volume	0.0086 m^3
Area	0.5547 m^2
Mass	67.4385 kg
Weight	661.3465 N

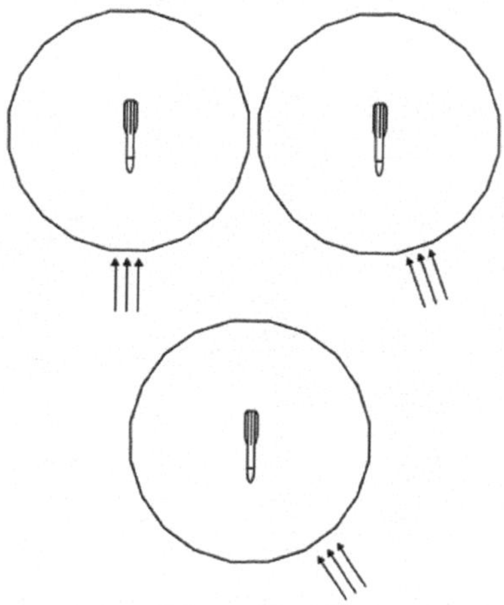

Fig. 5. Control volumen of the torpedo at differents angles.

The mesh of the body was performed in professional software, considering an Ortho-Trias elements. This mesh is specified in the next figure, where we can see a densification in the fins (Fig. 6).

To detect the viscous effects on the body surface, that has a high impact on the tangential drag forces, its necessary increased the elements close to surfaces. This process is summarize to compute the boundary layer thickness

$$\delta = \frac{0.377}{\sqrt[3]{\mathrm{Re}}} \tag{11}$$

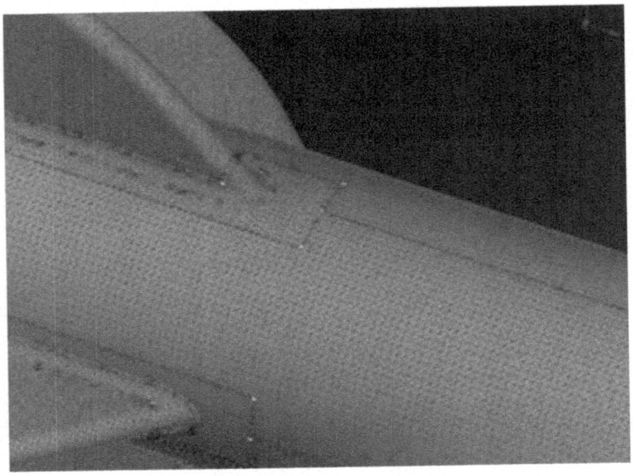

Fig. 6. Torpedo mesh view

Where Re is the Reynold's number. After, the boundary layer is divided by ten layers. The resulting volumetric mesh used in this analysis is shown in the Fig. 7.

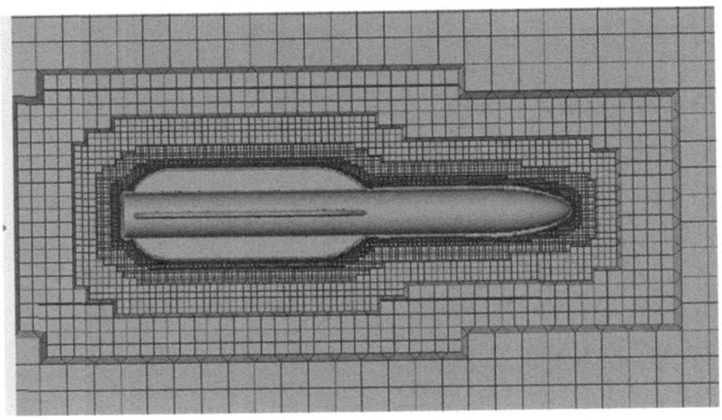

Fig. 7. Side view of the volumetric mesh

The Navier-Stokes equations are simplified to its incompressible form and solved in a classical Reynolds Averaged Navier-Stokes (RANS) coupling the moment and continuity equations [4]. The turbulent model is the Spalart Almaras (S-A), with one equation for the turbulent viscosity including the convective and transport elements to calculate the production and dissipation of turbulent viscosity [5].

For the torpedo, the Fig. 8 shows the contours of the velocity and pressure fields as a result of the numerical simulations.

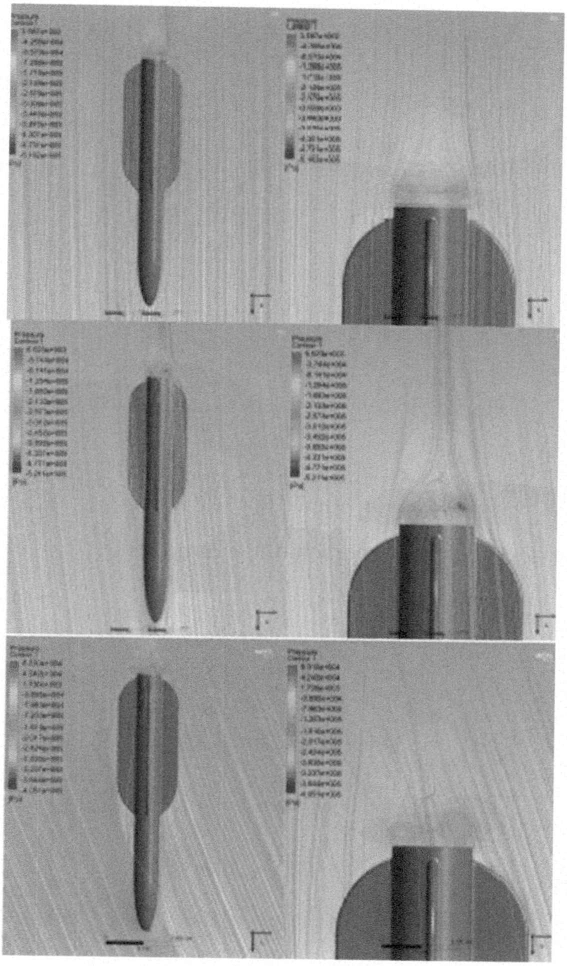

Fig. 8. Velocity and pressure contours

This figure shows the instabilities of the fluid at angles greater of 5, this impact directly on the stability of the torpedo. To show this behavior we made a graph of the moment coefficient and angle of attack (Fig. 9).

We can see that the torpedo has a high stability in angles of attack between −5 to + 5. With these results we can continue with the control.

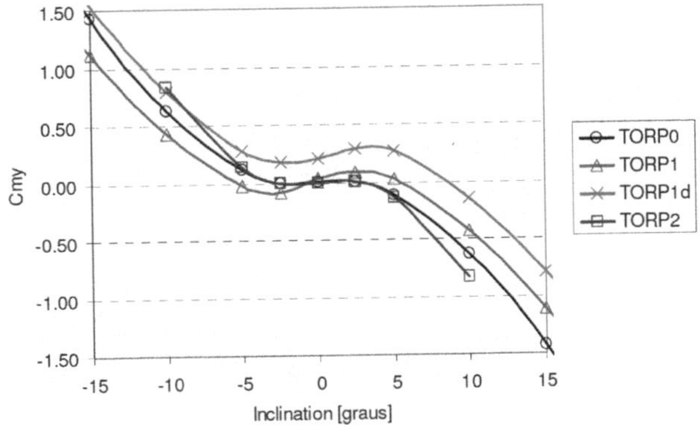

Fig. 9. Moment coefficient vs angle of attack

3 Control Design

The equations that describe the orientation of the torpedo are

$$\begin{bmatrix} M_{px} \\ M_{py} \\ 0 \end{bmatrix} + \begin{bmatrix} M_{hx} \\ M_{hy} \\ 0 \end{bmatrix} = \begin{bmatrix} I_x\dot{p} - (I_y - I_z)qr \\ I_y\dot{q} - (I_z - I_x)rp \\ I_z\dot{r} - (I_x - I_y)pq \end{bmatrix} \tag{12}$$

$$\begin{bmatrix} \dot{p} \\ \dot{q} \\ \dot{r} \end{bmatrix} = \begin{bmatrix} \frac{1}{I_x}[(I_y - I_z)qr + M_{hx} + M_{px}] \\ \frac{1}{I_y}[(I_z - I_x)rp + M_{hy} + M_{py} \\ \frac{1}{I_z}[(I_x - I_y)pq \end{bmatrix} \tag{13}$$

Or

$$\begin{aligned} \dot{x} &= f(x) + g(\pi) \\ y &= h(x) \end{aligned} \tag{14}$$

Where $f(x)$ is the state function and $g(\pi)$ the input function. Hence the equation that describe the output of the system in terms of the perturbation and propulsion force are

$$[\ddot{y}] = \begin{bmatrix} qr(I_x - I_z) + \frac{1}{I_x}M_{hx} \\ rp(I_z - I_y) + \frac{1}{I_y}M_{hy} \end{bmatrix} + \begin{bmatrix} \frac{1}{I_x}(F_p\cos\theta_p\cos\varphi_p)(l_p) \\ \frac{1}{I_y}(F_p\cos\theta_p\sin\varphi_p)(l_p) \end{bmatrix} \tag{15}$$

And

$$s_2\dot{s}_{\dot{s}} = -|s_2|D_2 - s_2 d_2 \le 0 \tag{16}$$

Using the input parameters obtained in the CFD analysis we can simulate the behaviour of the torpedo at different angles of attack (Figs. 10 and 11).

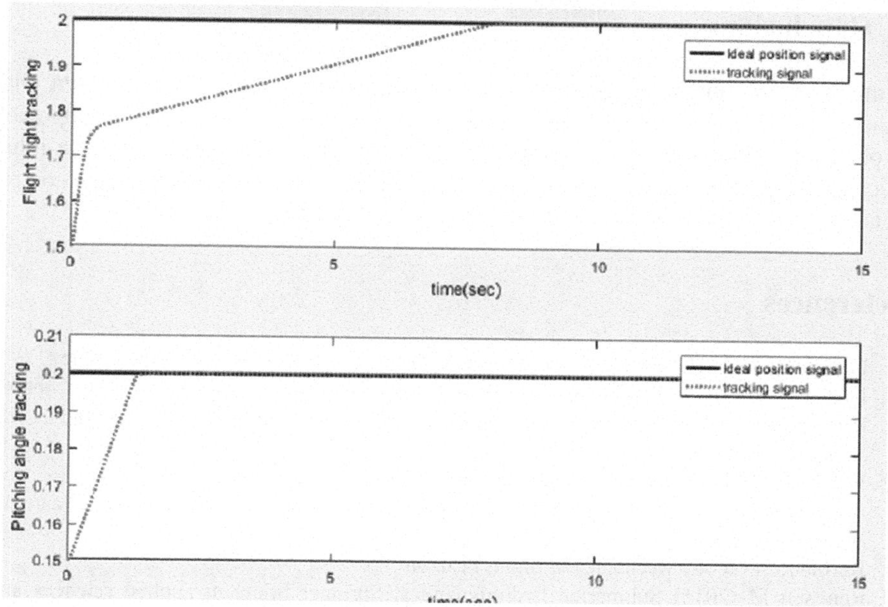

Fig. 10. Torpedo at 4 degrees.

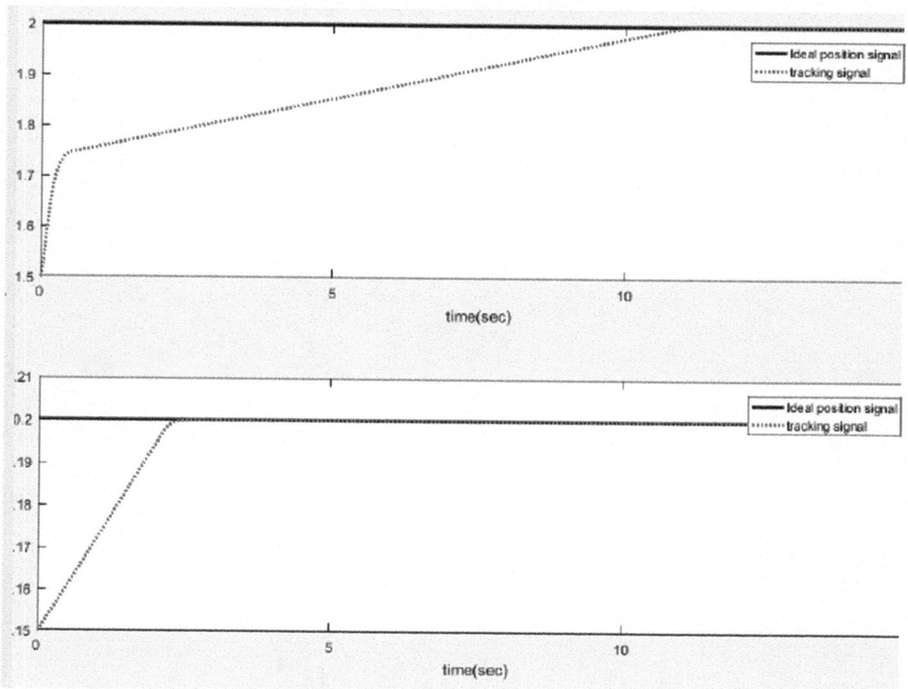

Fig. 11. Torpedo at 8 degrees

4 Conclusions

In this work, we obtain the dynamic equations that describe the motion of a DPA with a thrust, limiting only to a dynamic analysis. These equations are the first approach to obtain a control model to modify the trajectory. The CFD analysis gives a more precise data to complete the dynamic equations, the next step is implementing a control technique, and the design the devise to give the thrust force.

References

1. Fossen T (2011) Handbook of marine craft hydrodynamics and motion control. Wiley
2. Khac Duc Do JP (2009) Control of ships and underwater vehicles. Springer-Verlag London
3. Fernandes C (2011) Directional stability of the torpedo anchor pile during its installation. IES J Part A Civil Struct Eng 4:180–189
4. Wilcox D (1994) Turbulence modeling for CFD. DCW Industries, Inc
5. Alder B (1964) Methods in computational physics: advances in research and applications. Academic Press
6. Kermode A (2006) Mechanics of flight. Pearson
7. Renilson M (2015) Submarine hydrodynamics, Springer briefs in applied sciences and technology
8. Antonelli G (2006) Underwater robots, Springer tracts in advanced robotics
9. Stevens B (2003) Aircraft Control and Simulation. Wiley-Interscience Publication
10. Do D (2009) Control of ships and underwater vehicles. Springer

Geometric Parameters and Workspace Optimization of Sitting-Type Lower Limb Rehabilitation Robot

Askhat Diveev[1], Grigory Dubrovin[2], Dmitry Malyshev[3](✉), and Anna Nozdracheva[4]

[1] Federal Research Center "Computer Science and Control" of Russian Academy of Sciences, Moscow, Russia
[2] Kursk State Medical University, Kursk, Russia
[3] Belgorod State Technological University named after V.G. Shukhov, Belgorod, Russia
rlbgtu@gmail.com
[4] Gamaleya National Research Center for Epidemiology and Microbiology, Moscow, Russia

Abstract. A lower limb rehabilitation system based on a serial RRR manipulator and an active parallel 2PRP-2PPR manipulator is considered, effective numerical methods and algorithms are developed and tested that allow determining the optimal geometric parameters of the active manipulator mechanism that provides the required movements within the workspace. The influence of the range of rotation angles of the joints, which must be ensured during the rehabilitation process on the minimum overall dimensions of the 2PRP-2PPR manipulator, is analyzed.

Keywords: Workspace · Lower limb rehabilitation · Parallel robot · Optimization

1 Introduction

Worldwide, limb disability is one of the major problems faced in the medical sector and it is more common in the elderly age group. There are numerous reasons for these disabilities; one major reason is stroke which happens due to brain attack, spinal cord injury (SCI) which happens due to accidental injury to the spine, disability due to non-polio acute flaccid paralysis (NPAFP), etc. In addition, stroke is one of the common causes of mortality after coronary artery disease and it is the commonest cause of chronic adult disability as well. The lifetime risk of stroke after 55 years of age is 1 in 5 for women and 1 in 6 for men [1]. Besides, another kind of disease which is acute flaccid paralysis is common. Clinically indistinguishable from polio paralysis but acute flaccid paralysis is twice as deadly as polio paralysis and takes longer duration to recover from this disease [2]. These limb impairments need to go through rehabilitation process to regain normal functioning of limb activities. However, rehabilitation therapeutic treatments are limb specific and they are different from upper limbs to lower limbs. Upper limb treatments focus on regain of hand neural, muscular action ability and strength while lower limb treatments focus on various joint movements of the legs and their synchronized motions.

E. E. Hernandez et al. (Eds.): LASIRS 2019, MMS 86, pp. 279–289, 2020.
https://doi.org/10.1007/978-3-030-45402-9_27

Lower limb rehabilitation or therapeutic treatment has been a warm topic in recent years as mechanized systems have promised effective results and produced significant improvements in patients' recovery through the robotic physiotherapy [3]. In the lower limb treatments: every lower limb has a specific movement (body and limb specific) and it requires a specialized setup for the individual limb. These issues make the system development quite complex and require modular and robust mechanical systems for the lower limb treatments. However, lower limb treatments are very common in the rehabilitation domain and cover a wide range of patients.

It is widely accepted that human locomotion depends both on basic patterns generated at the spinal level, and the prediction and reflex-dependent fine control of these patterns at different levels [4]. These physiological motions recorded from healthy subjects are implemented on lower-limb impairment patients. These data sequence for lower-limb joints are called gait data (walking pattern). Other than gait data lower-limbs have movements like hip-flexion and extension, knee flexion and extension, ankle plantar flexion and dorsiflexion. Gait training rebuilds muscular action synchronization in the lower-limb and for strengthening each leg-joint is treated separately for regaining the motor actions.

For the rehabilitation of the lower limbs, mechanotherapy is widely used. For this purpose, such commercially available rehabilitation under development. Most rehabilitation systems have an active orthosis in their design that has a simple structure, but their dynamic characteristics are low due to the presence of drives in the lower limb orthosis system. Other types of rehabilitation systems have linear drives and high dynamic characteristics, but at the same time, they have a limited workspace with limited movement of linear drives or do not include supporting or orthopedic systems [5, 6]. The development of mechanisms for the rehabilitation of the lower limbs is also considered in the works [7–9]. An important task is the design of mechanisms with enhanced structural rigidity of the mechanical structure, ergonomics, compactness of the system, positioning accuracy, improved weight and size characteristics and reduced dynamic load on the drives to ensure stable and safe movement of the lower and upper limbs of patients, including using parallel mechanisms that are well studied in [10–12].

2 Formulation of the Problem

In this paper, a sitting/lying type lower limb rehabilitation robot is considered for the performance analysis which is providing an improved rehabilitation robotic system. The conceptual design of the system is given in Fig. 1. The device consists of two mechanisms namely a planar 2PRP-2PPR parallel manipulator and a planar RRR serial manipulator as a passive orthosis system.

Consider the system (Fig. 2), where in addition to the passive orthosis in the form of a four-link RRR mechanism, the active 2PRP-2PPR mechanism is used. An RRR mechanism is used to secure a person's legs. Its revolute joints of the RRR mechanism correspond to the patient's joints: P_1 - to the patient's hip joint, its angle of rotation relative to the horizontal axis X is indicated by α, the joint P_2 with the angle of rotation γ relative to the $1'$ link and β relative to the horizontal axis X corresponds to the knee joint, and P_3 with the angle rotation θ relative to the beam perpendicular to the link $2'$-ankle joint.

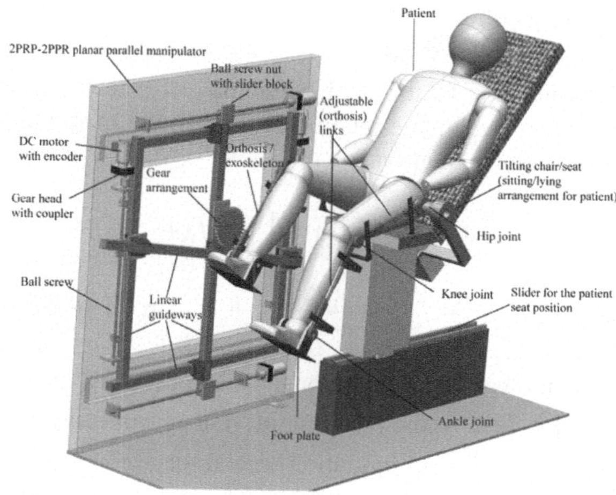

Fig. 1. Conceptual design of the lower limb rehabilitation system

The active 2PRP-2PPR mechanism provides the rehabilitation angles of rotation of all joints of the patient and coordinates x_{P3}, z_{P3} of the ankle joint P_3 of the patient in the XZ plane required for rehabilitation. In the designation 2PRP-2PPR, the letter P denotes prismatic joints, R - revolute joints, numerals 2 - two similar kinematic chains. Links 1, 3 and 4 move progressively relative to the fixed base 0. Link 2 rotates relative to link 1 due to the revolute joint A_3. Link 5 moves along the Z axis and along link 2, and also rotates relative to link 4 due to the revolute joint A_4. Link 6 moves along link 2. Link 8 is connected to link 6 through a gear reducer with gear ratio $i_{12} = \frac{\omega_1}{\omega_2} = 0,25$.

The article described the task of determining the optimal geometric parameters of a parallel 2PRP-2PPR mechanism that provides movement of a passive orthosis - RRR mechanism within the workspace in accordance with the required joint movements.

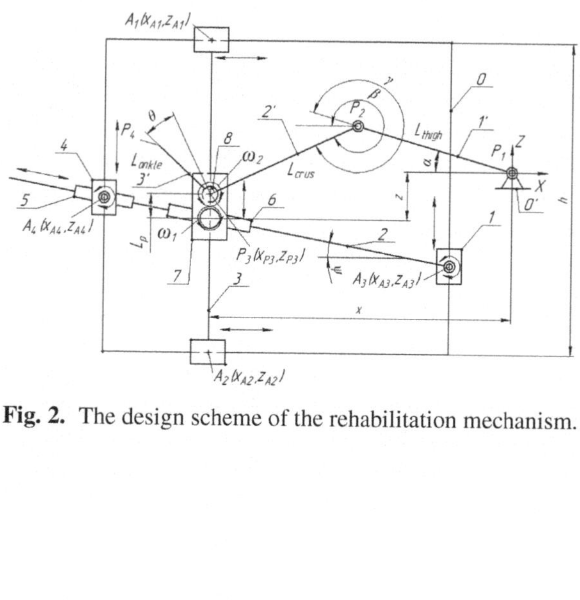

Fig. 2. The design scheme of the rehabilitation mechanism.

3 Kinematics Analysis

Consider the based kinematic relations of manipulators. The end-effector position is obtained from the forward kinematic relation of the parallel manipulator. The forward kinematic relation is given in equation:

$$\boldsymbol{\mu} = \left[x_{P3}\ z_{P3}\ \psi \right]^T = f(\boldsymbol{\eta}) \tag{1}$$

$$\boldsymbol{\mu} = \left[x_{A2}\ z_{A3} + \left(\frac{(x_{A3}-x_{A2})(z_{A4}-z_{A3})}{(x_{A3}-x_{A4})} \right) + L_p \tan^{-1}\left(\frac{(z_{A4}-z_{A3})}{(x_{A3}-x_{A4})} \right) \right]^T \tag{2}$$

where, $\boldsymbol{\mu}_p \in \mathcal{R}^{3\times 1}$ is the vector of operational-space positions, i.e., end-effector positions and orientation. $\boldsymbol{\eta} \in \mathcal{R}^{3\times 1}$ is the vector of linear displacements of the actuated linear joints i.e., $\boldsymbol{\eta} = \left[x_{A2}\ z_{A3}\ z_{A4} \right]^T$ (configuration or joint space positions or displacements).

We express z_{A3} and z_{A4} from formulas (1)–(2):

$$z_{A4} = z_{P3} - L_p + (x_{P3} - x_{A4}) \tan \psi \tag{3}$$

$$z_{A3} = z_{P3} - L_p - (x_{A3} - x_{P3}) \tan \psi \tag{4}$$

Similarly, the end- effector position is obtained from the forward kinematic relation of the serial manipulator (i.e., a passive orthosis) and is given in equation:

$$\boldsymbol{\mu}_s = \left[x\ z\ \psi \right]^T = f(\boldsymbol{\Phi}) \tag{5}$$

$$\boldsymbol{\mu}_s = \begin{bmatrix} x_1 - L_{thing} \cos\alpha - L_{crus} \cos\beta \\ y_1 + L_{thing} \sin\alpha + L_{crus} \sin\beta + L_p \\ G(\alpha + \beta + \theta) \end{bmatrix} \tag{6}$$

where $\boldsymbol{\Phi} \in \mathcal{R}^{3\times 1}$ is the vector of angular displacements of the passive joints of the orthosis i.e., $\boldsymbol{\Phi} = \left[\alpha\ \beta\ \theta \right]^T$.

For designing a rehabilitation system to ensure its compactness, it is important to determine the minimum possible overall dimensions at which it will perform its functions with the given parameters. In this case, these parameters are the sizes of the lower limbs of the patient and the range of rotation angles of the joints required for rehabilitation. Determination of the RRR mechanism workspace is used to calculate the optimal size of the 2PRP-2PPR mechanism.

4 Workspace Definition

Consider the problem of determining the boundaries of the workspace of the RRR mechanism when changing the parameters of the mechanism: angles α and γ in accordance with the movements of the passive orthosis, which is modeled by this mechanism. For rehabilitation, it is necessary to provide the following angles in the hip, knee and ankle joint when simulating walking: 1 step phase - flexion in the hip joint from 0° to 20°, flexion in the knee joint from 0° to 60°, extension in the ankle joint from 0° up to 10°; 2 phase - 0°, 0° and 90° respectively; 3 phase - extension in the hip joint from 0° to 10°, in the knee joint °, flexion in the ankle joint from 0° to 20°.

To determine the workspace, we will use the method of non-uniform coverings, which is described in detail in [13–15] for other types of parallel robots. For this purpose, we will use the methods of interval analysis described in detail in [16, 17].

We record the intervals that describe the ranges of changes in the rotation angles of the patient's joints. The lower boundary of the intervals corresponds to the minimum values of the angles necessary for rehabilitation, and the upper maximum:

$$\mathbf{A} := \left[\underline{\mathbf{A}}, \overline{\mathbf{\Gamma}}\right] = \left\{\underline{\mathbf{A}} \le \alpha \le \overline{\mathbf{A}}\right\}$$

$$\mathbf{\Gamma} := \left[\underline{\mathbf{\Gamma}}, \overline{\mathbf{\Gamma}}\right] = \left\{\underline{\mathbf{\Gamma}} \le \gamma \le \overline{\mathbf{\Gamma}}\right\}$$

$$\mathbf{\Theta} := \left[\underline{\mathbf{\Theta}}, \overline{\mathbf{\Theta}}\right] = \left\{\underline{\mathbf{\Theta}} \le \theta \le \overline{\mathbf{\Theta}}\right\}$$

$\beta = \alpha + \gamma$, that is, $\mathbf{B} := \left[\underline{\mathbf{B}}, \overline{\mathbf{B}}\right] = \left\{\underline{\mathbf{A}} + \underline{\mathbf{\Gamma}} \le \beta \le \overline{\mathbf{A}} + \overline{\mathbf{\Gamma}}\right\}$

To reduce the dimension of the computational problem, we propose using the following approach:

We express the angle α from (5)–(6):

$$\alpha = \sin^{-1}\frac{z_{P3} - L_{crus}\sin\beta}{L_{thigh}}$$

Denote: $f_1(z_{P3}, \beta) = \sin^{-1}\frac{z_{P3} - L_{crus}\sin\beta}{L_{thigh}}$, $f_2(a, \beta) = -L_{thigh}\cos\alpha - L_{crus}\cos\beta$.

Consider the intervals of values guaranteed to include the ranges of admissible values $\boldsymbol{Z}_{\boldsymbol{P}3}\!: = \left[\underline{\boldsymbol{Z}_{\boldsymbol{P}3}}, \overline{\boldsymbol{Z}_{\boldsymbol{P}3}}\right] = \left\{\underline{\boldsymbol{Z}_{\boldsymbol{P}3}} \le z_{P3} \le \overline{\boldsymbol{Z}_{\boldsymbol{P}3}}\right\}$, $\boldsymbol{X}_{\boldsymbol{P}3}\!: = \left[\underline{\boldsymbol{X}_{\boldsymbol{P}3}}, \overline{\boldsymbol{X}_{\boldsymbol{P}3}}\right] = \left\{\underline{\boldsymbol{X}_{\boldsymbol{P}3}} \le x_{P3} \le \overline{\boldsymbol{X}_{\boldsymbol{P}3}}\right\}$ and the interval **B**.

For the interval $\boldsymbol{Z}_{\boldsymbol{P}3}$ and the interval **B** we define the interval $\mathbf{A}'$ of the values of the angle α:

$$\mathbf{A}' := \left[\underline{\mathbf{A}'}, \overline{\mathbf{A}'}\right], \text{ where} \tag{7}$$

$\underline{\mathbf{A}'} = \min_{z_{P3} \in \boldsymbol{Z}_{\boldsymbol{P}3}, \beta \in \boldsymbol{B}} f_1(z_{P3}, \beta)$, $\overline{\mathbf{A}'} = \max_{z_{P3} \in \boldsymbol{Z}_{\boldsymbol{P}3}, \beta \in \boldsymbol{B}} f_1(z_{P3}, \beta)$

For the interval $\mathbf{A}'$ and the interval **B** we define the interval $\mathbf{\Gamma}'$ of the values of the angle γ:

$$\mathbf{\Gamma}' = \mathbf{B} - \mathbf{A}' \tag{8}$$

Then, for the intervals A′ and B, we calculate the interval $\boldsymbol{X'}_{\boldsymbol{P}3} := \left[\underline{\boldsymbol{X'}_{\boldsymbol{P}3}}, \overline{\boldsymbol{X'}_{\boldsymbol{P}3}}\right]$, where

$$\underline{\boldsymbol{X'}_{\boldsymbol{P}3}} = \min_{a \in \mathbf{A}', \beta \in \boldsymbol{B}} f_2(a, \beta),\ \overline{\boldsymbol{X'}_{\boldsymbol{P}3}} = \max_{a \in \mathbf{A}', \beta \in \boldsymbol{B}} f_2(a, \beta)$$

If the system is running

$$\begin{cases} \mathbf{A} \cap \mathbf{A}' \ne \emptyset \\ \boldsymbol{X}_{\boldsymbol{P}3} \cap \boldsymbol{X'}_{\boldsymbol{P}3} \ne \emptyset \\ \mathbf{\Gamma} \cap \mathbf{\Gamma}' \ne \emptyset \end{cases} \tag{9}$$

then for intervals $\boldsymbol{Z}_{P3}$, $\boldsymbol{X}_{P3}$ and $\mathbf{B}$ there is at least one point entering the workspace.

The interval constraints of the variables form an n-dimensional box, where n corresponds to the number of variables, i.e. $n = 3$. To determine the workspace, the initial and subsequent boxes, for which (9) holds, are divided into smaller boxes. For each box B obtained as a result of the partition, the verification is performed similarly. The work area is described by the set $\mathbb{P}_I$ of boxes B, for which $d(P_i) \leq \delta$, where δ is a given positive value that determines the accuracy of the approximation. For interval calculations, we use the Snowgoose interval analysis library [18]. Next, we determine the optimal geometric parameters of the 2PRP-2PPR mechanism.

5 Optimization of Geometric Parameters of the 2PRP-2PPR Mechanism

Consider the task of determining the optimal parameters of the 2PRP-2PPR mechanism, providing the movement of the output link of the RRR mechanism within a given work area.

Using a set of boxes that describe the workspace, we will choose the optimal overall dimensions along the Z axis:

$$x_{A3} = \max_{B \in \mathbb{P}_I} \overline{\boldsymbol{X}_{P3}} \tag{10}$$

$$x_{A4} = \min_{B \in \mathbb{P}_I} \underline{\boldsymbol{X}_{P3}} \tag{11}$$

We determine the optimal angle φ between the link P_3P_4 of the RRR mechanism and the link A_3A_4 of the 2PRP-2PPR mechanism with the coordinates $z_{A3} = z_{A4}$ of the joints A_3 and A_4 equal. As an optimization criterion, we choose a function that determines the distance h along the Z axis between the hinges A_1 and A_2, equal to the distance between the minimum and maximum values of z_{A3} and z_{A4}. Using formulas (3)–(4), we obtain:

$$\begin{aligned} f_3(\varphi) &= \min_{\varphi \in 0,\ldots,90^\circ} h \\ &= \min_{\varphi \in 0,\ldots,90^\circ} \left\{ \max_{B \in \mathbb{P}_I} \max_{i \in 1,\ldots,4} g_i - \min_{B \in \mathbb{P}_I} \min_{i \in 1,\ldots,4} g_i \right\}, \text{ where} \end{aligned} \tag{12}$$

$$g_1 = \frac{\underline{\boldsymbol{Z}_{P3}} + \overline{\boldsymbol{Z}_{P3}}}{2} - L_p + t_1\left(\frac{\underline{\boldsymbol{X}_{P3}} + \overline{\boldsymbol{X}_{P3}}}{2} - x_{A4}\right),$$

$$g_2 = \frac{\underline{\boldsymbol{Z}_{P3}} + \overline{\boldsymbol{Z}_{P3}}}{2} - L_p + t_2\left(\frac{\underline{\boldsymbol{X}_{P3}} + \overline{\boldsymbol{X}_{P3}}}{2} - x_{A4}\right),$$

$$g_3 = \frac{\underline{\boldsymbol{Z}_{P3}} + \overline{\boldsymbol{Z}_{P3}}}{2} - L_p - t_1\left(x_{A3} - \frac{\underline{\boldsymbol{X}_{P3}} + \overline{\boldsymbol{X}_{P3}}}{2}\right),$$

$$g_4 = \frac{\underline{\boldsymbol{Z}_{P3}} + \overline{\boldsymbol{Z}_{P3}}}{2} - L_p - t_2\left(x_{A3} - \frac{\underline{\boldsymbol{X}_{P3}} + \overline{\boldsymbol{X}_{P3}}}{2}\right),$$

where

$$t_1 = \tan\left(0,25\left(\frac{\underline{\boldsymbol{B}} + \overline{\boldsymbol{B}}}{2} + \frac{\pi}{2} + \underline{\boldsymbol{\Theta}} - \varphi\right)\right)$$

$$t_2 = \tan\left(0,25\left(\frac{\underline{\boldsymbol{B}} + \overline{\boldsymbol{B}}}{2} + \frac{\pi}{2} + \overline{\boldsymbol{\Theta}} - \varphi\right)\right)$$

The values of the functions g_i are determined for each of the boxes of the list $\mathbb{P}_I$. We define the minimum and maximum values of the coordinates of the joints A_1 and A_2 along the Z axis for the set of boxes of the list $\mathbb{P}_I$ obtained earlier and the angle φ. This procedure is performed for angles $\varphi \in [0°, 90°]$ with a step of 1°. The angle φ at which the distance h has a minimum value is assumed to be optimal:

$$h = \max_{B\in\mathbb{P}_I} \max_{i\in 1,..,4} g_i - \min_{B\in\mathbb{P}_I} \min_{i\in 1,..,4} g_i$$

Using these data, we will choose the optimal overall dimensions of the 2PRP-2PPR mechanism along the Z axis, which are determined by the following relationships:

$$z_{A1} = \max\nolimits_{B\in\mathbb{P}_I} \max\nolimits_{i\in 1,...,4} g_i \tag{13}$$

$$z_{A2} = \min\nolimits_{B\in\mathbb{P}_I} \min\nolimits_{i\in 1,...,4} g_i \tag{14}$$

6 Algorithm Synthesis

To solve the tasks of determining the workspace and choosing the optimal geometric relationships of the parallel 2PRP-2PPR mechanism, an algorithm is synthesized that works with two lists of three-dimensional boxes $\mathbb{P}$ and $\mathbb{P}_I$. The block diagram of the algorithm is shown in Fig. 3.

The algorithm works as follows:

1. Set the geometric parameters of the robot, the intervals $\mathbf{A}$, $\mathbf{\Gamma}$, $\mathbf{\Theta}$ and the accuracy of the approximation δ.
2. At the first step of the algorithm, the list of internal approximations $\mathbb{P}_I$ is empty, the list $\mathbb{P}$ consists of only one box Q, including the intervals $\boldsymbol{Z}_{P3}$, $\boldsymbol{X}_{P3}$ and $\mathbf{B}$: $\boldsymbol{Z}_{P3} = \boldsymbol{X}_{P3} = \left[-\left(L_{thigh} + L_{crus} + L_p\right), \left(L_{thigh} + L_{crus} + L_p\right)\right], \mathbf{B} = \left[\underline{\mathbf{A}} + \underline{\mathbf{\Gamma}}, \overline{\mathbf{A}} + \overline{\mathbf{\Gamma}}\right]$.
3. Remove the box B from the list $\mathbb{P}$.
4. If $d(B) < \delta$, then add B to the list $\mathbb{P}_I$ and go to step 13.
5. Calculate $\mathbf{A}'$ by the formula (7).
6. If $\underline{\mathbf{A}'} > 1$ or $\overline{\mathbf{A}'} < -1$, then go to step 13.
7. Calculate $\mathbf{\Gamma}'$ by the formula (8).
8. If $\mathbf{A} \cap \mathbf{A}' = \emptyset$ or $\mathbf{\Gamma} \cap \mathbf{\Gamma}' = \emptyset$, then go to step 13.
9. Calculate $\boldsymbol{X}'_{P3}$.

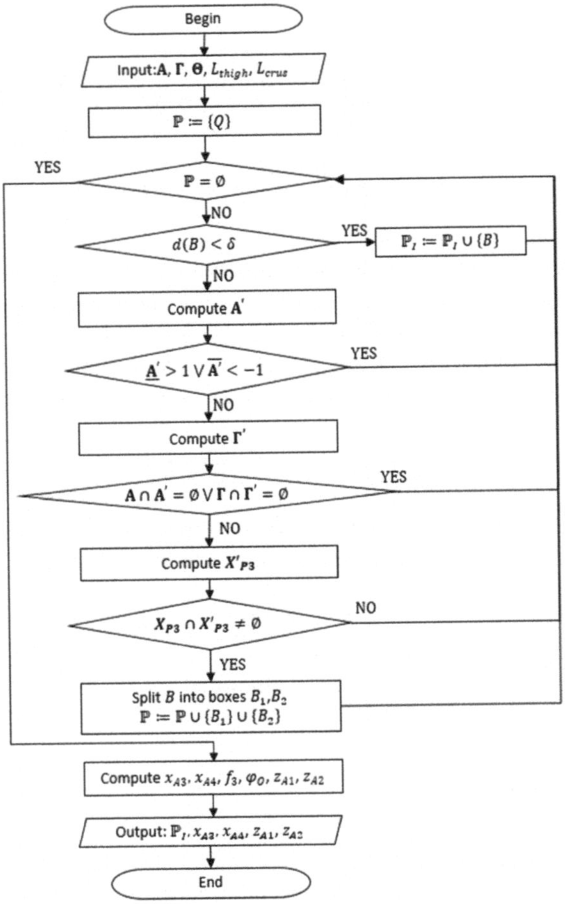

Fig. 3. The block diagram of the algorithm for determining the workspace and optimization of geometric parameters.

10. If $\boldsymbol{X}_{P3} \cap \boldsymbol{X}'_{P3} \neq \emptyset$, then B is divided into two equal boxes along the edge with the longest length. Insert these boxes at the end of the list $\mathbb{P}$.
11. If $\mathbb{P} \neq \emptyset$, then go to step 3.
12. Calculate x_{A3} and x_{A4} using formulas (10)–(11).
13. Calculate f_3 by the formula (12). The angle φ, for which h is minimal, is optimal.
14. Calculate z_{A1} by formula (13) and z_{A2} by formula (14).

To implement the synthesized algorithm, a program was developed in the C++ programming language with the ability to output results for visualization in the Gnuplot program.

7 Simulation Results

In accordance with the parameters required for rehabilitation: $\mathbf{A} = [-20°, 10°]$, $\mathbf{\Gamma} = [-60°, 0°]$, $\mathbf{\Theta} = [-10°, 20°]$, $\mathbf{B} = [-80°, 10°]$, $L_{thigh} = 450\,\text{mm}$, $L_{crus} = 450\,\text{mm}$, $L_P = 10\,\text{mm}$, a computational experiment was conducted, the simulation results are presented in Fig. 4 and 5. The calculation time for the approximation accuracy $\delta = 1$ mm on a personal computer was 27 s, of which 21 s were modeling the workspace, 6 s were optimizing the geometric parameters of the 2PRP-2PPR mechanism and determining the angle φ. It can be seen from Fig. 4 that the overall dimensions of the workspace are 401 × 754 mm. The optimal angle $\varphi = 0°$, the coordinates of the joints characterizing the overall dimensions and position of the 2PRP-2PPR mechanism were determined: $x_{A3} = -500$ mm, $x_{A4} = -901$ mm, $z_{A1} = 164$ mm, $z_{A2} = -598$ mm. In Fig. 5a shows the position of the joints of the 2PRP-2PPR mechanism, in which, to ensure the required positions of the RRR mechanism, the maximum value of the coordinate z_{A4} is reached at the optimum angle φ. In Fig. 5b shows the position of the mechanisms for the minimum value of the coordinate z_{A3}.

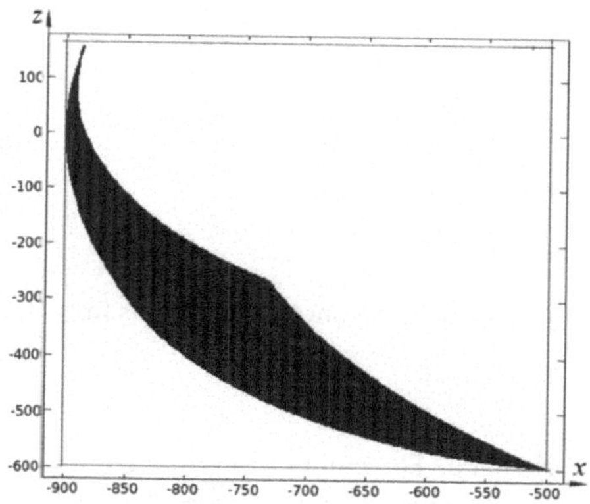

Fig. 4. The workspace of the RRR mechanism.

Note that the angle φ has a different value depending on the required rotation angles of the patient's joints. Defined and shown in Fig. 6 the dependence of the distance h on the angle φ at various intervals of the angle of rotation of the patient's hip joint. For each of the intervals, the angle φ was determined at which the distance h is minimal. For the interval $\mathbf{A} = [-20°, 10°]$ the optimal angle is $\varphi = 0°$, for $\mathbf{A} = [-10°, 20°]$ the angle is $\varphi = 10°$, for $\mathbf{A} = [0°, 30°]$ angle $\varphi = 21°$.

8 Conclusion

Thus, for the proposed lower limb rehabilitation system based on a serial RRR manipulator and an active parallel 2PRP-2PPR manipulator, effective numerical methods and algorithms were developed and tested that allowed us to determine the optimal geometric parameters of the mechanism of the active manipulator that provides the required

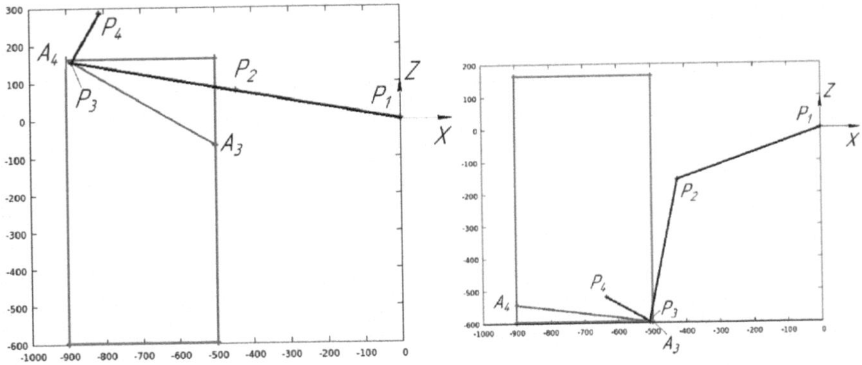

Fig. 5 The simulation results a - the extreme position of the joint A_4 for φ, b - the extreme position of the joint A_3 for φ.

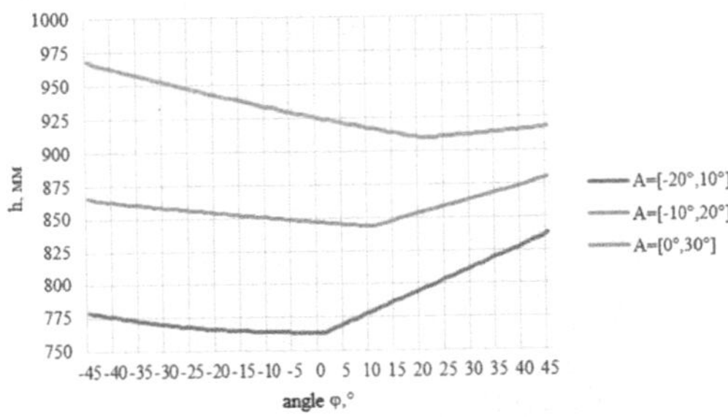

Fig. 6. Dependence of the distance h on the angle φ at various intervals of the angle of rotation of the patient's hip joint.

movements within the workspace. The calculation time for the approximation accuracy $\delta = 1$ mm on a personal computer was 27 s. In the simulation, the joints were considered as material points, and for a real model of the mechanism, the overall dimensions of the 2PRP-2PPR mechanism should be increased taking into account the actual dimensions of the drive elements to exclude their collision.

Acknowledgements. This work was supported by the Russian Science Foundation, the agreement number 19-19-00692.

References

1. Seshadri S, Beiser A, Kelly-Hayes M, Kase CS, Au R, Kannel WB (2006) The lifetime risk of stroke: estimates from the Framinham study. Stroke 37:345–350
2. Vashisht N, Puliyel J (2012) Polio programme: let us declare victory and move on. Indian J Med Ethics 9(2):114–117
3. Truelsen T, Bonita R (2009) The worldwide burden of stroke: current status and future projections. Handb Clin Neurol 92(3):327–336

4. Loeb GE (1989) Neural control of locomotion. BioSciences 39:800–804
5. Mirbagheri MM, Tsao C, Pelosin E (2005) Therapeutic effects of robotic-assisted locomotor training on neuromuscular properties. In: Proceedings of the IEEE 9th international conference on rehabilitation robotics (ICORR), Chicago, USA, pp 561–564
6. Hachisuka K (2010) Robot-aided training in rehabilitation IIJ. Brain Nerve 2:133–140
7. Hsieh CC, Kuo CH, Matsuura D, Takeda Y (2019) A 7R spatial linkage for ankle rehabilitation with an arbitrary ankle rotation axis. In: IFToMM world congress on mechanism and machine science, pp 1491–1498
8. Zhang C, Lan B, Matsuura D, Mougenot C, Sugahara Y, Takeda Y (2018) Kinematic design of a footplate drive mechanism using a 3-DOF parallel mechanism for walking rehabilitation device. J Adv Mech Des Syst Manuf 12:JAMDSM0017
9. Rios A, Hernandez E, Moreno JA, Keshtkar S, De la Garza R (2018) Kinematics analysis of a new 3DOF parallel manipulator as walking rehabilitation device a rios. In: 15th international conference on electrical engineering, computing science and automatic control (CCE)
10. Aleshin AK, Glazunov VA, Rashoyan GV, Shai O (2016) Analysis of kinematic screws that determine the topology of singular zones of parallel-structure robots. J Mach Manufact Reliab 45(4):291–296
11. Kong H, Gosselin CM (2007) Type synthesis of parallel mechanisms. Springer, Heidelberg
12. Rybak LA, Khalapyan SY, Gaponenko EV (2018) Issues of planning trajectory of parallel robots taking into account zones of singularity. IOP Conf Ser Mater Sci Eng 327:1–8 042092
13. Evtushenko YuG, Posypkin MA, Rybak LA, Turkin AV (2017) Finding sets of solutions to systems of nonlinear inequalities. Comput Math Math Phys 57(8):1241–1247
14. Evtushenko Yu, Posypkin M, Turkin A, Rybak L (2017) The non-uniform covering approach to manipulator workspace assessment. In: Proceedings of the 2017 IEEE Russia section young researchers in electrical and electronic engineering conference, ElConRus, pp 386–389. IEEE
15. Evtushenko Y, Posypkin M, Rybak L, Turkin A (2018) Approximating a solution set of non-linear inequalities. J Global Optim 7:129–145
16. Jaulin L (2001) Applied interval analysis: with examples in parameter and state estimation, robust control and robotics. Springer, New York
17. Moore RE (1966) Interval analysis. Prentice-Hall, Englewood Cliffs
18. Posypkin M, Usov A (2019) Basic numerical routines. https://github.com/mposypkin/snowgoose. Accessed 18 Nov 2019

Comparison of Parallel Versions of SA and GA for Optimizing the Performance of a Robotic Manipulator

Adán H. Baena[1,3](✉), Sergio I. Valdez[2,3](✉), Felipe de Jesús Trujillo Romero[1,3](✉), and Moisés M. Montes[1,3](✉)

[1] División de Ingenierías, Universidad de Guanajuato, Carr. Salamanca - Valle de Santiago km 3.5 + 1.8, Palo Blanco, 36885 Salamanca, GTO, Mexico
{a.hernandezbaena,fdj.trujillo,mm.montesvillanueva}@ugto.mx

[2] Centro de Investigación en Ciencias de Información Geoespacial, CENTROGEO A.C., Contoy 137, Lomas de Padierna, 14240 Ciudad de México, Mexico
sergio.valdez@conacyt.mx

[3] Cátedra CONACYT, Ciudad de México, Mexico

Abstract. The objective of this work is to optimize a robotic manipulator with five links and two degrees of freedom. Our optimization criterion is the lowest energy consumption subject to delivering an adequate performance, this means that the mechanism should follow the defined trajectory with a maximum bounded error. For this purpose, we used the Simulated annealing (SA). which is a probabilistic technique for approximating the global optimum of a given function. In our case, the given function is the integral of the torque returned by the simulation of the mechanism. The SA evaluates the function with random configurations and according to the returned value, the configuration is accepted or rejected in agreement with an acceptance criterion. The stopping criterion is that three consecutive optimization loops do not improve the current best solution. Energy consumption is directly proportional to the torques signals $\tau 1$ and $\tau 2$, so we propose a criterion which considers energy consumption as well as the position error, as follows:

$$E(l_1, l_2, l_4, k_1^p, k_2^p, k_1^d, k_2^d) = \lambda \sum_i^n (\tau_{1_i} + \tau_{2i}) + \sum_i^n (ep_1 + ep_2) \tag{1}$$

Where τ_1 is the torque of link l_1 and τ_2 is the torque of link l_2 as it is shown in Fig. 1, λ is a penalty value that indicates the importance of the energy with respect to the position error. It should be mentioned that the larger the λ, the greater tracking error in the trajectory. The second part of Eq. (1) indicates the numerical integral of position error, where ep_1 is the error caused by the angle φ_1 and ep_2 is the error caused by

S. Ivvan Valdez is supported by grant Cátedra CONACYT 7795.

E. E. Hernandez et al. (Eds.): LASIRS 2019, MMS 86, pp. 290–303, 2020.
https://doi.org/10.1007/978-3-030-45402-9_28

the angle φ_2. The SA step with the greatest computational cost is the simulation and evaluation of candidate configurations since each of them is independent of each other, this part is highly parallelizable, that is to say, several configurations are evaluated in parallel. Each one of the threads simulates a configuration with its own memory and only share specific information, such as the best configuration found so far. The results demonstrate the time reduction as well as the high quality of the results. In addition, we compare two versions of Parallel SA and a simple GA. The first SA is completely equivalent to the sequential algorithm but the computing time is reduced by parallelizing a set of simultaneous evaluations, even though the synchronized acceptance/rejection step introduces overhead in the algorithm. On the other hand, the second SA eliminates the synchronization, so it is not equivalent to the sequential version, but in contrast, it the overhead is reduced. Finally, both versions are contrasted with a simple GA, in the objective function of the best solution as well as in computational time.

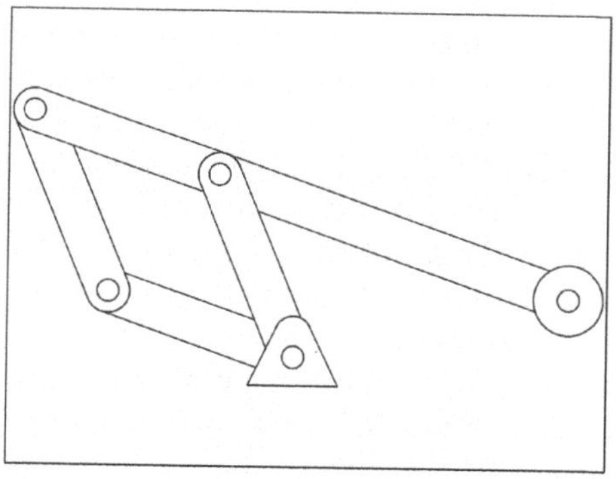

Fig. 1. Robotic manipulator

Keywords: Parallel · Optimization · Simulated annealing · Genetic Algorithm

1 Introduction

In all areas of engineering, optimization is a way of improving designs, it has served as a tool to save material, time, effort, energy, etc. When working in a new design, mechanical engineers need to analyze and test each candidate configuration, usually, in a test-error cycle they improve an initial design with respect to a given criterion. To test a specific configuration of a mechanism, a computer simulation is convenient in order to save resources, and reducing costs and time. In addition, an adequate computer program could test much more candidate configurations than a human designer, nevertheless, it must do it in an adequate an efficient way.

The Simulated Annealing Algorithm (SA) [1] is a global optimization algorithm based on Gibbs probability distribution. It generates random configurations from the Gibbs distribution using the metropolis sampling method, the better a configuration is the greater the probability to be sampled. Nevertheless, a drawback of the SA is the need of a huge amount of objective function evaluations, in this case an evaluation is a numerical simulation of a particular robot configuration. The SA requires lots of simulations under different conditions to know when a configuration is better than another. Due to this fact, parallel computation is used to reduce the time require to find an approximation to the global optimal mechanism. So, SA returns the best configuration of a mechanism, selected among a huge amount of randomly generated configurations, having in mind that this random process samples intensively the most promising regions of the search space.

Another optimization method is the Genetic Algorithm (GA) [2]. GA imitates the evolution of the living beings, described by Charles Darwin, and it is being part of the group of Evolutionary Algorithms (EA). GA are directly, parallel, stochastic method for global search and optimization. GA [3] uses genetic operations such as selection, crossover and mutation to generate new individuals. Such individuals can be the configurations for the robotic manipulator. As the species' evolution in nature, GA tries the development of new and better across the generations. Since this method is heuristic and not have a defined behavior and according to the specific problem the number of generations needed to discover a solution that solves the problem can be lots, the development in parallel could have an interesting performance comparing it to a sequential GA.

These both methods have been developed previously by several researchers, e.g. Paradalos et al. [4] describe the implementation of these methods sequential and parallel. Parallel GA and parallel SA have had better performance mostly in the execution time comparing to sequential versions working in optimization problems of several fields (e.g. Traveling Salemen Problem [5], Combinatorial Optimization [4], Shape optimization [6]).

The method proposed in this article can be used as design tool, although, it can not consider some features such as manufacturability, standardization, etc., it intends to approximate the optimum configuration regarding energy consumption subject to a bounded error, then a designer can use this preliminary configuration in an initial stage of the design process.

1.1 4-Bar Manipulator

The mechanism has four bars and two active joints which angles are denoted by φ_1 and φ_2, as it is shown in Fig. 2. A parallelogram is formed with the four elements, one of them is larger in order to be the end effector, which must follow a given trajectory. The position of the end effector is defined by the direct kinematics in Eq. (2), the literals L_1 and L_4 as well as φ_1 and φ_2 are those shown in Fig. 2.

$$\begin{bmatrix} x \\ y \end{bmatrix} = \begin{bmatrix} L1\cos\varphi_1 - L4\cos\varphi_2 \\ L1\sin\varphi_1 - L4\sin\varphi_2 \end{bmatrix} \tag{2}$$

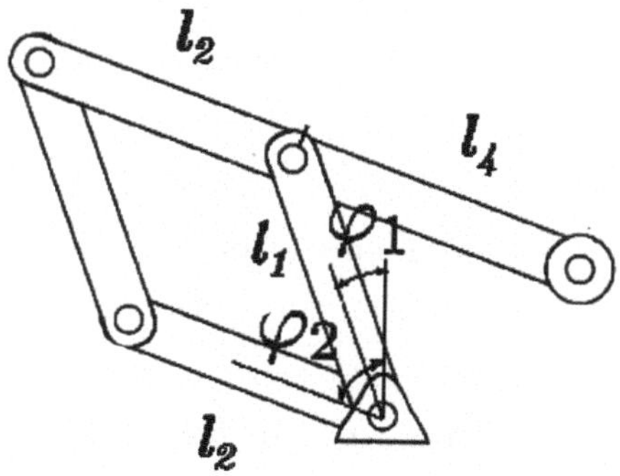

Fig. 2. Four-bar manipulator.

The direct kinematics is used to find the position of the end effector given the angles φ_1 and φ_2, while the inverse kinematics is used to compute the angles given the end effector position, it is given by Eqs. (3) and (5). Both of them are used to numerically simulate the manipulator movement to following a given trajectory.

$$\varphi_1 = \arctan(\frac{y}{x}) - \arctan(\frac{L_4\phi}{L_1 + L_4\phi}) \tag{3}$$

$$\phi = \arctan(\frac{-\sqrt{1-d^2}}{d}) \tag{4}$$

$$\varphi_2 = \varphi_1 + (\phi + \pi) \tag{5}$$

$$d = cos\phi = \frac{x^2 + y^2 - L_1^2 - L_4^2}{1L_1L_4} \tag{6}$$

The manipulator velocity is computed via the differential kinematics, shown in Eqs. (7) to (10).

$$\dot{\mathbf{x}} = J \cdot \dot{\mathbf{q}} \tag{7}$$

$$\dot{\mathbf{x}} = [\dot{x}\dot{y}]^T \tag{8}$$

$$\dot{\mathbf{q}} = [\dot{\varphi_1}\dot{\varphi_2}]^T \tag{9}$$

$$J = \begin{bmatrix} -L_1\sin\varphi_1 & L4\sin\varphi_2 \\ L_1\cos\varphi_1 & -L4\cos\varphi_2 \end{bmatrix} \tag{10}$$

Table 1. Inertial data used in the manipulator numerical simulation.

Masses	
$m_1 = 2L_1/0.3$	$m_3 = 0.5L_2/0.15$
$m_3 = 2L_3/0.3$	$m_4 = 2L_4/0.4$
Mass centers	
$l_{c1} = L_1/2$	$l_{c3} = L_3/2$
$l_{c3} = L_3/2$	$l_{c4} = L_4/2$
Moments of inertia	
$I_1 = 0.0226$	$I_2 = 0.00641$
$I_3 = 0.0226$	$I_4 = 0.0226$
Proportional control	
$u_i(t) = K_{pi} \cdot (\varphi_{id}(t) - \varphi_i(t))$	
Proportional-Derivative control	
$u_i(t) = K_{pi} \cdot (\varphi_{id}(t) - \varphi_i(t)) + K_{pi} \cdot (\dot{\varphi_{id}}(t) - \dot{\varphi}_i(t))$	

Finally, the dynamic model considers inertial forces and torques, hence, it is necessary to calculate the effects of the gravity, masses, lengths, mass centers, and inertial moments, which are shown in Table 1.

The physics of the mechanism is governed by the Ordinary Differential Equations (11).

$$\begin{aligned} d_{11}\ddot{\varphi_1} + \phi_1(\varphi_1) = \tau_1 \\ d_{22}\ddot{\varphi_2} + \phi_2 2(\varphi_2) = \tau_2 \end{aligned} \tag{11}$$

Where the functions ϕ_i are shown in Eq. (12)

$$\begin{aligned} \phi_1(\varphi_1) = g\cos\varphi(m_1 l_{c1} + m_3 l_{c3} + m_4 L_1) \\ \phi_2(\varphi_2) = g\cos\varphi(m_2 l_{c2} + m_4 l_{c4} + m_3 L_2) \end{aligned} \tag{12}$$

Where d_{ii} are constants related with the masses of the elements. They are defined as shown in Eq. (13).

$$\begin{aligned} d_{11} = m_1 l_{c1}^2 + m_3 l_{c3}^2 + m_4 L_1^2 + I_1 + I_3 \\ d_{22} = m_2 l_{c2}^2 + m_4 l_{c4}^2 + m_3 L_2^2 + I_2 + I_4 \end{aligned} \tag{13}$$

The proportional-integral-derivative control is used to compute the torque τ_i, associated with each active join.

$$\tau_i(t) = k_p e_i(t) + k_i \int e_i(t)dt + k_d \dot{e_i}(t), \quad e_i(t) = \varphi_{id} - \varphi_i \tag{14}$$

- The integral and derivatives are solved with numerical methods, in this case, the Euler method for differential equations for an interval T.
- The integrals and derivatives are calculated in each step t, with the value obtained of the instant torque computed in the previous step.
- The direct and inverse kinematics are solve each step according to the new positions and the required intermediate steps in the method.

Finally, we propose a trajectory to test the method. In this experiment we used a circular path:

$$\begin{aligned} X &= \delta x + R\cos(\omega t) \\ Y &= \delta y + R\sin(\omega t) \\ R &= 0.05 \\ \delta x &= 0.4 \\ \delta y &= 0.3 \\ \omega &= 1 \end{aligned} \tag{15}$$

2 Simulated Annealing (SA)

The simulated annealing is a method for approximating the global minimum of an objective function over a search space, usually in presence of non-linearities and discontinuities. The SA algorithm is based on the analogy between the simulation of the annealing of solids and the problem of solving large combinatorial optimization problems [7].

We use an SA inspired on the Coranna's version [1] which is shown in Algorithm 1. This algorithm requires two inputs: initial temperature and candidate solutions which are randomly generated in an internal loop. The initial temperature T_0 starts with a great value and it is reduced by multiplying by a constant coefficient $0 < C < 1$ (line 10) to decrease the temperature until it reach a minimum target value.

The evaluation step is the most computationally expensive in the SA, in consequence, the parallelization efforts must be focused in this step. The general workflow of the parallel algorithm works as follows: The sequential SA uses one thread, given an initial configuration X the SA perturbs N times this configuration during the optimizing process. In the sequential algorithm the N perturbations are made and evaluated by a single core, in this parallel proposal we use all cores of the microprocessor. This is the unique modification to the sequential algorithm, nevertheless, the implications are deeper, considering that the sequential SA works as follows: it perturbs the initial configuration in one variable, then, the perturbed configuration is evaluated and compared with the initial configuration, using the metropolis criterion the perturbed configuration could be accepted or rejected for replacing the initial configuration, the new initial configuration is going to be perturbed in the next iteration of this loop and so on. Notice, that in the sequential algorithm there is an stochastic dependency between the current perturbed configuration and the next one, on the other hand, in the parallel version this dependency is removed.

Algorithm 1: Simulated annealing

```
   Input  : T0, Points
   Output: XB
 1 X ← RAND(lim_inf, lim_sup);
 2 f(X) ← EVALUATION(X);
 3 f(XB) ← f(X);
 4 XB ← X;
 5 while STOP CRITERION do
 6 |  for temp ← T0 to temp < 0.005 do
 7 |  |  for t ← 1 to Points do
 8 |  |  |  PARALLEL EVALUATIONS
 9 |  |  end
10 |  |  temp ← temp * C
11 |  end
12 |  XB ← X
13 |  f(XB) ← f(X)
14 end
```

Our SA algorithm is shown in Algorithm 1, it is similar to the Corona's Algorithm but with the parallel evaluation step (line 8).

Given the fact that each evaluation of the perturbed solutions is independent of the rest, each thread can carry out the evaluation of some of these solutions, such as it is shown in Algorithm 2. First, temporary variables of $\mathbf{X}$, $\hat{\mathbf{X}}$, $f(\mathbf{X})$ and $f(\mathbf{XB})$ are declared, as well as an array of disturbed solutions $\hat{\mathbf{X}}$ and its function $f(\hat{\mathbf{X}})$ (lines 1 to 6). From here, each thread works with its memory and they just compare (line 21) the current best solution to the best global one, that is, threads don't have access to shared memory and, this way, we expect to reduce lost time doing access to memory in common. When the evaluations of perturbed solutions end (lines 7 to 20) where all changes and improvements are stored in temporary variables, the best solution of each thread are compared with the best global solution and it is selected the best among of all them.

2.1 Stop Criterion

Stop criterion (line 5) is flexible. It is recommended to repeat the SA a few times until the value of the objective function converges to, approximately, the same value, this value is stored in $f(\mathbf{XB})$ and is associated with the solution $\mathbf{XB}$.

3 Genetic Algorithm (GA)

Genetic Algorithms (GAs) [8] are optimization techniques based on the Darwinian principle of evolution through genetic selection. These kinds of algorithms have been successfully applied to an ample variety of real-world problems of different complexity. A GA uses an abstraction of the natural selection, using a

Algorithm 2: PARALLEL EVALUATIONS

input : $\mathbf{X}$, $\mathbf{XB}$, $f(\mathbf{X})$, $f(\mathbf{XB})$, N
output: $\mathbf{XB}$, $f(\mathbf{XB})$

$\hat{\mathbf{X}} \leftarrow \mathbf{X}_{tmp} \leftarrow \mathbf{X}$
$\mathbf{XB}_{tmp} \leftarrow \mathbf{XB}$
$f(\hat{\mathbf{X}}) \leftarrow 0$
$f_{tmp}(\mathbf{XB}) \leftarrow f(\mathbf{XB})$
$f_{tmp}(\mathbf{X}) \leftarrow f(\mathbf{X})$
for $i \leftarrow 1$ **to** N **do**
    $e_i \leftarrow \text{RAND}(lim\_inf_i, lim\_sup_i)$
    $\hat{\mathbf{X}} \leftarrow \hat{\mathbf{X}} + [0, 0, ..., e_i, ..., 0, 0]$
    $f(\hat{\mathbf{X}}) \leftarrow \text{EVALUATION}(\hat{\mathbf{X}})$
    if $f(\hat{X}) < f_{tmp}(\mathbf{XB})$ **then**
        $\mathbf{XB}_{tmp} \leftarrow \hat{\mathbf{X}}$
        $f_{tmp}(\mathbf{XB}) \leftarrow f(\hat{\mathbf{X}})$
    end
    /* ACCEPTANCE OF METROPOLIS CRITERIA */
    $\Delta f \leftarrow f(\hat{\mathbf{X}}\mathbf{G}) - f(\mathbf{X}_{tmp})$
    if $\Delta f \leq 0$ ***OR*** $\{\Delta f \geq 0$ ***AND*** $RAND(0,\ 1) < \exp \frac{-\Delta f}{tmperature}\}$ **then**
        $\mathbf{X}_{tmp} \leftarrow \hat{\mathbf{X}}$
        $f_{tmp}(\mathbf{X}) \leftarrow f(\hat{\mathbf{X}})$
    end
end
if $f_{tmp}(\boldsymbol{XB}) < f(\boldsymbol{XB})$ **then**
    $\mathbf{XB} \leftarrow \mathbf{XB}_{tmp}$
    $f(\mathbf{XB}) \leftarrow f_{tmp}(\mathbf{XB})$
end

population of artificial chromosomes. Chromosomes are usually chains of data that provide information (a real number) about how the individual behaves in a particular habitat (the problem). The chromosome is evaluated in order to determine how fitted an individual is, with the purpose of preserving or discarding his genes in the following generations. So, chromosomes are solutions of the particular problem and the fitness is related or equal to its objective function evaluation. GAs start with a random population (of chromosomes) which is evaluated to know what individuals are convenient for the next generation. GA carries out a process of chromosomes recombination with the best current chromosomes (parents) to produce a successor population (children), then children are mutated, evaluated and used to replace the current population, and so on. This process is iterated for a finite number of times (generations). Each iteration, the solution is expected to improve.

For this GA (Algorithm 3), we propose as an individual of the population a chain of 7 genes, which represent a candidate configuration for the robotic manipulator. Therefore, the population POP which the GA works is an array of the mentioned chains. The optimizing begins with an initial POP generated randomly of possibles solutions and then, the GA carries out the process of minimization (lines 3 to 11).

Algorithm 3: Genetic algorithm

input : *generations*, *population*, *genes*
1 **POP** ← RANDOM(lim_inf, lim_sup);
2 **F** ← [0, 0, ..., 0] ;
3 **for** $i \leftarrow 1$ **to** *generations* **do**
4 | **F**← PARALLEL EVALUATIONS;
5 | **POP**← SORT(**POP, F**);
6 | **CHILDREN** ← SELECTION(**POP**);
7 | **CHILDREN** ← RECOMBINE(**CHILDREN**);
8 | **CHILDREN** ← MUTATION(**CHILDREN**);
9 | **POP** ← NEW_GEN(**CHILDREN**);
10 **end**

Like the SA, the evaluation step is parallelized. As it is shown in Algorithm 4, each thread works with POP and F (lines 1 and 2) temporary arrays. After each thread carries out a part of the all evaluations, the results are inserted in the global functions array F, this way, each thread has access to shared memory only one time.

Algorithm 4: Parallel evaluations

1 **Parallel for** $i = 1, 2..|POP|$ **do**
2 | $F_i = EVALUATION(POP_i)$
3 **end**

4 Results

In this section, the results of our experiment are described. Since our purpose is comparing parallel versions of the SA and the GA, and as each of both algorithms works with parameters which determine how they carry out the process, we must use conditions where the algorithms behave of an adequate way. The conditions which govern the performance of our SA are initial temperature, amount of points, times the SA works and, of course, the number of cores used.

'Times' is the value to which the stop criterion mentioned in previous sections is subject. To find out which are the adequate conditions for the experiment, we carry out preliminary tests with different parameters which are shown in Table 2. Analyzing the data, It can be noted that the configuration where values are 125, 50 and 10 (for *temperature*, *points* and *times*, respectively) has the lowest average for the objective function, as well as the lowest standard deviation. Furthermore, the execution time that the SA takes with these parameters is below of the arithmetic mean for recorded times. Thus, we opted to take the aforesaid configuration like the conditions for true tests.

Table 2. Objective function varying the conditions for the SA

Points	Temps	Times	Mean	SD	Max	Min	Time (Mean)
25	25	5	6594.514	578.2811	7241.312	6127.422	67.79439
25	25	10	7403.964	919.7598	8450.774	6725.321	135.4295
25	75	5	6118.325	187.3511	6333.709	5993.086	80.83644
25	75	10	5948.943	63.07227	6018.22	5894.847	155.4485
25	125	5	6449.567	473.9563	6865.875	5933.76	80.895
25	125	10	8100.752	1291.439	8890.017	6610.396	690.752
25	175	5	6661.187	678.9761	7390.485	6047.335	90.40371
25	175	10	7655.782	237.4511	7929.832	7511.314	196.5836
50	25	5	596.5836	11.1195	206.8913	184.7998	141.0356
50	25	10	6992.395	1590.787	8816.63	5893.924	265.6407
50	75	5	7140.305	604.4596	7826.031	6684.72	151.7968
50	75	10	6725.831	330.8201	7099.958	6471.954	310.9205
50	125	5	6338.199	504.8088	6883.95	5887.985	156.6621
50	125	10	5903.71	18.63697	5925.194	5891.894	312.5321
50	175	5	6126.922	253.421	6393.968	5889.779	162.3664
50	175	10	6290.447	411.5418	6704.64	5881.609	330.7269
75	25	5	6167.833	359.2555	6571.708	5883.871	229.0864
75	25	10	6023.295	116.8177	6113.512	5891.342	392.8536
75	75	5	6368.638	490.9635	6893.742	5921.034	224.4166
75	75	10	6523.281	855.9222	7509.494	5974.127	459.3084
75	125	5	6485.036	185.7312	6612.42	6271.925	240.5058
75	125	10	6853.661	536.2561	7472.461	6524.623	480.7757
75	175	5	6122.928	363.733	6542.868	5906.702	239.6924
75	175	10	6447.443	3359.526	6845.913	607.7692	500.4325

Moreover, for our GA, the conditions which govern its performance are the amount of population and generations, as well as the number of cores used. For the GA, tests were executed with different combinations of *generations* and *population* which are shown in Table 5. Like the SA tests (Table 2), we considerate as the better parameters for the true tests, the conditions where the

GA has the lowest average, lowest standard deviation and where the average time is below of the arithmetic mean. The configuration, that describes the previous description, is the GA working with 300 generations and a population of 175 individuals. Even though the configuration previously described fulfills the necessary criteria, we consider more feasible to use 75 individual for the population through 400 generations given that it has almost the same statistics and by considerably less time.

Table 3. Objective function statistics for the SA

Threads	Points	Temp	Times	Mean	SD	Max	Min
8	50	125	10	5968.21375	106.0896113	6224.180244	5899.072774
1	50	125	10	5975.654339	128.6481023	6301.167460	5898.078654

Table 4. Execution time statistics for the SA

Threads	Points	Temp	Times	Mean	SD	Max	Min
8	50	125	10	312.4205	7.325753659	319.9205	302.9205
1	50	125	10	775.2191731	75.24748331	807.231196	698.236

The results of the true simulations were summarized in statistics concerning the minimization of the objective function and the execution time. These statistics are shown in Tables 3 and 4 for the SA and they are shown in Tables 6 and 7 for the GA.

Additionally, in Fig. 3 (For the SA) and Fig. 4 (For the GA) can be shown how the objective function behaves through the minimize process for both algorithms. These graphs show how GA tends slightly faster to a global minimum, while SA tends to have a less steep slope. Something equally interesting is how the SA behaves when the temperature is very close to zero, how in these temperatures the minimum decreases a little more, something that in higher temperatures it carries out in a small scale.

Table 5. Objective function varying the conditions for the GA

Gens	Population	Mean	SD	Max	Min	Time (Mean)
100	25	5948.81	87.4562956	6123.066	5896.562	3.064973
100	75	5896.105	11.6481354	5916.619	5882.258	8.988048
100	125	5887.878	4.318659675	5896.305	5883.955	14.15521
100	175	5886.633	3.91596454	5893.639	5883.16	19.70656
100	225	5884.212	1.550943486	5886.635	5882.294	26.12352
200	25	5940.722	50.01127077	6030.216	5892.125	5.847065
200	75	5884.957	1.111173328	5886.36	5883.617	16.99983
200	125	5886.275	4.328700825	5892.569	5882.312	28.5828
200	175	5884.772	2.843259643	5889.056	5881.904	39.30059
200	225	5885.181	3.090062037	5890.908	5882.93	51.91625
300	25	5913.216	14.55270843	5933.787	5897.383	8.731338
300	75	5885.806	5.310659254	5895.842	5881.749	25.53023
300	125	5883.863	2.416952464	5888.347	5881.573	42.11146
300	175	5881.653	0.360897382	5882.147	5881.161	58.56627
300	225	5882.314	0.881892134	5883.541	5881.429	79.4524
400	25	5904.799	18.5366548	5927.339	5884.802	11.71828
400	75	5882.036	0.98095959	5883.272	5880.977	34.01984
400	125	5882.242	1.057190353	5883.391	5880.835	56.1136
400	175	5882.643	0.612921241	5883.162	5881.497	78.35479
400	225	5881.919	1.237379279	5883.766	5880.778	104.978
500	25	5917.184	40.0366268	5992.548	5887.238	14.46798
500	75	5883.117	1.694247849	5885.302	5881.585	42.54183
500	125	5883.581	2.075564146	5886.204	5880.886	69.93083
500	175	5882.511	1.804717969	5885.561	5880.928	97.76304
500	225	5881.592	0.521225604	5882.31	5880.967	134.5352

Table 6. Objective statistics for the GA

Threads	Gens	Population	Mean	SD	Max	Min
8	400	75	5881.103	2.19020944	5888.697	5880.895
1	400	75	5882.243	1.315935506	5881.456	5880.88

Table 7. Execution time statistics for the GA

Threads	Gens	Population	Mean	SD	Max	Min
8	400	75	34.37404	0.398747021	35.54596	33.65045
1	400	75	155.5839	0.883990483	156.2735	154.5873

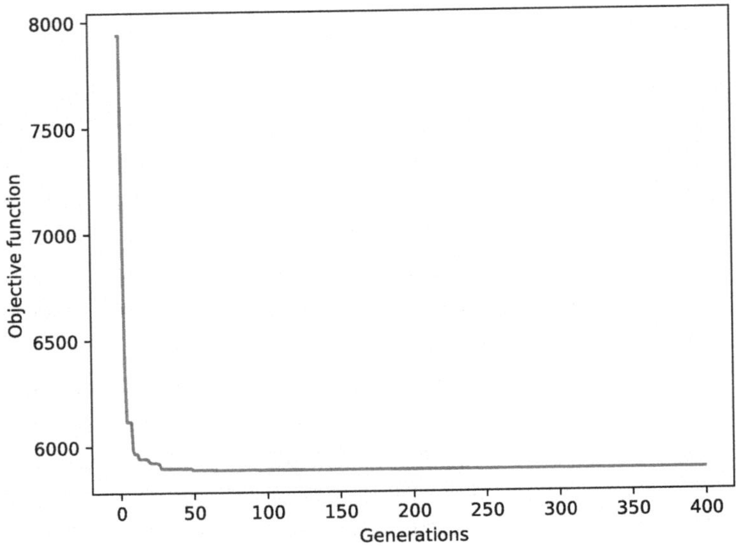

Fig. 3. Objective function behavior with the GA.

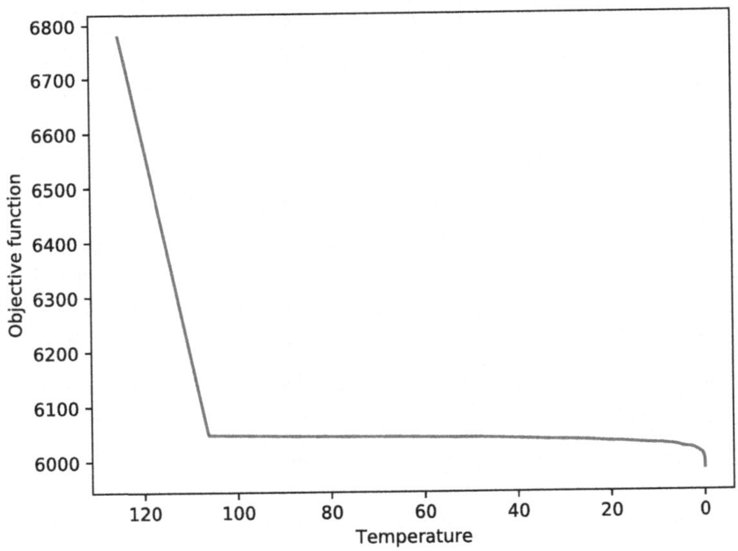

Fig. 4. Objective function behavior with the SA.

5 Conclusions

Firstly, the computation time, as expected, is lower when various threads carry out the optimization process in contrast with a single one. Nevertheless, there is a significant variance in the computation times among different executions with the same parallelization, in all cases the parallel version shows a significant time reduction. This is due to the adequate implementation and selection of the parallel strategy. We parallelized the most time-consuming task, additionally, we take care of using a low amount of memory, by reusing variables and avoiding to preserve results that are no longer needed inside the loops with the purpose of maximizing the cache use and minimizing the request of memory updating to the RAM.

The analysis of the data from the previous section provides enough information to assert that GA is a more convenient algorithm for the optimization of the robotic manipulator. Comparing the SA with the GA, the second converges faster to the best solution. Besides, the total processing time of the GA to is significantly less than that used by the SA. Another important consideration about the comparison is that the SA solutions present a huge variance among executions, while the GA reports more similar objective function values of the best solutions from different executions, that indicates that the GA is a more confident algorithm. Future work considers the comparison of other optimization algorithms for this problem.

References

1. Martini C, Ridella S, Corana A, Marchesi M (1987) Minimizing multimodal functions of continuous variables with the “simulated annealing” algorithm. ACM Trans Math Softw 13:262–280
2. Popov A (2005) Genetic algorithms for optimization. User Manual, Hamburg 2013
3. Chen Z, Zhang Z, Xie J, Guo Q, Tao Y (2019) Metaheuristic optimization method for compact reactor radiation shielding design based on genetic algorithm. Ann Nucl Energy 134:318–329
4. Pardalos PM, Pitsoulis L, Mavridou T, Resende MGC (1995) Parallel search for combinatorial optimization: genetic algorithms, simulated annealing, tabu search and GRASP. In: Parallel algorithms for irregularly structured problems, pp 317–331. Springer, Heidelberg
5. Suh JY, Van Gucht D, Prasanna J (1991) Parallel genetic algorithms applied to the traveling salesman problem. SIAM J Optim 1:515–529
6. Brauer H, Ziolkowski M (2005) Shape optimization with adaptive simulated annealing and genetic algorithms. In: Wiak S, Krawczyk A, Trlep M (eds) Computer engineering in applied electromagnetism. Springer, Dordrecht, pp 25–30
7. van Laarhoven PJ, Aarts EH (1987) Simulated Annealing: Theory and Applications. Springer, Netherlands
8. McCall J (2005) Genetic algorithms for modelling and optimisation. J Comput Appl Math 184:205–222

Dynamic Modeling and Control of a Pivoting Nozzle

Christopher Virgen-Vázquez(✉), Eusebio Eduardo Hernández, and Armando Oropeza Osornio

ESIME Ticomán, Instituto Politécnico Nacional, Ciudad de México, Mexico
tablaton@gmail.com

Abstract. This proposal is based on the dynamic analysis of the orientation mechanism of a pivoting nozzle, that is composed of a bell-type nozzle and three actuators mounted and positioned at 120° respectively, for the orientation of the output thrust of a propelled system. The behavior of the elements together will be studied to obtain the dynamic model, generating the equations that describe the movement of the actuators, and thus evaluate a preliminary law of control that allows the correct interaction of the entire system.

Keywords: Propulsion · Vector · Nonlinear system · Law of control · Nozzle · Pivotal · Actuator stabilization

1 Introduction

Originally the idea of designing a control law for a pivoting nozzle is the need to be able to implement orientation control in propelled systems, which allows us to orient the exit thrust. Recently, in thrust nozzle vectoring area, there was a proposal made by Naguil, Jorge L. Pedroni, et al. [1], which sets out the dynamic considerations on the control of a pivoting nozzle, however, in this case they analyzed a system with only two actuators that are positioned perpendicularly. Other references have mentioned the aerodynamics and evaluation of performance of this kind of systems, Schaefermeyer [2] and Berrier and Taylor [4], for rockets mainly, and the simulation of the mass flow control of a common system, Ruoyu Deng, Fanshi Kong and Heuy Dong Kim [3].

For the development of the proposal, initially it was necessary to obtain the dynamic equations of the pivoting nozzle system. Subsequently, we proposed and evaluated a control scheme based on the dynamic model. Lagrange mathematical modeling has been used to obtain dynamic equations. For the control law it was expected to obtain a non-linear model and use the approximate linearization method and thus propose a control law.

At the beginning of the mechanical design, three symmetrically located actuators were used, which improve the movement in the nozzle space and address the outflow (Fig. 1). The system is composed of a bell-type nozzle and three actuators mounted and positioned at 120° respectively, for the orientation of the output thrust of a propelled

E. E. Hernandez et al. (Eds.): LASIRS 2019, MMS 86, pp. 304–309, 2020.
https://doi.org/10.1007/978-3-030-45402-9_29

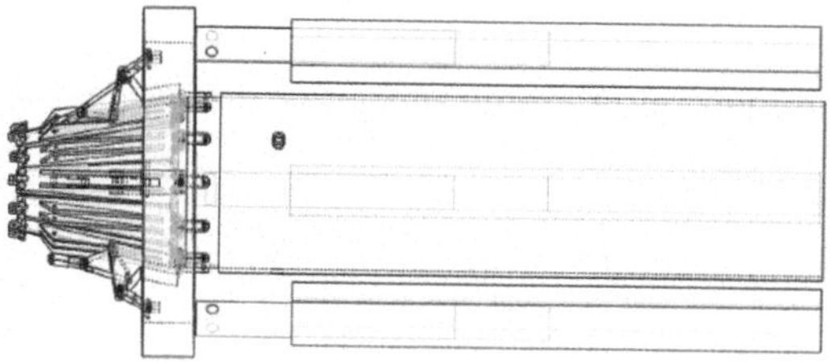

Fig. 1. Proposed model.

system. The deeper analysis of oscillations and vibrations will also be left for the future, with the corresponding control of these phenomena.

Therefore, a mechanical design and its control law are proposed, which allow to address the flow of the nozzle, considering the dynamic and kinematic constraints.

2 Governing Equations

Considering the mechanism of Fig. 1, we should analyze an over-actuated mechanical system, which has external forces acting on it. We must find the mathematical model that describes the behavior of the system. After, it is necessary to design control for the mechanism that will help to maintain the system variables in their nominal operating values, for existing or in the presence of disturbances. The Lagrange method has been used to obtain a dynamic representation of the mechanism. In Fig. 2, reference and rotated frames are presented.

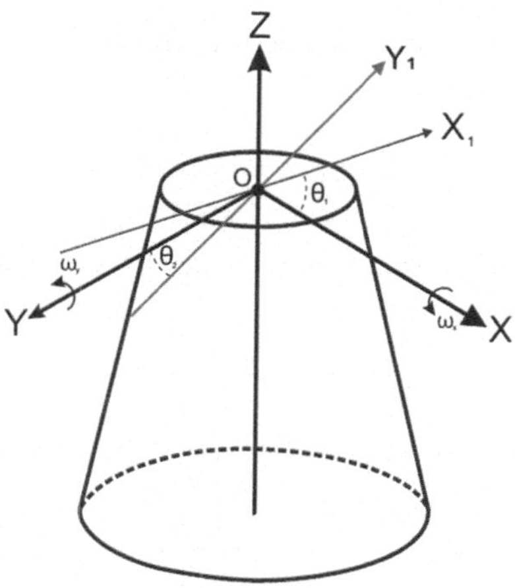

Fig. 2. Origin point, angles and related axes.

Kinetic Energy. The equation of kinetic energy for the system can be written as next equation [5]:

$$T = \frac{1}{2}\left[I_{xx}\left(\dot{\theta}_1^2 + \dot{\theta}_2^2\right)\right] \tag{1}$$

where:

I_{xx} represents the moment of inertia coefficient.
θ_1 is the first angle to rotate.
θ_2 is the second angle to rotate.

Potential Energy. Potential energy only involves the force of gravity, mass and displacement at two angles, and can be defined by next equation:

$$V = -mgl\cos\theta_1\cos\theta_2 \tag{2}$$

where:

m represents the mass.
g is the gravity constant.
l is the distance from the origin to the point of analysis.

Lagrangian Equations. In the next, we formulate Lagrange's dynamic equations for the system, where $Q = 0$. By defining $q_1 = \theta_1$, $q_2 = \theta_2$, we obtain next equations.

$$\begin{aligned} I_{xx}\ddot{q}_1 + mgl\sin q_1\cos q_2 &= Q_1 \\ I_{xx}\ddot{q}_2 + mgl\cos q_1\sin q_2 &= Q_2 \end{aligned} \tag{3}$$

Generalized Forces. The complete equations of the generalized forces are given by a set of equations as follows:

$$\begin{aligned} I_{xx}\ddot{q}_1 + mgl\sin q_1\cos q_2 - f_1\left[\left(l_{xy}\sin q_2 - l\cos q_2\right)\sin q_1\right] \\ + f_2\left[-\sin q_2\left(\sin q_2\tfrac{l_{xy}}{2} + l\cos q_2\right) + \cos q_1\tfrac{l_{xy}\sqrt{3}}{2}\right] \\ + f_3\left[-\sin q_1\left(\sin q_2\tfrac{l_{xy}}{2} + l\cos q_2\right) + \cos q_1\tfrac{l_{xy}\sqrt{3}}{2}\right] \\ I_{xx}\ddot{q}_2 + mgl\cos q_1\sin q_2 - f_1\left[-\cos q_1\left(l_{xy}\cos q_2 - l\sin q_2\right)\right] \\ + f_2\left[\left(\tfrac{l_{xy}}{2}\cos q_2 - l\sin q_2\right)\cos q_1\right] \\ + f_3\left[\left(\tfrac{l_{xy}}{2}\cos q_2 - l\sin q_2\right)\cos q_1\right] \end{aligned} \tag{4}$$

3 Control

The problem of controlling a pivoting nozzle is presented, where the behavior of a non-linear model of the system around a given operating point is analyzed (in our case, where the corresponding angles are zero, which is the vertical position).

In the case of a highly non-linear system, with more outputs that input variables, the modeling in the state space could be suitable [6, 7]. This implementation can be obtained as next equations.

$$
\begin{aligned}
x_1 &= q_1 \qquad \dot{x}_1 = x_2 \\
x_2 = \dot{q}_1 \quad \dot{x}_2 &= -\tfrac{mgl}{I_{xx}} \sin x_1 \cos x_3 + \tfrac{1}{I_{xx}} u_1((\sin x_3 - l \cos x_3)\sin x_1) + \\
&\quad \frac{1}{I_{xx}} u_2 \left(-\sin x_3 \left(\sin x_3 \frac{l_{xy}}{2} + l \cos x_3 \right) + \cos x_1 \frac{l_{xy}\sqrt{3}}{2} \right) \\
&\quad + \frac{1}{I_{xx}} u_3 \left(-\sin x_1 \left(\sin x_3 \frac{l_{xy}}{2} + l \cos x_3 \right) + \cos x_1 \frac{l_{xy}\sqrt{3}}{2} \right) \\
x_3 &= q_2 \qquad \dot{x}_3 = \dot{x}_4 \\
&\quad -\frac{mgl}{I_{xx}} \cos x_1 \sin x_3 + \frac{1}{I_{xx}} u_1(-\cos x_1 (l_{xy} \cos x_3 + l \sin x_3)) + \\
x_4 &= \dot{q}_2 \qquad \dot{x}_4 \\
&= -\frac{mgl}{I_{xx}} \cos x_1 \sin x_3 + \frac{1}{I_{xx}} u_1(-\cos x_1 (l_{xy} \cos x_3 + l \sin x_3)) \\
&\quad + \frac{1}{I_{xx}} u_2 \left(\left(\frac{l_{xy}}{2} \cos x_3 - l \sin x_3 \right) \cos x_1 \right) \\
&\quad + \frac{1}{I_{xx}} u_3 \left(\left(\frac{l_{xy}}{2} \cos x_3 - l \sin x_3 \right) \cos x_1 \right)
\end{aligned} \tag{5}
$$

Where, the displacement angles θ_1 and θ_2 or also called q_1 and q_2, are defined as state variables, as well as their respective angular velocities. The input variables u_i represents the force that each actuator exerts on the ring, in order to move it and control the plane as is shown in Fig. 1.

Considering the model of Eq. (5) and taking into account that all the state variables are measurable, we can use them for feedback control. We define the disturbed variables around the nominal equilibrium values. For the initial evaluation of the proposed control scheme of the pivoting nozzle, the approximated linearization procedure around the equilibrium point are presented in Eqs. (6) and (7), this procedure is based in references [9–11].

$$
A = \frac{\partial f}{\partial x}|_{x_i, u_i = (0,0)} = \begin{bmatrix} 0 & 1 & 0 & 0 \\ -\frac{mgl}{I_{xx}} & 0 & 0 & 0 \\ 0 & 0 & 0 & 1 \\ 0 & 0 & \frac{mgl}{I_{xx}} & 0 \end{bmatrix} \tag{6}
$$

$$B = \frac{\partial f}{\partial u}|_{x_i, u_i=(0,0)} = \begin{bmatrix} 0 & 0 & 0 \\ \frac{l}{I_{xx}} & \frac{l_{xy}\frac{\sqrt{3}}{2}}{I_{xx}} & -\frac{l_{xy}\frac{\sqrt{3}}{2}}{I_{xx}} \\ 0 & 0 & 0 \\ -\frac{l_{xy}}{I_{xx}} & \frac{\frac{l_{xy}}{2}}{I_{xx}} & \frac{\frac{l_{xy}}{2}}{I_{xx}} \end{bmatrix} \tag{7}$$

Physically, this means that we can take the nozzle to the horizontal position, as long as we bring the nozzle to the ideal equilibrium position, in that case zero for both angles, which is its ideal operating position. We propose and evaluate a control law for trajectory tracking [6, 8], with $y_1 = x(1)$, $y_2 = x(3)$, $y_d = \sin(t)$ and $\dot{y}_d = \cos(t)$, and signals of control in the form $u_{1,2,3} = (K_1 * e_x) + (K_2 * e_y)$, where e_x and e_y are signals of error between state variables and trajectory variables.

4 Results

The system was compared in three different scenarios, the open loop system (nonlinear), closed loop system (with the proposed control scheme) and the linearized system. Initially, the simulation of the open loop system was carried out, without considering external forces affecting the system. Subsequently, the simulation of the linearized system was performed, and the response of the system was compared according to the definition, characteristics and advantages of the approximate linearization. Finally, the simulation of the controlled system was carried out, where one can see the advantages of using the proposed control law (Fig. 3).

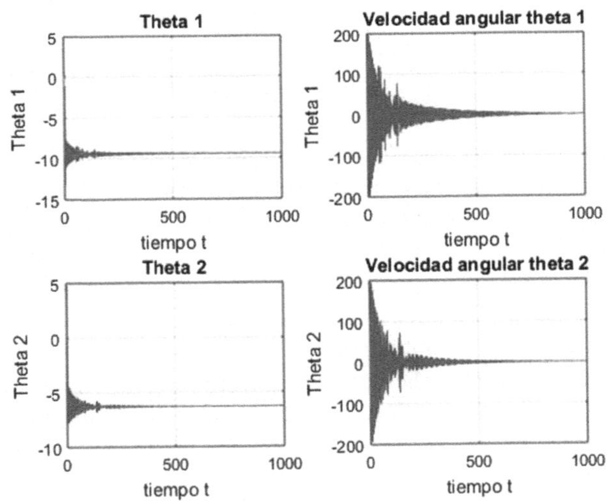

Fig. 3. Stabilized system with 10-degree disturbance.

5 Conclusions

Once the mathematical model by using Lagrange method has been obtained, an analysis of the generalized forces was required, this process was somewhat complex because the forces acting were determined and it was analyzed how these forces interact with the system. Thus, the mathematical model was completed and finally it was modeled in the form of state variables, a simple process that requires a complete and conscious prior analysis for the system. Then, we approached a pivoting nozzle that represents a non-linear model and its behavior at its point of operation, that for our use is around zero. Since the model is highly non-linear, an approximate linearization method was presented, which is useful near the point of equilibrium and low disturbances. Finally, a control scheme was evaluated for trajectory tracking, where rapid stabilization was obtained. This result was compared with them from the open loop and closed loop system, as well as the linearized system, and it was showed that the control law stabilizes the system in a faster and more efficient way.

References

1. Naguil JL, Pedroni JP, Cova, WJD, Consideraciones dinámicas sobre el control de una tobera pivotante
2. Schaefermeyer, MR (2011) Aerodynamic thrust vectoring for attitude control of a vertically thrusting jet engine. Utah State University Logan, Utah
3. Deng R, Kong F, Kim HD, Numerical simulation of fluidic thrust vectoring in an axisymmetric supersonic nozzle. Department of Mechanical Engineering, Andong National University, Andong, Korea, pp 760–749
4. Berrier BL, Taylor JG (1990) internal performance of two nozzles utilizing gimbal concepts for thrust vectoring, NASA TP-2991
5. Taylor JR, Díaz JID, Grau AC (2013) Mecánica clásica. Reverté, ISBN 9788429143126
6. Slotine J-JE, Li W, et al (1991) Applied nonlinear control, vol 199, Prentice Hall Englewood Cliffs, NJ
7. Khalil HK (1996) Nonlinear Systems. Prentice-Hall, New Jersey, vol 2(5), pp 5–1
8. Ogata K, Yang Y (1970) Modern control engineering. Prentice-Hall Englewood Cliffs, NJ
9. Sira-Ramirez H, Agrawal SK (2004) Differentially flat systems. CRC Press
10. Sira-Ramírez, H, Márquez, R, Rivas-Echevarría, F, Llanes-Santiago R (2005) Control de sistemas no lineales: linealización aproximada, extendida, exacta. Pearson Prentice Hall
11. Cortés-Romero J, Sira-Ramirez, H, Luviano-Juárez A (2011) Control lineal robusto de sistemas no lineales diferencialmente planos. RIAI Revista Iberoamericana de Automatica e Informatica Industrial

A Closed Loop Controller Design Which Can Minimize the Wind Effects on a Parabolic Solar Tracker

J. Díaz-Salgado[1], O. M. Huerta-Chávez[2(✉)], E. A. Villegas Jimenez[1], E. Enciso-Contreras[1], L. A. Cortés-Marín[1], and C. Ortiz-Cil[3]

[1] Tecnológico de Estudios Superiores de Ecatepec, Edo de Méx Ecatepec de Morelos, Mexico
jorgesalgado@tese.edu.mx, llanus_2@yahoo.com, eenciso@hotmail.com, lualcoma1@gmail.com

[2] Cátedras CONACyT, Tecnológico de Estudios Superiores de Ecatepec, Edo de Méx Ecatepec de Morelos, Mexico
lecram@hotmail.com

[3] Tecnológico de Apan Hidalgo, Apan, Mexico
ceoc_ceoc@hotmail.com

Abstract. In the present work, it was conducted a steady-state fluid analysis to obtain the aerodynamic coefficients of a Parabolic Solar Tracker (PST) for different azimuth and elevation angles. A computational fluid dynamic (CFD) wind flow analysis technique was applied to model an extreme weather condition called North Wind (NW), which can severely affect the PST under operational conditions. The North Wind case simulated was validated using experimental data without bottom effects in order to verify the degree of approximation between the results obtained with CFD approach. The obtained aerodynamic coefficients were used to: (i) establish both analytical and numerical models for each PST degree of freedom for dynamic analysis and simulation purposes, (ii) design different closed loop controllers which can minimize the wind effects and (iii) to establish a correct dimensioning of the control actuators.

Keywords: Parabolic Solar Tracker · CFD · Solar energy and control

1 Introduction

Solar energy is one the most attractive renewable energy because is unlimited, clean and free and its development permits a sustainable growing of alternate countries. Among the different forms of high-temperature solar energy harnessing Disc-Stirling systems present the highest solar-electric conversion efficiency thus having the potential of coming through the cheaper sun energy systems [1]. In a Disc-Stirling system the Parabolic Solar Tracker (PST) is the subsystem tasked to concentrate the solar radiation energy into a receiver that generally transfers such energy to a heat engine [2]. To achieve the high temperature a PST must have a two degree of freedom sun tracking system (azimuth, elevation) along the dairy operation.

E. E. Hernandez et al. (Eds.): LASIRS 2019, MMS 86, pp. 310–326, 2020.
https://doi.org/10.1007/978-3-030-45402-9_30

The presence of extreme weather conditions in areas with important solar energy resource is common. As PST are usually installed in open areas they suffer commonly this extreme wind conditions that might affect the efficiency and/or the integrity of the PST. Solar power companies have recognized that the wind is a fundamental element in the operation of a PST in three different forms: (i) the efficiency reduction due to the misalignment of the desired trajectory caused by forces due to the wind loads, (ii) temperature losses caused by the convection phenomenon due to the wind speed and (iii) the severe damage over the PST and/or its mechanical structure as a result of a critical wind condition [3–5].

In this work, the effect over a PST of an extreme wind condition named North Wind (NW) that is originated from a wind blowing from the north to the south, very common on winter in the mexican coast that surrounds the Gulf of Mexico, reaching wind speeds up to 22 m/s is studied [6]. A steady-state computational fluid dynamic (CFD) wind flow analysis technique is applied to obtain the aerodynamic coefficients of a PST for different azimuth and elevation angles along a test day with the NW condition. Experimental data were used to validate the simulated results obtained with CFD approach. Finally, the obtained aerodynamic coefficients were used to: (i) establish both analytical and numerical models for dynamic analysis and simulation purposes, (ii) to design a closed loop controller which can minimize the wind effects, and (iii) establish a correct dimensioning of the control actuators.

As the azimuth and elevation controllers were designed in a decoupled form, three different controllers were designed and compared to select the most appropriated for each case. The proposed controllers were designed in terms of the physical parameters of the PST. This feature permits to apply the same control law to another PST with different dimensions but with the same geometrical form.

This work is organized as follows: first an adequate numerical and analytical model is proposed for the PST, and its unknown parameters are identified; next the aerodynamic momentum coefficients are determined by the application of a (CFD) wind flow analysis technique under a NW weather condition; in the next section are designed different sun tracking controllers for the PST, and compared in nominal and perturbed conditions simulations in the subsequent section; finally some conclusions are drawn.

2 Parabolic Solar Tracker Modeling

In this section the mathematical and numerical models of the PST are established. The physical parameters (Table 1) of the PST considered (Fig. 1) makes it suitable for medium temperature applications [7].

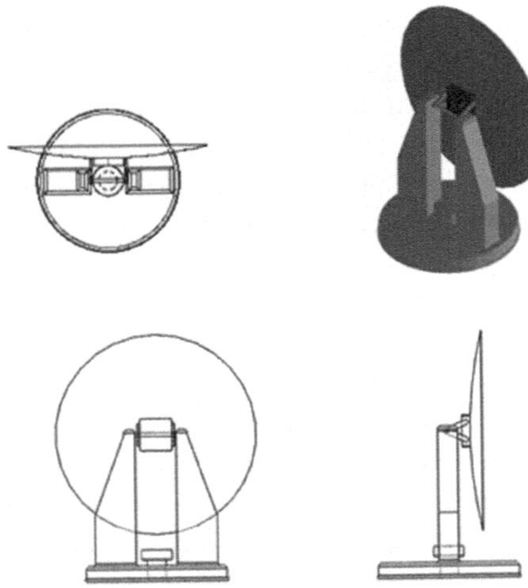

Fig. 1. Parabolic Solar Tracker

The PST was designed and manufactured with aluminum alloy 1060 and has two DC motors that generates two torques (τ_1, τ_2) to control the azimuth θ angle of the base and the elevation ϕ angle of the parabolic plate, respectively.

Table 1. Physical parameters of the PST considered.

Parameter	Value	Symbol	Units
Reflecting area	10.10	A_T	m^2
Focal distance	1.67	l_f	m
Projected area	4.59	A_P	m^2
Total mass	2.32×10^3	m_T	kg
Focal to diameter ratio	0.47	FD	
Concentrating ratio	10.00	η	

For simplicity two decoupled models are proposed one for the base and other for the parabolic plate of the PST.

2.1 Mathematical Model

Base Model
The rotation of the circular PST base produces the azimuth angle γ. Since in the considered PST the base is three times heavier that the parabolic dish and in closed loop it is expected a slow motion of the dish with respect to the base this second one is modelled as an over-damped rotational inertia system with a linear second order space state model

(1) that arises when a DC motor is coupled with a rotational inertia load.

$$\dot{x}_1 = x_2 \tag{1a}$$

$$\dot{x}_2 = -px_1 + k\tau_1 \tag{1b}$$

$$y_\gamma = x_1 \tag{1c}$$

$$x_1 = \gamma, x_2 = \dot{\gamma}, u_1 = \tau_1 \tag{1d}$$

where k is the input gain, and p is the not dominant pole.

Dish Model

The rotational movement of the parabolic dish produces the elevation angle θ. Because of the symmetry of the dish this is considered as a pendulum with a defined center of mass and so a non-linear second order model in space state is used as the mathematical model of the dish (2).

$$\dot{x}_3 = x_4 \tag{2a}$$

$$\dot{x}_4 = -a_1 x_4 - a_2 sin(x_3) + a_3 u_2 \tag{2b}$$

$$y_{\phi\theta} = x_3 \tag{2c}$$

$$x_3 = \theta, x_4 = \dot{\theta}, u_2 = \tau_2 \tag{2d}$$

where a_1, a_2, a_3 can be expressed in terms of physical parameters γ

$$a_1 = \frac{b}{ml_f^2}, a_2 = \frac{g}{l_f}, a_3 = \frac{1}{ml_f^2}$$

where m is the dish mass, l_f is the parabolic dish focal length, b is the viscous damping coefficient given as the sum of the bearings, gears and motor viscous friction, and g is the gravitational constant.

2.2 Numerical Model

The numerical model of the PST is obtained exploiting the fact that many CAD specialized mechanical design software can formulate and solve the movement equations of a mechanical system only from the geometry and dimensions of the physical elements and the type of material it is composed. In particular the methodological design followed was: (i) the drawing, assembly, and configuration of the materials of the mechanical elements of the system in a CAD software, (ii) the use of a simulation mechanics toolkit to export the mechanical model, including the joints, body elements, frames, and transforms to a numerical simulation software, and (iii) the inclusion of the gravitational forces in the numerical model. Different dynamic open and closed loop response simulations can be done with this numerical model and its outputs are considered as the actual outputs of the system.

2.3 Parameter Identification

The parameters (k, p) of the base and (a_1, a_2, a_3) of the dish are identified by using the numerical model outputs as the measured signals of a recursive least squares parameter identification method. This can be done when the mathematical model can be expressed as a linear regressor (3) in the unknown parameters:

$$y(k) = \Psi^{\mathrm{T}}(k)\Theta \tag{3}$$

The measured discrete time outputs generated by the numerical model $\dot{y}_\gamma(k)$, $y_\theta(k)$ and $\dot{y}_\theta(k)$ are used to express (1b) and (2b) as linear regressors (4) and (5):

$$\ddot{y}_\gamma(k) = \left[-\dot{y}_\gamma(k)\; u_1(k) \right] \begin{bmatrix} p \\ k \end{bmatrix} = \Psi_1(k)\Theta_1 \tag{4}$$

$$\ddot{y}_\theta(k) = \left[-\dot{y}_\theta(k)\; -sin[y_\theta(k)]\; u_2(k) \right] \begin{bmatrix} a_1 \\ a_2 \\ a_3 \end{bmatrix} = \Psi_2(k)\Theta_2 \tag{5}$$

Where $y(k)$ is the measured signal vector of the system, $\Psi(k)$ is the regression matrix constituted by known function observations, and Θ is the unknown parameter vector. The variable k denotes the discrete time index.

The scalar least square method has the next form [8]:

$$\hat{\Theta}(k) = \hat{\Theta}(k-1) + \frac{P(k-1)\Psi(k)e(k)}{1 + \Psi(k-1)^T P(k-1)\Psi(k)} \tag{6}$$

$$P(k) = P(k-1) - \frac{P(k-1)\Psi(k)\Psi(k)^T P(k-1)}{1 + \Psi(k-1)^T P(k-1)\Psi(k)} \tag{7}$$

$$e(k) = y(k) - \Psi(k)^T \hat{\Theta}(k-1) \tag{8}$$

Where $P(k)$ is a positive definite covariance matrix, $e(k)$ is the prediction error, and $\hat{\Theta}(k)$ is the estimated parameter vector. The followed methodology considers in applying two different tests: (i) the parabolic dish free response from the upper position and zero velocity as the initial condition, and (ii) the unit step response of the base from zero position and velocity initial condition.

The simulated free response (Fig. 2) for the parabolic dish $(\gamma, \dot{\gamma})$ and unit step forced response for the base $\left(\theta, \dot{\theta}\right)$ shows good agreement between the proposed numerical and analytical models, in both transient and steady state responses, and makes the proposed model (1-2) suitable for control design purposes.

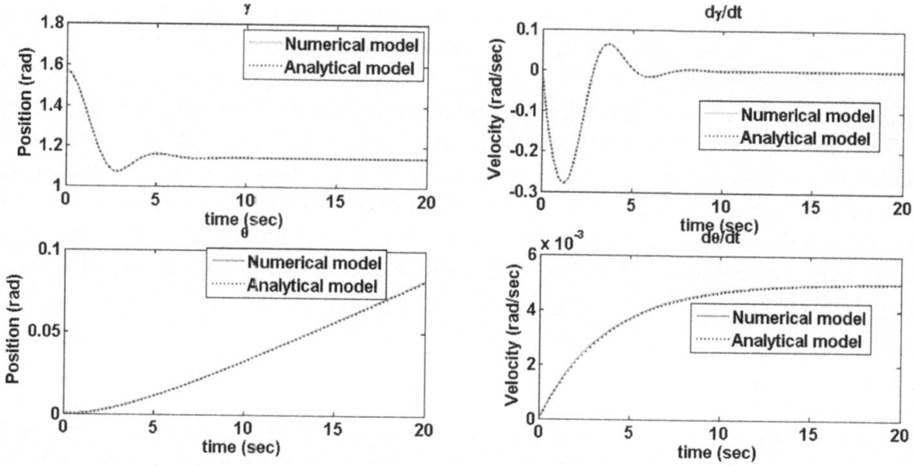

Fig. 2. Comparison between the numerical and analytical model responses.

3 CFD Wind Flow Analysis

A computational fluid dynamic (CFD) wind flow analysis technique is applied to the PST CAD model to determine the aerodynamic momentum coefficients under an NW extreme weather condition [9]. As mentioned above, the CFD Fluent software was used to perform the fluid flow simulations of the wind loads over the PST. The incompressible Navier-Stokes equations were solved in a classical model based on Reynolds Average formulation, better known as RANS. Therefore, momentum and continuity equations were coupled to solve the velocity field. The Spalart Allmaras (S-A) viscous model was applied to solve a single partial differential equation (PDE). In the simulations, the S-A turbulent model was considered which as most turbulence models follow Boussinesq's assumption and the Reynolds stresses are considered proportional to the fluid shear stresses (see Čedomir Kostić 2015).

The S-A viscous model is not typically applied in the literature for PST problems, because is a special model for aeronautical applications in the aerospace industry, however, it has been identified that the S-A model works correctly under the following two scenarios: 1. For a fine enough resolution to completely resolve viscous sub-layer (but not so fixable for Re~10E6) and 2. Putting the first grid point in the log-region (z+~30-50) which need high grid intensity near wall (see Gerasimov, A.V., 2003).

3.1 CFD Results

The residual behavior computed from the S-A model Δ8, where 8(Δ8 is the convergence criteria used to stop the solver, e.g. Δ10_8 → 1x10^8, for the most significant case there were simulated 18 h. All data are based on absolute value and the same process was applied in all cases from 6 to 18 h. Also, there were conducted simulations with S-A model Δ4 which have not sufficient accuracy in results.

To obtain better agreements in the results, the CFD team has been followed professional recommendations with a new set of simulation with S-A Δ10.

Plots in Fig. 5 from a to c shows not discrepancy between S-A Δ8 and S-A Δ10, its mean no more than 3% error for each data point. Actually, S-A Δ10 behavior tend to RKE Δ8 results and the differences between plots are not significant in results, both models have a good quality, quantity and great convergence. However, the computational cost are more expensive for S-A Δ10, the number of cells needed increase from 5 to 123 million and the computational time from 6 to 28.4 h with more than 18 000 iterations each case (working with 256 GB of RAM and 48 processors in a Professional Server from TESE Institute).

3.2 Experimental Test

In order to validate the simulation results, it was performed laboratory experimental work. Two different numerical methods were compared with the experimental data which was obtained with a wind channel (Andre, Mier-torrecilla, & Wüchner, 2015). In this work only the results of the S-A Δ8 method were compared with experimental data. : (i) not significant differences were founded with respect the RKE Δ8 method and (ii) the first one is as twice as fast than the second one.The PST-prototype used in this study was made of ABS-HI-100 plastic material (see Fig. 6). l. The geometric similarity was scaled by a constant factor CGF 1:4.51 and the Reynold Number (Re~2.54E6) to scale all forces (e.g., viscous and inertial forces) in the prototype was used as a dynamic similarity parameter.

3.2.1 Results

Comparisons between the total forces from experimental measurements vs numerical results were qualitative and quantitative favourable in most cases, with predictions usually within 77% of the measured data as clearly shown in Fig. 3a–b.

During the analyses, there were identified two particular and visible behaviours over the curves obtained from a sixth order polynomial approximation. First, Fig. 3a shows a phase shift between the experimental measurements and the numerical data curves and second, Fig. 3b, shows more emphasized behavior in the valleys and crests of the experimental measurement over the curve. Quantitative differences were found at some data points in the graph due to the effects of significant blockages during the rotation of the dish within the wind test section, since the area of obstruction of the prototype within the wind channel varies. However, qualitatively the behavior between numerical vs analytical results is very similar, which gives a great validity to the numerical results.

In addition, during experiments, three particular altitude and azimuth angle positions from the PST at 11 h (50°, 135°), 12 h (63°, 140°) and 18 h (45°, 320°) presented oscillation forces around -3 to 3% of medium effective velocity. In fluids mechanics, the effect is called vortex shedding, where the vortices are created at the back of the dish and detach periodically from either side of the dish due alternating low pressure on the downstream side.

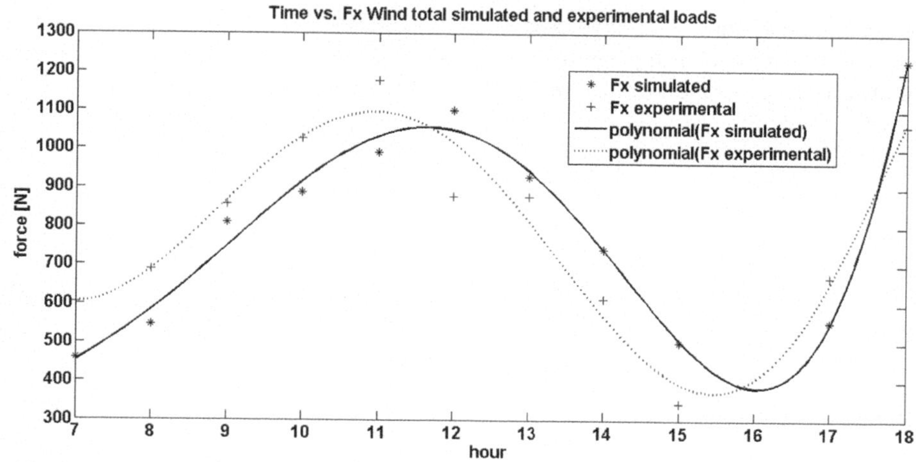

(a)

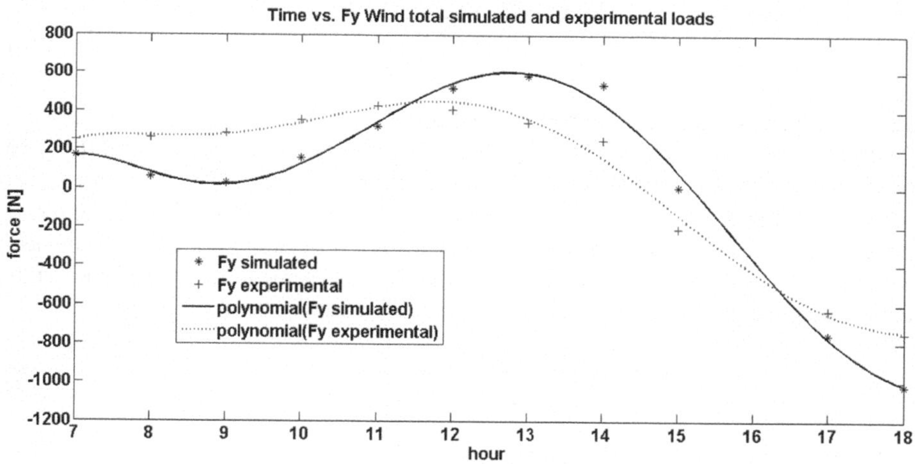

(b)

Fig. 3. a.–b. Represent time vs Fx and Fy wind total loads simulated compared with experimental total loads over the PST.

4 Control Algorithms Design

Three different control algorithms are designed for each PST angle. Departing from the static and dynamic performance in closed-loop response one of the three algorithms is selected for each angle.

4.1 Parabolic Dish Controllers

Since the mathematical model of the dish is non-linear, two of the proposed control algorithms are designed by standard linearization method around the upward dish position, meanwhile the last is anon linear one.

Model Linearization

The upward parabolic dish position corresponds to an elevation angle $\theta = 0$. To obtain the linearized model the Jacobian matrices A and B are defined as:

$$A = \left.\begin{bmatrix} \frac{\partial f_1}{\partial x_3} & \frac{\partial f_1}{\partial x_4} \\ \frac{\partial f_2}{\partial x_3} & \frac{\partial f_2}{\partial x_4} \end{bmatrix}\right|_{\theta=0}, \quad B = \left.\begin{bmatrix} \frac{\partial f_1}{\partial u_2} \\ \frac{\partial f_1}{\partial u_2} \end{bmatrix}\right|_{\theta=0} \tag{9}$$

Where:

$$f_1 = x_4 \tag{10a}$$

$$f_2 = -a_1 x_4 - a_2 sin(x_3) + a_3 u_2 \tag{10b}$$

yielding the linearized model

$$\dot{x} = \begin{bmatrix} 0 & 1 \\ -a_2 & -a_1 \end{bmatrix} x + \begin{bmatrix} 0 \\ a_3 \end{bmatrix} u \tag{11a}$$

$$y = \begin{bmatrix} 1 & 0 \end{bmatrix} x \tag{11b}$$

State Feedback + Integral Control Algorithm

The first control designed is a state feedback + integral control algorithm [10]. The design consists in determining a gain feedback vector K by proposing a control signal as $u = r - Kx$ and forcing a desired dynamic response. The addition of the integral part permits to eliminate the tracking error simultaneously to a desired dynamic performance. The first stage consist in the definition of an additional element x_N of the space state vector and an integral gain k_i yielding:

$$\begin{bmatrix} \dot{\boldsymbol{x}} \\ \dot{x}_N \end{bmatrix} = \begin{bmatrix} (\boldsymbol{A} - \boldsymbol{B}\boldsymbol{K}) & \boldsymbol{B}k_i \\ -\boldsymbol{C} & 0 \end{bmatrix} \begin{bmatrix} \boldsymbol{x} \\ x_N \end{bmatrix} + \begin{bmatrix} 0 \\ 1 \end{bmatrix} r \tag{12}$$

With the substitution of A, B and C gives:

$$\begin{bmatrix} \dot{\boldsymbol{x}} \\ \dot{x}_N \end{bmatrix} = \begin{bmatrix} \left(\begin{bmatrix} 0 & 1 \\ -a_2 & -a_1 \end{bmatrix} - \begin{bmatrix} 0 \\ a_3 \end{bmatrix} \boldsymbol{K} \right) & \begin{bmatrix} 0 \\ a_3 \end{bmatrix} k_i \\ -\begin{bmatrix} 1 & 0 \end{bmatrix} & 0 \end{bmatrix} \begin{bmatrix} \boldsymbol{x} \\ x_N \end{bmatrix} + \begin{bmatrix} 0 \\ 1 \end{bmatrix} r \tag{13}$$

The new space state matrix A_n is:

$$A_n = \begin{bmatrix} 0 & 1 & 0 \\ -a_2 - a_3 k_1 & -a_1 - a_3 k_2 & a_3 k_i \\ -1 & 0 & 0 \end{bmatrix} \tag{14}$$

With this matrix the proposed closed loop third grade polynomial is stated:

$$|sI - A_n| = \left| \begin{bmatrix} s & 0 & 0 \\ 0 & s & 0 \\ 0 & 0 & s \end{bmatrix} - \begin{bmatrix} 0 & 1 & 0 \\ -a_2 - a_3 k_1 & -a_1 - a_3 k_2 & a_3 k_i \\ -1 & 0 & 0 \end{bmatrix} \right|$$
$$= \begin{vmatrix} s & -1 & 0 \\ a_2 + a_3 k_1 & s + a_3 k_2 + a_1 & -a_3 k_i \\ 1 & 0 & s \end{vmatrix} = s^3 + (a_3 k_2 + a_1) s^2 + (a_3 k_1 + a_2) s + a_3 k_i \tag{15}$$

A desired closed loop dynamic is proposed

$$s^3 + 90.06s^2 + 95.48s + 23.85 \tag{16}$$

Finally the desired and proposed closed-loop dynamics are equated (15) = (16) and the gains determined

$$k_1 = \frac{95.48 - a_2}{a_3}, k_2 = \frac{90.06 - a_1}{a_3}, k_i = \frac{23.85}{a_3} \tag{17}$$

Optimal Control

This algorithm starts with a similar control signal proposition as in the state feedback case: $u = -Kx$. But in this case the gain vector is determined by solving a cost function that minimizes the state trajectory and at the same time the control effort:

$$J = \int_0^\infty \left(x^* Qx + u^* Ru\right)dt \tag{18}$$

Where Q is a symmetric semi-definite positive matrix and R is a symmetric definite positive one. The gain vector is determined from

$$K = R^{-1}B^* P \tag{19}$$

Where P is a symmetric definite positive matrix calculated by solving:

$$(A - BK)^* P + P(A - BK) = -\left(Q + K^* RK\right) \tag{20}$$

By choosing

$$Q = \begin{bmatrix} 1 & 0 \\ 0 & 1 \end{bmatrix}, R = 1e - 7 \tag{21}$$

is obtained the gain vector

$$K = \left[9.1627e3 \; 9.9472e3\right] \tag{22}$$

Non-linear Control Algorithm Design

This design consists in the definition of the position error (**23**), the proposition of a pole assignable second order desired dynamic (**24**), and finally the substitution of this dynamic in the non-linear model yielding the control signal (**25**).

$$e = y_\phi - \bar{y}_\phi \tag{23}$$

$$\ddot{e}(t) + k_1\dot{e}(t) + k_2 e(t) + k_3 \int e(t)dt = 0 \tag{24}$$

$$u = \left(\frac{1}{a_3}\right)\left(-k_1\dot{e}(t) - k_2 e(t) - k_3 \int e(t)dt + a_1\dot{y} + a_2 sen(y) + \ddot{\bar{y}}_\phi\right) \tag{25}$$

The proposed dynamic (24) is stable if

$$k_1 > 0, k_2 > 0, k_3 > 0 \tag{26}$$

So the control gains are chosen in terms of the model parameters

$$k_1 = 3a_1, k_2 = 10a_2, k_3 = 0.5a_1 \tag{27}$$

4.2 Base Controllers

State Feedback

The state feedback controller proposes a control input as:

$$u = -Kx + k_i r \tag{28}$$

This control applied in the system dynamics yields:

$$\dot{x} = Ax + Bu = (A - BK)x + Bk_i r \tag{29}$$

The desired dynamics is proposed in terms of the not dominant pole p as:

$$(s + 25p)(s + 250p) \tag{30}$$

The proposed dynamics is given by:

$$\begin{vmatrix} s + p + k_1 & k_2 \\ -1 & s \end{vmatrix} = s^2 + (k_1 + p)s + k_2 \tag{31}$$

Optimal Control

The methodology design is the same of the base optimal control, the only difference is the proposition of the R value.

$$R = [9e - 6] \tag{32}$$

Proportional + Integral + Derivative Control

The PID transfer function commonly established in terms of proportional, integral and derivative gains here is written in terms of the arbitrary position of two zeroes $[z_1, z_2]$ and a global gain k_d.

$$\frac{k_d\left(s^2 + \frac{k_p}{k_d}s + \frac{k_i}{k_d}\right)}{s} = \frac{k_d(s + z_1)(s + z_2)}{s} \tag{33}$$

The proposed methodology considers the design in two stages [10]: (i) the PD component design to achieve the desired transient response (that fixes one of the zeroes) and (ii) the PI component design to achieve the desired steady state response (that fixes the other zero and the global gain).

For the PD component a zero z_1 is proposed one hundred times quicklier than the not dominant pole in open loop response.

$$z_1 = 100p \tag{34}$$

Meanwhile for the PI component a zero z_2 is proposed to minimize the transient response effect of the PD component.

$$z_2 = p - \frac{p}{10} \tag{35}$$

The described method yields the (xx) control gains.

$$k_d = 1e5, \; k_i = 2e5, \; k_p = 3e5 \tag{36}$$

5 Simulation Results

Two different scenarios are considered for the simulations: nominal and perturbed conditions. Nominal simulations considers the operation of the PST without perturbations while perturbed simulations considers the NW condition with a wind velocity of 22.2 m/s. The solar azimuth and elevation angles are calculated using a well-known sun position algorithm [11] for Juchitán de Zaragoza, Oaxaca (16.5988373, −95.043291) on august 5 (Fig. 4).

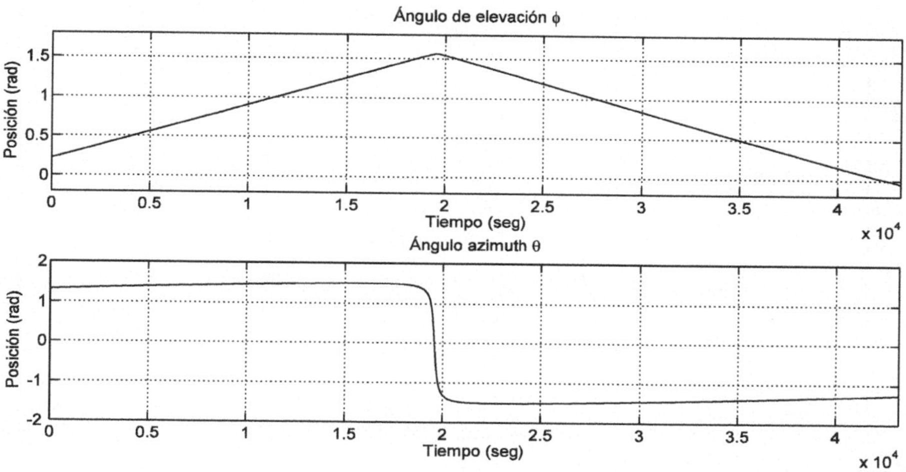

Fig. 4. Solar azimuth and elevation angles calculated with [11] sun position algorithm.

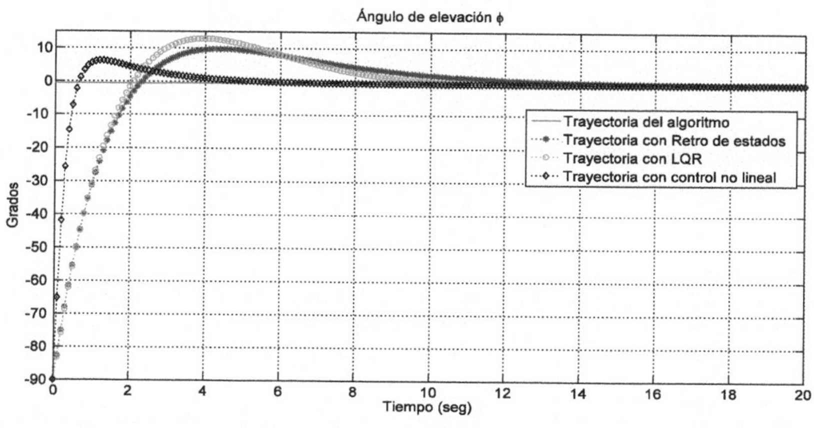

Fig. 5. PST dish (elevation angle) transient response with proposed controllers.

5.1 Nominal Conditions

Transient and steady state responses are simulated for base and parabolic dish proposed controllers. In Table 2 the performance parameters of the transient response of the parabolic dish (elevation angle) (Fig. 5) and base (azimuth angle) (Fig. 6) are presented.

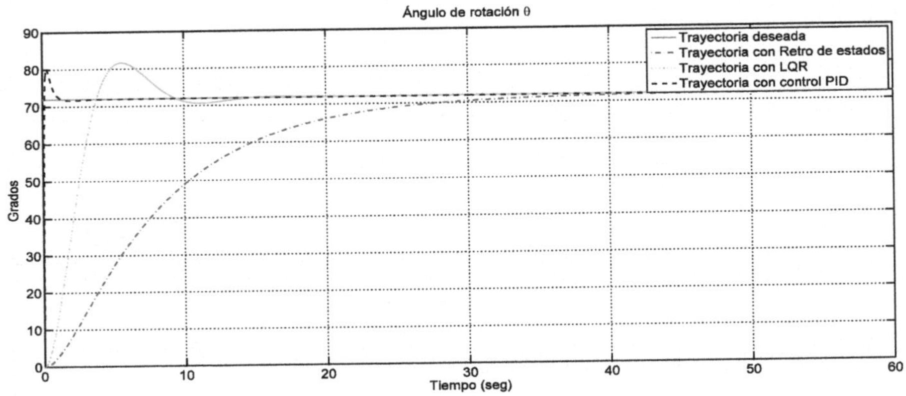

Fig. 6. PST base (azimuth angle) transient response with proposed controllers.

The comparison shows a better performance of non-linear and PID control for dish and base controllers respectively.

Table 2. Performance of dish and base transient responses.

	Parameter	$t_s[s]$	$\%OS[/]$	$e_{ss}[°]$
Dish (elevation angle) control	State feedback + Integral control	15	10	0.22
	Optimal control	11	13	0.23
	Non-linear control	5	7	0.15
Base (azimuth angle) control	State feedback	45	0	11
	Optimal control	18	12	2
	PID control	2	11	0

The steady state trajectory integral error of dish (Fig. 7) and base (Fig. 8) controllers shows that non-linear and PID controllers respectively present the lower steady state error.

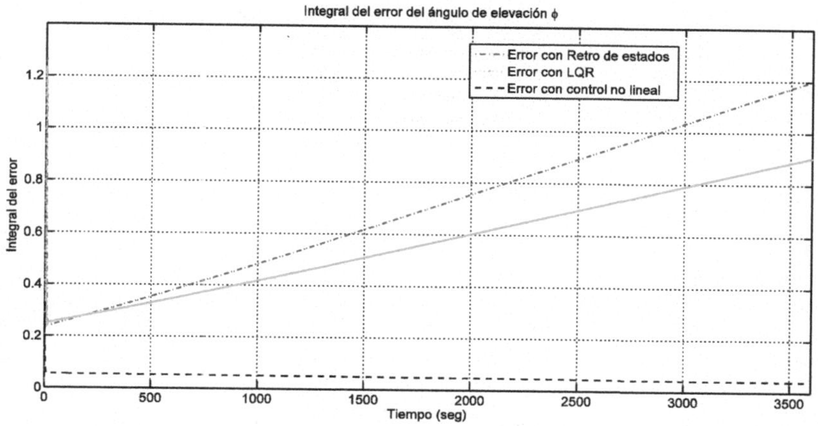

Fig. 7. PST dish (elevation angle) steady state integral error response with proposed controllers.

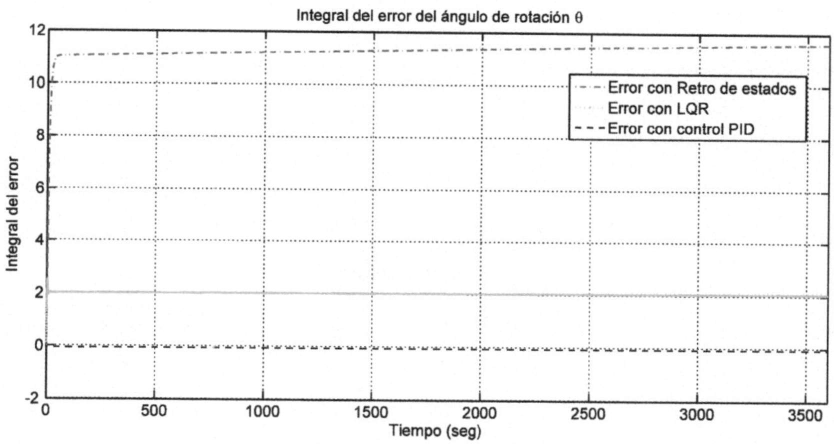

Fig. 8. PST base (azimuth angle) steady state integral error response with proposed controllers.

5.2 Perturbed Conditions

As non-linear and PID controllers for the dish and base controllers respectively shown the better transient and steady state performance, the perturbed conditions are only proved with this selected controllers.

The perturbed input signals (Fig. 9) are obtained with a 6th order polynomial approximation of the hourly moment coefficients generated in the CFD simulations.

The sun trajectory tracking along the test day (Fig. 10) shows good agreement between desired and actual elevation and azimuth angles with the selected designed controllers with perturbed conditions, as the steady state tracking errors (Fig. 11) stays in an order of 10^{-5} in both cases.

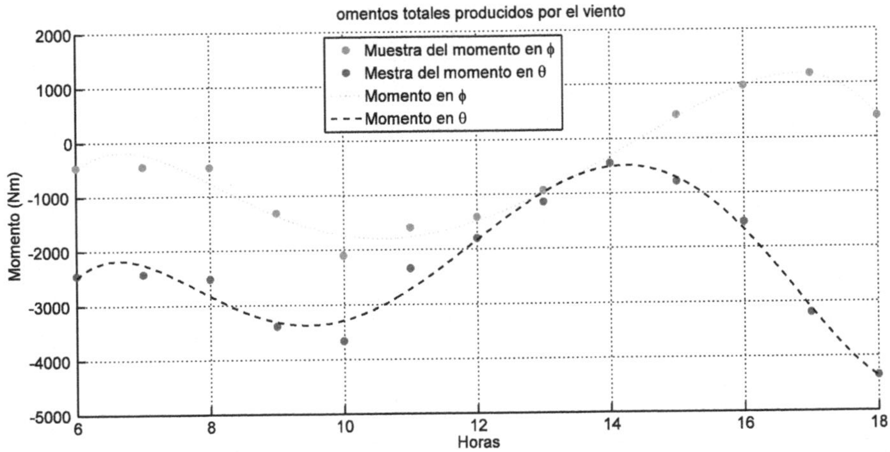

Fig. 9. PST moment dish and base perturbations considered for perturbed conditions.

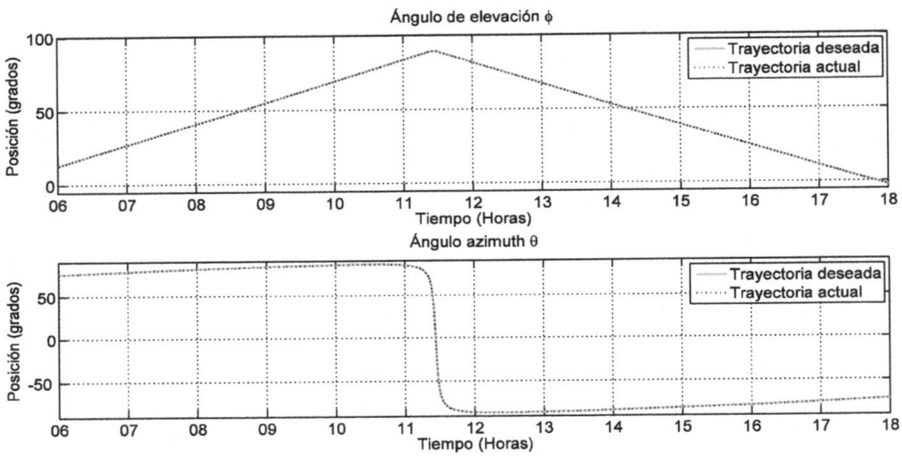

Fig. 10. Trajectory tracking of dish and base controllers with perturbed conditions

Finally the comparison between nominal and perturbed control signals (Fig. 12) shows a huge difference between the nominal and perturbed control signals which justifies: (i) the need to incorporate a closed loop control scheme in order to minimize the wind load deviation, and (ii) gives an insight of the correct dimensioning of the system actuators under perturbed NW condition.

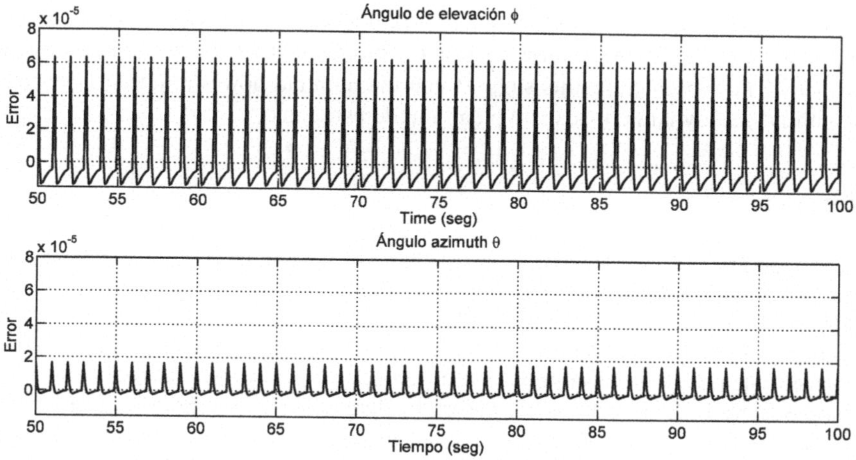

Fig. 11. Trajectory tracking errors of dish and base controllers with perturbed conditions

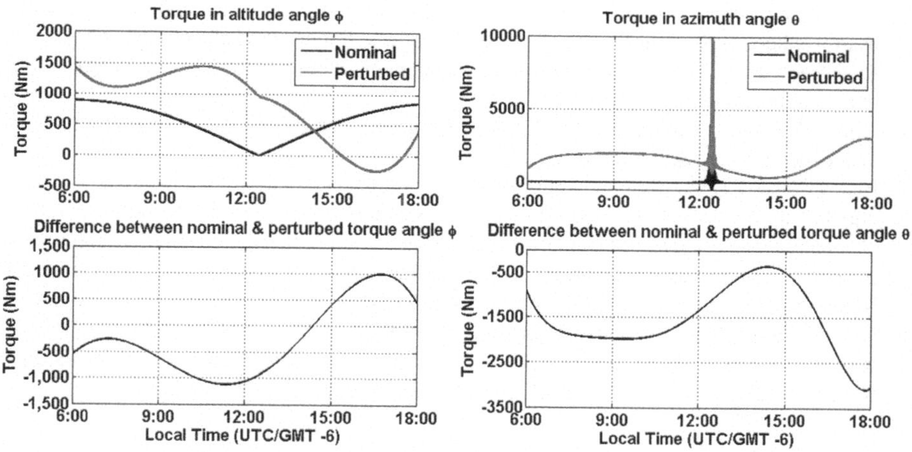

Fig. 12. Comparison between nominal and perturbed control signals

6 Conclusions

In this work, a (CFD) wind flow analysis technique was applied to model a North Wind weather condition to obtain the aerodynamic momentum coefficients of a Parabolic Solar Tracker for different azimuth and elevation angles along a test day. The proposition of a numerical and analytical model of the PST in conjunction with the aerodynamic momentum coefficients, obtained in the CFD analysis, permits to (i) establish a dynamic analysis and (ii) design a closed loop controller to minimize the effect of the wind. The closed loop responses show that the control torque in worst cases must increase two to three times the nominal torque (without wind) to counteract the aerodynamic forces. The results of the present work can be used to (i) correct sizing of the control system actuators for a specific wind condition and location and (ii) design a lighter weighted structure because a great part of the disturbance rejection of the wind load can be done by the actuator of the control system, instead of an increment of the mechanical structure robustness. The presented results suggest that the nominal power required to control a PST without aerodynamic loads is typically underestimated.

References

1. Baharoon DA, Rahman HA, Wan Omar WZ, y Fadhl SO (2015) Historical development of concentrating solar power technologies to generate clean electricity efficiently a review. Renew Sustain Energy Rev 41:996–1027
2. Schiel W, Keck T (2012) Parabolic dish concentrating solar power (CSP) systems, Schlaich Bergermann und Partner, Germany, pp 284–322
3. Ruelas Ruiz JE, Cota A, Ochoa F, Lucero B, Burgos T, Delfin J, Soto A, Tzab J, Lopez J, Nañez J (2014) Structural design methodology for solar concentrators subjected to wind loads. Energy Procedia 57:2872–2878
4. Roschke EJ (1984) Wind loading on solar concentrators: some general considerations. Wiley, New Jersey
5. Paetzold J, Cochard S, Fletcher D, Vassallo A (2015) Wind engineering analysis of parabolic trough collectors to optimise wind loads and heat loss. Energy Procedia 69:168–177
6. Meteored Homepage. https://www.meteored.mx/noticias/actualidad/evento-de-norte-en-mexico.html. Accessed 10 Jan 2020
7. Hijazi H, Ossama Mokhiamar OE (2016) Mechanical design of a low cost parabolic solar dish concentrator. Alexandria Eng J 55:1–11
8. Reyes Cortés F (2012) Matlab Aplicado a Robótica y Mecatrónica. Alfaomega
9. Huerta Chavez O, Díaz Salgado J, Ortiz Cil C, Jimenez Escalona J, Torres Cedillo S (2019) Cuamatzi R (2019) 3D CFD wind flow analysis technique applied to a parabolic solar tracker for two extreme weather conditions with experimental results and a controller proposition. Renew Sustain Energy 11:023702
10. Nise N (2011) Control systems engineering. John Wiley & Sons
11. Reda I, Andreas A (2003) Solar position algorithm for solar radiation application. National Renewable Energy Laboratory (NREL) Technical report NREL/TP-560-34302

Author Index

E. E. Hernandez et al. (Eds.): LASIRS 2019, MMS 86, pp. 327–328, 2020.
https://doi.org/10.1007/978-3-030-45402-9

The manufacturer's authorised representative in the EU is Springer Nature Customer Service Centre GmbH, Europaplatz 3, 69115 Heidelberg, Germany. If you have any concerns regarding our products, please contact ProductSafety@springernature.com

Printed and bound by CPI Group (UK) Ltd, Croydon, CR0 4YY
23/09/2026
02221528-0013